Renewable Energy

A Global Review of Technologies, Policies and Markets

Edited by
Dirk Aßmann, Ulrich Laumanns and Dieter Uh

Routledge
Taylor & Francis Group

LONDON AND NEW YORK

First published by Earthscan in the UK and US in 2006

For a full list of publications please contact:
Earthscan
2 Park Square, Milton Park, Abingdon, Oxon OX14 4RN
711 Third Avenue, New York, NY 10017

Routledge is an imprint of the Taylor & Francis Group, an informa business

First issued in paperback 2016

Typeset by RefineCatch Limited, Bungay, Suffolk
Cover design by Susanne Harris

A catalogue record for this book is available from the British Library

Library of Congress Cataloging-in-Publication Data
Renewable energy : a global view of technologies, policies and markets / edited by Dirk Aßmann, Ulrich Laumanns and Dieter Uh

    p. cm.

1. Renewable energy sources—Research.   I. Aßmann, Dirk, 1968–   II. Laumanns, Ulrich, 1977–   III. Uh, Dieter.
TJ808.2R45   2006
333.79′4—dc22

                                                        2006000305

    ISBN13: 978-1-84407-261-3 (hbk)
    ISBN13: 978 1 138 98512 4 (pbk)

Renewable Energy

Contents

PART FOUR – DEVELOPING COUNTRIES

PART FIVE – CAPACITIES

List of Figures, Tables and Boxes

FIGURES

TABLES

BOXES

List of Contributors

Carsten Agert is head of the fuel cell systems group at the Fraunhofer Institute for Solar Energy Systems (ISE) in Freiburg, Germany. His previous assignments include a post as research analyst for the German government's Advisory Council on Global Change (Wissenschaftlicher Beirat Globale Umweltveränderungen, WBGU), strategic planning for Fraunhofer ISE's business areas and projects on renewable energy policy as a freelance consultant in South Africa. His key areas of research are fuel cell technology, photovoltaics and renewable energy policy.

Dirk Aßmann is project manager and energy expert at the Deutsche Gesellschaft für Technische Zusammenarbeit (GTZ, German Technical Cooperation), Germany. He worked on the preparation of the renewable 2004 conference, where he was in charge of the Thematic Background Papers, the Conference Issue Paper and the Policy Recommendations. Dirk has studied energy technologies and system analysis, and holds a doctorate in economics and social sciences. From 1996 to 2003 he worked at the Wuppertal Institute for Climate, Environment and Energy, latterly as scientific coordinator for 'Non-European Projects'. His previous assignment areas include research, business development and development aid.

Adrian Bradbrook is the Bonython Professor of Law at the University of Adelaide, Australia. He is a Fellow and Member of the Board of Directors of the International Energy Foundation; Fellow of the Australian Institute of Energy; and Deputy Chair of the Working Group on Climate Change and Energy Law of the World Conservation Union (IUCN).

Sonja Butzengeiger is an environmental engineer and partner of Perspectives consultancy. She has worked extensively on emissions trading and initial allocation of allowances, especially in the Working Group on Emissions Trading of the German Ministry of the Environment.

John M. Christensen is head of the secretariat for the Global Network on Energy for Sustainable Development (GNESD) – a partnership initiative established at the World Summit on Sustainable Development in 2002. He also heads the United Nations Environment Programme (UNEP) Risoe Centre on

Energy, Climate and Sustainable Development, which he has been managing since its establishment in 1990. John has worked on energy and development issues in more than 20 countries over the past two decades; he has also been a Bureau member of the Intergovernmental Panel on Climate Change (IPCC) and a lead author in both the IPCC second and third assessments.

Joy Clancy is an associate professor in technology transfer with the Technology and Development Group, the University of Twente, the Netherlands. She studied chemistry in London and her PhD is about alcohol fuels in stationary engines. She joined the University of Twente in 1989. Since then her research has focused on small-scale energy systems for developing countries. Gender and energy has been an important factor addressed in this research. Dr Clancy is a founder member and a technical adviser to ENERGIA.

Christopher Flavin is President of the Worldwatch Institute and was a member of the International Steering Committee of the International Conference for Renewable Energies (Renewables 2004). He is co-author of *Power Surge: Guide to the Coming Energy Revolution*. Chris was a founding director of the Business Council for Sustainable Energy and is a member of the board of the American Wind Energy Association.

Franco Fugazza is a consultant at MacroConsulting, a consultancy in public utilities regulation economics. He holds a degree in economics awarded by the University of Buenos Aires and completed a Master's degree in economics and philosophy at the London School of Economics. Besides his work on different projects for MacroConsulting, mainly in the electricity and water sectors in Latin America and Africa, his career has evolved around the following areas of competence: research on international competition and foreign trade, disloyal commercial practices (dumping, subsidies) case following up and analysis, and research and analysis on fiscal matters.

José Goldemberg earned his PhD in physical science from the University de São Paulo, of which he was Rector. He has served as President of the Energy Company of the State of São Paulo (CESP), Minister for Science and Technology and Minister of Education of the Federal Government of Brazil. Presently he is Secretary for the Environment of the State of São Paulo.

Thomas B. Johansson is professor and director, International Institute for Industrial Environmental Economics (IIIEE), Lund University, Sweden. He was director of the Energy and Atmosphere Programme of the United Nations Development Programme, 1994–2001, where he, inter alia, developed and launched the World Energy Assessment. Since 1992 he has been the international co-chair of the Working Group on Energy Strategies and Technologies of the China Council on International Cooperation for Environment and Development (CCICED). He received the Volvo Environment Prize in 2000.

Stephen Karekezi is the director of the African Energy Policy Research Network (AFREPREN) as well as the executive secretary of the Foundation for Woodstove Dissemination (FWD), Nairobi. In 1995 he was appointed member of the Scientific and Technical Advisory Panel (STAP) of the Global Environment Facility (GEF) co-managed by the World Bank, the United Nations Development Programme (UNDP) and the UNEP. He has written, co-authored and edited some 87 publications on sustainable energy development. In 1990 he received the Development of Association Award in Stockholm, Sweden, in recognition of his work on the development and dissemination of the Kenya Ceramic Jiko.

Matthias Krey is an economic engineer at the University of Flensburg. He has specialized in Clean Development Mechanism (CDM) transaction cost analysis and has carried out a thorough case study for India.

Ricardo Külheim is studying production engineering focused on environmental issues such as renewable energies at the University of Freiberg, Germany. He has done several internships related to renewables, including one with a Green Party member of the German Bundestag and another in GTZ headquarters. There he worked on an international study on the use of biofuels for transportation.

Ulrich Laumanns has studied political science at the Free University in Berlin, Germany. He was working for the Secretariat of the International Conference for Renewable Energies (renewables 2004). After that, he worked on issues of energy and development in the Division Environment and Infrastructure of Deutsche Gesellschaft für Technische Zusammenarbeit (GTZ). At the moment, Ulrich is employed as a GTZ energy expert in the Energy Advisory Project (EAP) in Uganda.

Kusum Lata is with The Energy Research Institute (TERI) in India. Kusum has been involved in extensive research work on biomass energy in Asia. TERI is one of the apex institutions on biomass energy issues in Asia. Kusum has participated in numerous projects, and has co-authored several research papers on the energy sector in Asia.

Joachim Luther is director of the Fraunhofer ISE and professor of solid-state physics and solar energy at the University of Freiburg, Germany. His key areas of research and development are solar buildings, photovoltaics, off-grid electrification and fuel cell technology. Joachim acts as coordinator of the energy activities of the Global Research Alliance (GRA).

Kes McCormick is a research associate and PhD candidate at IIIEE at Lund University, Sweden. He holds a Master's degree in environmental science from Monash University, Australia, and a Bachelor of Arts with Honours from Melbourne University, Australia. His research concentrates on overcoming barriers to bioenergy implementation and the role of local governance.

Susan McDade is the UN Resident Coordinator and the United Nations Development Programme (UNDP) Resident Representative in Cuba. Prior to moving to Cuba in 2005, she was the Manager of the Sustainable Energy Programme in UNDP, based in New York. While in New York, she worked on the definition of UNDP's overall programme approach and policy framework linking energy activities to sustainable development. From 1990–96, Ms McDade worked in UNDP country offices in Guatemala and China, managing capacity-building activities in the social sectors and subsequently in energy and the environment. Ms McDade is a development economist with a Master's of Development Studies in Economic Policy and Planning from the Institute of Social Studies in the Netherlands.

Jim Mackenzie is a senior associate in the World Resources Institute's (WRI's) Programme in Climate, Energy and Pollution. Dr MacKenzie is a Professorial Lecturer in the School of Advanced International Studies of the Johns Hopkins University and a Visiting Fellow in the Woodrow Wilson National Fellowship Foundation Programme.

Axel Michaelowa is head of the International Climate Policy Programme at the Hamburg Institute of International Economics, Germany, and Chief Executive Officer of the climate policy consultancy Perspectives. He has been analysing the Kyoto Mechanisms since their inception and consulting international organizations, governments and businesses on the design of rules and programmes to harness the benefits from the Clean Development Mechanism, Joint Implementation and emissions trading.

Lena Neij is associate professor at the IIIEE at Lund University, Sweden. Her research is focused on the dynamics of energy systems, methods for analysing technical change, renewable energy, energy efficiency and energy policy.

Sheila Oparaocha is the coordinator of ENERGIA, the International Network on Gender and Sustainable Energy. Her first degree is in veterinary medicine (University of Zambia) and her Master's degree is in gender and rural and regional development planning (Asian Institute of Technology). Sheila joined ENERGIA in 1999. Since then she has supported the programme's activities on advocacy, information exchange, regionalization and research.

Jonathan Pershing is the director of the Climate, Energy and Pollution Programme at the WRI, an independent non-profit organization providing technical analyses, policy recommendations and practical solutions to a variety of environmental problems. Prior to joining the WRI, he was the head of the Energy and Environment Division at the International Energy Agency in Paris.

Jem Porcaro is an Energy Policy Analyst with the Sustainable Energy Programme in UNDP. He has co-authored several publications on energy for sustainable development, including *Achieving the Millennium Development Goals: The Role of Energy Services* and *Energizing the MDGs: A Guide to Energy's Role in Reducing Poverty*. Prior to joining UNDP, Mr Porcaro worked as an environmental engineer for several international energy and environmental firms. He holds a Master's degree in International Energy Management and Policy from Columbia University and a Bachelor of Science in Civil and Environmental Engineering from Tufts University.

Ulrike Roehr is a civil engineer and sociologist. She has worked for many years as a teacher and project leader in solar energy training for women. Since 1994 she has been Director of the Frankfurt am Main branch of LIFE eV, which focuses on the integration of gender issues in environmental/ sustainability policy. One of the main fields of activity is research and information on gender aspects in sustainable energy and climate change in industrialized countries.

Janet L. Sawin, PhD, is director of the Worldwatch Institute's Energy and Climate Change Programme. She has earned Master's and Doctoral degrees in energy and environmental policy from the Fletcher School of Law and Diplomacy, Tufts University. Her doctoral thesis, completed in September 2001, examined the impact of government policy on the development and diffusion of renewable energy technologies.

Richard Schlirf is a partner at Macroconsulting, a consultancy based in Buenos Aires and Brussels, specializing in utility economic regulation. He studied economics at the Université Libre de Bruxelles in Belgium and then worked for several years in the fields of financial engineering and new technologies in Europe and the US. At Macroconsulting he specializes in financial and economic modelling, contract negotiation and in the design and implementation of regulatory information.

Frederik Schutyser (LLM (University of Stellenbosch, South Africa) and MSc (University of Warwick, UK)) is Project Manager, Biodiversity Analysis and Indicators at the European Environment Agency in Copenhagen, Denmark. At the time he contributed to this publication, he was a Programme Officer with

IUCN – The World Conservation Union. Previously he has worked with the Energy Charter Secretariat in Brussels, and at the University of Leuven (Belgium).

Niklas Sieber, PhD, is a transport economist and regional planner with more than 15 years of professional experience in spatial planning and transport economics in industrialized and developing countries. He worked for seven years as a research assistant at the University of Karlsruhe, where he was involved in many commercial studies on transport. Later he was nominated to be deputy executive secretary of the International Forum for Rural Transport and Development in London, where he worked on rural transport in developing countries. Returning to Germany, Niklas became technical adviser for transport with the German Technical Co-operation GTZ. Presently he is a self-employed consultant.

Virginia Sonntag-O'Brien is managing director of Basel Agency for Sustainable Energy (BASE), a non-profit foundation and UNEP Collaborating Centre that aims to mobilize investment in energy efficiency and renewable energy in both industrialized and developing countries (www.energy-base.org). BASE provides a platform to bring together members of the business and finance communities, and develops and promotes practical tools to facilitate investment.

Achim Steiner is Executive Director of the United Nations Environment Programme (UNEP). Mr Steiner has a BA from the University of Oxford (Philosophy, Politics and Economics) and a MA from the University of London (Economics and Regional Planning). He also studied at the German Development Institute in Berlin as well as the Harvard Business School. Previously he has acted as Secretary General of the World Commission on Dams, and Director General of IUCN – The World Conservation Union.

Minoru Takada is the acting Sustainable Energy Programme Manager in the Energy and Environment Group of UNDP in New York, where he has been posted since 1998. From 1996 to 1998, Minoru was posted with UNDP in Angola where he was the focal point for energy and environment issues. Before joining UNDP, Minoru spent three years, from 1990 to 1993, in Ghana as a volunteer teacher, where he also assisted in youth mobilization and community development projects. He holds a PhD in Renewable Energy Applications from the University of Mie (Japan) and a Master's in Nuclear Engineering from the University of Hokkaido (Japan).

Suani Teixeira Coelho is a professor and thesis adviser at the Energy Graduation Programme of the University of São Paulo. She is also the Deputy Secretary for the Environment of São Paulo State and also Executive Secretary of CENBIO –

the Brazilian Reference Centre on Biomass. Suani has coordinated technical, economic, environmental and institutional studies on biomass for energy subjects with Brazilian and foreign institutions, including Federal and State governments. She has participated in the development of the Brazilian Energy Initiative, and published several papers and attended national and international conferences.

Wim C. Turkenburg, trained as a physicist, is professor and head of the Department of Science, Technology and Society (STS) at Utrecht University, the Netherlands. He is also the director of the Copernicus Institute for Sustainable Development and Innovation at Utrecht University and scientific director of the Utrecht Centre for Energy research (UCE). Recently he was also chairman of the Sub-Committee on Energy of the UN Committee on Energy and Natural Resources for Development (UN-CENRD) and convening lead author of the chapter on Renewable Energy Technologies of the World Energy Assessment (WEA), published in 2000.

Dieter Uh has studied electrical engineering in Karlsruhe and Berlin. He has worked as a teacher in mathematics and physics, and later as a planning engineer for energy concepts. From 1999 to 2003 he worked as scientific consultant to the German Parliament (Deutscher Bundestag); in 2003 he joined the team for the renewable 2004 conference. Since 2005 he has been involved in the follow-up process of this conference with the establishment of the 'Renewable Energy Policy Network for the 21st Century' (REN21).

Eric Usher is head of the Renewable Energy and Finance Unit, and deputy coordinator of the Energy Branch within UNEP's Paris-based Division of Technology, Industry and Economics (www.uneptie.org/energy). As part of the organization's overall effort to link the global energy and environment agendas his work focuses on new approaches to financing renewable energy and energy efficiency enterprises, and new means for delivering clean energy services in regions without access.

Thomas Wälde (Dr iur (Frankfurt); LLM (Harvard)) is Jean Monnet Chair for EU Economics and Energy Law (Centre for Energy, Petroleum and Mineral Law and Policy, University of Dundee, UK). From 1980 to 1990 he was the principal UN adviser on international investment, natural resources and energy.

Foreword

The Bonn Conference for Renewable Energies was a watershed event, bringing together a diverse range of technical and policy experts, political representatives and members of civil society who wanted to share their knowledge and discover the path to a sustainable energy future. More than 3600 participants saw how far some technologies and countries have come and how far we still have to go to reach the goal of a global economy running on clean energy.

We have also seen how far some prices can go. At the time of the conference a barrel of petroleum cost US$50, but traded a year later for up to $70 on the back of steep demand from India and China and one of the most severe natural disasters the US has ever experienced – Hurricane Katrina.

Since the conference – and hopefully in some way because of it – there is a growing acceptance that our addiction to fossil fuels makes our economies less secure and increases the likelihood that economic costs and human misery will only increase from global climate change – change that is predicted to include more frequent and more intense extreme weather events such as Katrina.

And yet for me, the most impressive part of the conference was the realization that the shift to a clean energy economy comes with enormous *opportunity* – new industries, new jobs, new wealth and greater national and international security. This shift and these opportunities can also help lift the burden of poverty from the 2 billion of our fellow citizens who survive today on less than $2 per day and must burn wood and wastes for poor and polluting energy services.

From generating clean electricity in large, offshore windfarms to small biogas digesters for a single house, the conference delegates left few stones unturned to answer the question 'Why not sustainable energy?' Some of the answers were surprising: nearly 20,000 Indian families have solar lighting because they were able to borrow money from an innovative loan programme to buy them. Some are incredibly simple: hand-cranked radios and cellphones that dramatically 'empower' local communities, and biodiesel made from local plants. And some were just impressive: the annual 30 per cent growth of the wind energy industry to 50,000 megawatts (MW) in just three decades.

Of course, some answers will only be known when the right policy meets the right technology. Getting those 'rights' together is now not just a desirable outcome, but economically imperative and environmentally urgent. For this, one outcome of the conference can help. REN21 is a global policy network designed to help rapidly expand the use of renewable energy in developing and industrial countries by bolstering the development of the best policies and decision-making at the local, national and international levels.

In these chapters I hope you find added knowledge and inspiration to continue on this necessary and intriguing path. Although we have made substantial progress, we can and must remind ourselves that sustainability is not simply a process of doing things better. It is instead a *destination* – a place where our energy demands are satisfied through resources and technology that do not leave our children and theirs with a polluted world that cannot adapt to climate change.

I hope we meet again on this exciting adventure!

Klaus Toepfer
Ex-Executive Director
United Nations Environmental Programme

List of Acronyms and Abbreviations

ADEME	Environment and Energy Management Agency (France)
AIJ	'Activities Implemented Jointly'
AIM	Alternative Investments Market (London Stock Exchange)
APEC	Asia-Pacific Economic Cooperation
ASEAN	Association of Southeast Asian Nations
bbl	barrel (of oil)
BEDP	Bagasse Energy Development Programme (Mauritius)
BEERECL	Bulgarian Energy Efficiency and Renewable Energy Credit Line
BtL	biomass to liquids
CABEI	Central American Bank of Economic Integration
CAES	compressed air energy storage
CAREC	Central American Renewable Energy and Cleaner Production Facility
CBO	Congressional Budget Office (US)
CD	capacity development
CDM	Clean Development Mechanism (Kyoto Protocol)
CEO	chief executive officer
CER	Certified Emission Reduction
CHP	combined heat and power
CNG	compressed natural gas
CO_2	carbon dioxide
CRW	combustibles and renewable waste
DFI	development finance institution
DME	dimethyl ether
DNA	Designated National Authority
DNES	Department of Non-conventional Energy Sources (India)
DOE	Designated Operational Entity
EB	Executive Board (of the Clean Development Mechanism)
EBRD	European Bank for Reconstruction and Development
ECA	export credit agency
EGAT	Electricity Generating Authority of Thailand
EIT	economy in transition
EJ	exajoule
ERU	emission reduction unit
ESMAP	Energy Sector Management Programme (World Bank)

ETBE	ethyl tertiary-butyl ether
ETEF	Empowerment Through Energy Fund
EU	European Union
EVA	Austrian Energy Agency
FAME	fatty acid methyl ester
FI	financing institution
FSU	former Soviet Union
GATT	General Agreement on Tariffs and Trade
GDB	Global Development Bond
GEF	Global Environment Facility (United Nations Environment Programme)
GEIX	Global Energy Innovation Index
GHG	greenhouse gas
GNESD	Global Network on Energy for Sustainable Development
GREFF	Global Renewable Energy Fund of Funds
Gtoe	gigatonnes of oil equivalent
GTZ	Deutsche Gesellschaft für Technische Zusammenarbeit (German Technical Cooperation)
GVEP	Global Village Energy Partnership
GW	gigawatt
HVAC	heating, ventilation, air conditioning
HVDC	high-voltage direct current
IAEA	International Atomic Energy Agency
IBT	improved biomass technology
ICT	information and communication technologies
IEA	International Energy Agency
IET	International Emissions Trading (Kyoto Protocol)
IIASA	International Institute for Applied Systems Analysis
IPCC	Intergovernmental Panel on Climate Change
IPP	independent power producer
IPSE	Intergovernmental Panel on Sustainable Energy (proposed)
IREDA	Indian Renewable Energy Development Agency
ITDG	Intermediate Technology Development Group
IUCN	World Conservation Union
JI	Joint Implementation (Kyoto Protocol)
JREC	Johannesburg Coalition for Renewable Energy
KCJ	Kenya Ceramic Jiko
Ksh	Kenyan shilling
LAC	Latin America and the Caribbean
LDC	less developed country
LPG	liquefied petroleum gas
MBT	modern biomass energy technology
MDG	Millennium Development Goal

MMSD	Mining, Minerals and Sustainable Development Project
MNES	Ministry of Non-conventional Energy Sources (India)
MRET	Mandatory Renewable Energy Target (Australia)
MTBE	methyl tertiary-butyl ether
Mtoe	megatonnes of oil equivalent
MW	megawatt
MWe	megawatt electrical
NFFO	Non-Fossil Fuel Obligation (United Kingdom)
NGO	non-governmental organization
NO_x	oxides of nitrogen
NREA	New and Renewable Energy Agency (Egypt)
NREL	National Renewable Energy Laboratory (US)
ODA	Official Development Aid
OE	Operational Entity
OECD	Organisation for Economic Co-operation and Development
Ofgem	Office of Gas and Electricity Markets (United Kingdom)
OLADE	Organización Latinoamericana de Energía
OPEC	Organization of the Petroleum Exporting Countries
OPIC	Overseas Private Investment Corporation
OTEC	ocean thermal energy conversion
OWC	oscillating water column
PCF	Prototype Carbon Fund
PDD	Project Design Document
PES	public electricity supplier(s)
PPA	power purchase agreement
ppmv	parts per million by volume
PPO	pure plant oil
PRSP	Poverty Reduction Strategy Paper
PSE	producer subsidy equivalent
PTC	production tax credit
PURPA	Public Utilities Regulatory Act 1978 (US)
PV	photovoltaic(s)
R&D	research and development
RD&D	research, development and demonstration
RE	renewable energy
REC	renewable energy certificate
REEEP	Renewable Energy and Energy Efficiency Programme (REEP)
REILP	Renewable Energy and International Law Project
REN21	Renewable Energy Network for the 21st Century
RES	renewable energy source(s)
RESCO	rural energy service company
RES-E	Electricity from Renewable Energy Source
RET	renewable energy technology

RIGES	Renewables-Intensive Global Energy Scenario
RO	Renewables Obligation (England and Wales)
ROC	Renewable Obligation Certificate
ROE	return on investment
RPS	Renewables Portfolio Standard
RTE	renewable technology
SDHW	solar domestic hot water
SEFI	Sustainable Energy Finance Initiative
SEWA	Self-Employed Women's Association (India)
SHS	solar home system
SHW	solar hot water
SI	Shell International
SMEs	small and medium-size enterprises
SNG	substitute natural gas
STE	solar thermal electricity
SWERA	Solar and Wind Resource Assessment
TBE	traditional biomass energy technology
TBP	Thematic Background Paper
TREC	tradable renewable energy certificates
TWh	terawatt-hour
UNDP	United Nations Development Programme
UNEP	United Nations Environment Programme
UNFCCC	United Nations Framework Convention on Climate Change
VAT	value added tax
WBGU	German Advisory Council on Global Change
WCD	World Commission on Dams
WEC	World Energy Council
WEHAB	Water, Energy, Health, Agriculture and Biodiversity
WSSD	World Summit for Sustainable Development
WTO	World Trade Organization

Introduction

The International Conference for Renewable Energies (Renewables 2004) that took place in Bonn, Germany, in June 2004 marked a major highlight in the growing worldwide discussion on the role of renewable energies in the future global energy mix. In many countries all over the world there is a growing awareness of the opportunities and potentials that renewable energy (RE) might have in the short and in the long term. Nevertheless, tapping these potentials needs political will and action. This will have to include the establishment of favourable framework conditions, the involvement of the finance and business sectors in order to enable investment in RE, and the development of capacities both in industrialized and in developing countries.

Taking a look at the history of energy systems, the 'renewable energy age' is not a new phenomenon. It already existed before fossil fuels were discovered and was used on an increasingly large scale in the context of industrialization. Over the subsequent decades, different types of coal were discovered; then oil was added to the mix as a very user-friendly type of energy source, especially for mobile purposes. In the 20th century, natural gas became an increasingly important source of energy for the modern economies of the world. Nuclear energy, which was introduced in the 1950s, is, in contrast, a very complex technology that requires large investments. It can therefore only be a technology option for industrialized and politically stable countries if the risks are accepted by societies.

Now, some centuries after the starting point of the fossil fuel age, the world is faced with the negative consequences of the present energy system: carbon dioxide (CO_2) concentration in the atmosphere is constantly increasing and the effects of global warming are becoming ever more evident. As the Intergovernmental Panel on Climate Change (IPCC) points out in its third Assessment Report, there is an urgent need for action in order to achieve a low-carbon global energy system.

After decades of growth in energy consumption and with the rapidly increasing energy demand in transition and developing countries, the 'mid-depletion point' of oil (and gas) resources is near. Given the relevance of energy for economic development and welfare, the following facts underline the necessity for transition of the global energy system.

One-third of humankind still lives in poverty – which also means energy poverty. Around 2 billion people have no access to modern forms of energy. The lack of a sufficient and affordable modern energy supply limits the opportunities for the poor to improve their quality of life and contributes to the increase of migration from rural areas to the cities.

The global implementation of RE technology can contribute significantly to addressing these problems. However, there are some key issues that need to be considered. Generally, the introduction of any new, innovative technology in an existing market requires some preconditions:

- There has to be a demand for the new product through an added value in terms of comfort or economic, social and environmental benefits.
- The product must be affordable in relation to income or investment.
- A structure of support for the product is necessary: information, awareness, maintenance, quality standards, improvement by research and development (R&D).

Therefore, the introduction of new technologies rests on three pillars:

1 The policy framework must be favourable, as policy not only has to consider today's markets but also needs to take into account long-term developments, as much on a national as on an international level.
2 Financing for the new technology must be available; it is therefore essential that investments are commercially attractive in order to mobilize resources from private companies and financing institutions.
3 A supporting environment is crucial, including capacities for information, awareness-raising, R&D, after-sales service and know-how on the technologies, from primary education up to universities.

This book constitutes a review of the different determining factors that drive the worldwide dissemination of renewable energy technologies. With a clear emphasis on policy and action, contributions from outstanding international experts on renewable energy combine to form a holistic picture of the current status, impacts and future potential of renewable energies. Each chapter deals with a different issue, including detailed information on social, environmental, political, economic and technological aspects.

To start with, José Goldemberg makes the case for REs by outlining their benefits and emphasizing the need to increase RE worldwide. This chapter is followed by Thomas Johansson, Kes McCormick, Lena Neij and Wim Turkenburg's contribution in which the authors demonstrate the huge potential of RE and explain the state of the art of the various renewable energy technologies. Dirk Aßmann, Niklas Sieber and Ricardo Külheim describe the situation of the transport sector and analyse the potential role of biofuels in view of high oil prices and the relevance of transport for climate change. In this sense the chapter represents a critical contribution to the debate on sustainability in the transport sector – in terms not only of renewable but also of 'conventional' options.

In Part 2 of the book, on policies, Janet Sawin and Christopher Flavin describe and compare strategies and instruments for the promotion of RE on the national policy level: feed-in tariffs versus quotas, tax relief, subsidies, loans and renewable energy portfolio standards. An increasing number of countries have developed a range of valuable experience in the utilization of these instruments, allowing the authors to give some interesting examples for best practices. Franco Fugazza and Richard Schlirf on their part reflect on the importance of a functioning and favourable regulation for the deployment of renewable energies, supporting their arguments with case studies. An important matter for concern, and not only for RE, is the subsidies for fossil fuels that impede the creation of a level playing field. Jonathan Pershing and Jim Mackenzie scrutinize the rationale for subsidies and consider its effects, outlining the need for policy reform.

With the contribution from Achim Steiner, Thomas Wälde, Adrian Bradbrook and Frederik Schutyser we touch on another key aspect of energy policy: the question of how to organize an international driving force for the promotion of renewable energies. Against the background of the proposal of an International Renewable Energy Agency (IRENA) from the World Council for Renewable Energy (WCRE), the chapter discusses the state of international cooperation and potential future developments.

The third part of the book focuses on financing aspects of RE. Virginia Sonntag-O'Brien and Eric Usher start off by describing the investment conditions in the energy sector in general, then elaborating on the specific investment conditions for RE. Their proposal on how to develop financing solutions for RE constitutes a very interesting new approach. The Kyoto Protocol opens a window of opportunity for additional investments in RE: Axel Michaelowa, Matthias Krey and Sonja Butzengeiger explore the possibilities for RE projects deriving from Joint Implementation (JI) and the Clean Development Mechanism (CDM).

Part 4 places the emphasis on developing countries. In their contribution on RE, Susan McDade, Minoru Takada and Jem Porcaro assess its benefits of for developing countries and the role of development cooperation in the promotion of RE. Stephen Karekezi, Kusum Lata and Suani Teixeira Coelho offer some insights on a very difficult and often forgotten issue: the traditional use of biomass and its corresponding challenges, such as the need to improve its use and move to modern energy. Joy Clancy, Sheila Oparaocha and Ulrike Roehr examine the impact of energy issues on gender equity. Their chapter addresses gender aspects of household energy, and reflects on how different renewable energy technologies can contribute to reduce workload and save time, particularly for women.

The final part of the book looks into the issue of capacity development for RE. Since RE technology is still relatively new and underdeveloped, the need for capacity development in this area (including training, education, awareness creation and R&D) is considerable. In his contribution, John Christensen describes the history of capacity development in the energy sector with regard to the situation both in developed and in developing countries. Carsten Agert and Joachim Luther highlight the role of R&D in improving renewable energy technologies and reducing their costs. Although some of these technologies are not competitive today, they will be in the future. We know that the cost curve of fossil fuels will rise, whereas the specific costs of renewable energy technologies are continuously declining.

Most of the chapters have been written as Thematic Background Papers in preparation for the Renewables 2004 conference. For this publication they have been updated and revised in order to reflect recent developments and provide actual data. The editors, who have all been involved in the thematic preparation of renewable 2004, would like to thank the authors and the publisher, Earthscan, for their contribution to this collection, and for the good and constructive cooperation. We hope to have compiled an interesting text for the whole 'renewable energy community' and beyond. We wish all readers new insights, new ideas and interesting further discussions.

Dirk Aßmann
Ulrich Laumanns
Dieter Uh

PART ONE

BASICS

1

The Case for Renewable Energies

José Goldemberg

Introduction

All energy used by humankind originates from one of the following sources: radiant energy emitted by the sun (solar energy and its derivates); geothermal energy from the interior of earth; tidal energy originating from the gravitational pull of the moon; and nuclear energy.

By far the largest source is solar energy, thousands of times larger than all the others and inexhaustible for as long as the sun shines (approximately 4.5 billion years). Table 1.1 compares the energy available on earth with present energy consumption and fossil fuel reserves.

Long availability is not the only criterion by which to judge an energy source. The way it is converted into forms that meet our needs, environment and health issues – at the local, regional and global level – the problem of guaranteeing

Table 1.1 *Energy available on earth*

Type of source	× 10¹² watts
Solar	174,000
Geothermal	32
Tidal	3
Present world energy consumption*	12.7
Photosynthesis	40
Winds, waves, convection and currents	370
Fossil fuel reserves (mainly coal)	≈ 2000

Note: The world's average primary energy consumption (2001) is 2.1 kW per capita.
Source: Hubbert (1971)

energy security, and the overriding social issues such as the connection between energy and poverty, employment generation and gender also have to be addressed.

We will show in this chapter that the present energy system, essentially based on the use of fossil fuels (not renewable), cannot handle these problems very well and that increasing the share of renewable energies is one of the best ways of addressing them. We will specifically address the following aspects:

1   the contribution of renewable energy to mitigating climate change;
2   innovation, local markets and employment generation;
3   diversification of energy supply, energy security and prevention of conflicts over natural energy resources;
4   poverty reduction through improved energy access and gender aspects;
5   health-related impacts (local air pollution, indoor air pollution);
6   positive spill-over effects to other sectors and further benefits.

In the concluding section some recommendations will be made.

RENEWABLE ENERGY: SOME CHARACTERISTICS

Solar energy manifests itself as low-temperature solar heat, high-temperature solar heat, wind electricity and photovoltaics. Low-temperature solar heat is produced by the absorption of sunlight by darkened surfaces that convert it into heat, which can be used for warming water or other fluids. High-temperature solar heat can be obtained by focusing sunlight and heating fluids to a high temperature so that they can be used to generate electricity. The warming of the atmosphere by solar heating leads to turbulence manifested as wind, which can be used to generate electricity. Photovoltaics is the direct conversion of the ultraviolet component of sunlight into electricity in appropriate surfaces. These forms of energy are all, by definition, renewable.

Geothermal energy manifests itself in the form of hot water or vapour and can be used for heat or electricity production in some specific regions. It is generally considered to be a renewable energy source. Tidal energy can be used to generate electricity in some coastal areas and is also a renewable source of energy.

Hydropower is indirectly linked to sunlight, which evaporates the water in the oceans, which then precipitates on land masses as rain, and forms rivers. Dams are built on the resulting rivers to create reservoirs, which guarantee a steady supply of water for electricity generation. A large dam is a dam with a height of 15 metres or more, measured from the foundation. Small dams are smaller than that or have no storage reservoir. Usually they produce less than 10 megawatts (MW) of electricity. Large hydropower plants flooding large areas might displace people and have undesirable ecological or social impacts, so they are considered to be a non-renewable resource by some. Small hydropower plants are usually not affected by such problems.

A small part of the solar energy reaching the earth is converted by photosynthesis into biomass (organic matter). Some of this matter was buried in the distant past (hundreds of millions of years) by sedimentation and earthquakes and transformed by bacterial action into coal, oil and gas, which constitute the present fossil fuel resources (which are not renewable). Biomass is usually regarded as a renewable energy except when leading to deforestation.

BOX 1.1 BIOMASS

Biomass can be used in two ways:

- Traditional or non-commercial biomass is unprocessed biomass-based fuels, such as crop residues, fuelwood and animal dung. Such types of biomass are used frequently for cooking and heating in many developing countries, but they are of very low efficiency. Although traditional energy sources can be used renewably, they frequently lead to deforestation. This is why programmes to develop and disseminate improved biomass stoves in many African countries, China and India are so important.
- Modern biomass is biomass produced in a sustainable way and used for electricity generation, heat production and transportation (liquid fuels). It includes wood/forest residues from reforestation and/or sustainable management, energy crops, rural (animal and agricultural) and urban residues (including solid waste and liquid effluents), excluding the traditional uses of fuelwood in inefficient and polluting conversion systems. Most of the biomass used in Organisation for Economic Co-operation and Development (OECD) countries falls in this category.

The most important energy source in many developing countries is renewable only if realistically replaced. Much biomass use in developing countries, either domestic small-scale use or large-scale use for industrial purposes, is leading to deforestation. Moreover, biomass use for cooking and heating in developing countries is a major cause of serious indoor pollution, particularly for women, small children and the elderly. On the other hand, in OECD countries most biomass used is 'modern biomass', from wood plantations, wood, or urban or rural residues.

THE ADVANTAGES OF RENEWABLES

The contribution of renewable energy to mitigate climate change

The present energy system is heavily dependent on the use of fossil fuels. Worldwide, coal, oil and gas accounted for 78 per cent of primary energy consumption in 2002. Fossil fuel combustion is the prime source of carbon dioxide (CO_2) emissions, which are growing at the rate of 0.5 per cent per year. Present levels have reached 377 parts per million by volume (ppmv) (CDIAC, 2005), up from 278 ppmv at the dawn of the industrial revolution two centuries ago.

Emissions of anthropogenic greenhouse gases, mostly from the production and use of energy, are altering the atmosphere in ways that are affecting the climate. As stated in the Third Assessment Report of the Intergovernmental Panel on Climate Change (IPCC), there is new and stronger evidence that most of the warming observed over the past 50 years is attributable to human activities, and that significant climate change would result if 21st-century energy needs were met without a major reduction in the carbon emissions of the global energy system during this century. Current CO_2 emission trends, if not controlled, will lead to more than a doubling of atmospheric concentrations before 2050, relative to pre-industrial levels. Changes have already been observed in climate patterns that correspond to scientific projections based on increasing concentrations of greenhouse gases.

This is a serious challenge to sustainable development, and the main strategies to prevent it are:

* more efficient use of energy, especially at the point of end use in buildings, transportation and production processes;
* increased reliance on renewable energy sources;
* accelerated development and deployment of new and advanced energy technologies, including next-generation fossil fuel technologies that produce near-zero harmful emissions.

The relative importance of these options and the order in which they become relevant depends on the stage of development of the region as well as availability of natural resources and technology. There are, however, important differences in the energy systems of OECD countries as compared with those of developing countries, as indicated in Table 1.2.

In the OECD countries, which have reached a very high level of development, gains in energy efficiency have been the main strategy followed. In developing countries, where renewables (mainly biomass) are already very important (20.0 per cent), albeit used in inefficient ways, modernization of the way they are used seems the better strategy to follow. In these countries, energy consumption is growing 2.3 times more rapidly than in OECD countries, so

Table 1.2 *Characteristics of energy systems, 2002*

	Fossil fuels (%)	Renewable energy (%)	Nuclear (%)	Growth rate, all sources (%/year, 1971–2000)
OECD	83.0	5.9	11.1	1.48
Developing countries	78.1	20.0	1.9	2.81

Source: IEA (2005)

there is ample space for innovation as the energy system grows. In all cases the increased use of renewables, which are carbon-free (or neutral), will contribute to reduction in CO_2 emissions and thus mitigate climate change.

Innovations, local market and employment generation

The rapidly growing renewable energy industries and service sectors in many countries show clear evidence that the systematic promotion of such new technologies offers great opportunities for innovation, for the development of energy markets with locally or regionally oriented value chains, and thereby for the creation of new jobs with very different qualification requirements. While the development and deployment of new state-of-the-art renewable energy technologies, such as wind or photovoltaic energy, require highly skilled, knowledge-intensive workforces in industrialized countries, developing countries can, for instance, benefit economically from an increased use of improved biomass-based energy generation, both in terms of better availability of energy for productive use and through the provision of energy services as such. Examples are the widespread use of improved wood and charcoal cooking stoves in Kenya and other African countries as well as the production of ethanol – an excellent substitute for gasoline in the Otto cycle in internal combustion engines – from sugar cane in Brazil.

Generally speaking, renewable energies are important for local employment and income generation, which result from manufacturing, project development, servicing and, in the case of biomass, rural jobs for the biomass production.

Usually renewable energy devices are decentralized, are modular in size and have low operating costs in addition to involving short construction times, which means much greater flexibility in energy planning and investment. Table 1.3 provides an idea of the number of jobs per unit of energy generated from different sources.

These numbers were obtained from a variety of sources and include jobs involved in operating the generating stations as well as the jobs involved in producing and maintaining the equipment. Photovoltaic energy is usually generated (and used) in small modules of 100 watts, and the generation of 1 terawatt-hour (TWh) would require typically 10 million modules to be installed and maintained. This is the reason why a large number of jobs are created. Ethanol production involves large plantations of sugar cane, which again explains why numerous jobs are generated.

The main beneficiaries of the adoption of renewable sources of energy will be the developing countries, where biomass, and particularly fuelwood, is used widely with very inefficient and wasteful technologies for cooking and heating. In such countries the modernization of the use of biomass could bring great benefits, including – among others – a reduction in deforestation.

Table 1.3 *Direct jobs in energy production*

Sector	Job-years/Mtoe (fuel production)	Job-years/terawatt-hour (fuel production + power generation)
Petroleum[a]	396	260
Offshore oil[a]	450	265
Natural gas[a]	428	250
Coal[a]	925	370
Nuclear[b]	100	75
Wood energy[c]		733–1067
Hydro[d]		250
Minihydro[e]		120
Wind		918[e]–2400[f]
Photovoltaics		29,580–107,000[e]
Bioenergy (from sugar cane)		3711–22,5392

Notes:
a The staff level for operation of a 1350-megawatt (MW) nuclear power plant in the US, producing 9.45 TWh/year (or 2.138 million tonnes of oil equivalent (Mtoe)/year at an efficiency of 38 per cent, was 500 people.
b Electric generation based on herbaceous crops (5.5 direct jobs/megawatt electrical (MWe)) and on forestry crops (8 direct jobs/MWe), utilization 7500 h/year.
c World installed capacity for wind 17,300 MW, utilization 2000 h/year and 4.8 jobs/MW.
d Including 12 different activities to construct, transport, install and service 1 MW of photovoltaics (not including economies of scale between 2 kW and 1 MW), world installed photovoltaics capacity is 800 MW.
e Utilization of 1200 h/year; 35.5 jobs/MW (including 15 different activities to manufacture, transport, install and service 1 MW of wind power).
f Ethanol industry provides 33 direct jobs/million litres in Brazil, where ethanol production in the 1992–2001 period ranged between 10.6 and 15.4 billion litres/year (LHV of ethanol 6500 kcal/kg and density 0.8 kg/litre); energy production comprised 7 Mtoe of ethanol fuel, plus 9.6 TWh/year of cogeneration (installed capacity 2000 MW, utilization of 4800 hours/year).
Sources: [a] Grassi (1996); [b] Electric Power International (1995), Grassi (1996); [c] Grassi (1996); [d] Carvalho and Szwarcz (2001); [e] Perez (2001); [f] IEA (2002b)

Diversification of energy supply, energy security and prevention of conflicts about natural resources

Maintaining energy security in today's industrialized countries comes at high, but usually hidden, costs that find expression in military and security spending. The volatile world market prices for conventional energy sources, in particular oil, pose great risks for large parts of the world's economic and political stability, with sometimes dramatic effects on energy-importing developing countries. In this context, renewable energies can help to diversify energy supply and to increase energy security. They should increase the economic benefits that result from transformations in energy trading patterns. Additionally, in the mid- and long-term perspective renewable energies prolong the availability of most fossil fuels for the satisfaction of both energy needs and numerous other non-energy needs.

The potential for conflict, sabotage, disruption of production and trade of fossil fuels and fissionable materials cannot be dismissed. As far as electricity supply is concerned, this is dramatized by recent 'blackouts' in the eastern US, the United Kingdom and Italy, probably due to accidents, which are difficult to eliminate in highly centralized production and distribution systems. As far as oil is concerned, potential threats lead to sudden transient price increases (price spikes) that cause economic problems in many countries, and disrupt global economic growth.

The present energy system found in industrialized countries is heavily dependent on fossil fuels, which are geographically concentrated in a few regions of the world. Dependence on imported fuels leaves many countries vulnerable to disruption in supply, which might pose physical hardships and an economic burden for others; the effect of fossil fuels imports on their balance of payments is unsustainable for some countries (Box 1.2).

To reduce such dependence is a high priority in many countries, particularly as regards oil imports in developing countries, which frequently spend a large fraction of their foreign currency earnings on oil imports. Just to give an example: oil imports consume half of all export earnings in Barbados, and a similar situation is widespread among oil-poor countries. To increase the share of indigenous renewable energy in their system is an important step in solving this problem.

Poverty reduction through improved energy access and gender aspects

The enhanced use of renewables is closely linked to poverty reduction and elimination, since energy services can:

- improve access to pumped clean water for drinking and for cooking food to reduce hunger (95 per cent of food needs cooking);
- reduce the time spent by women and children on basic survival activities (gathering firewood, fetching water, cooking, etc.);
- provide lighting that permits home study, increases security and enables the use of educational media and communication in school; and
- reduce deforestation.

More than 2 billion people cannot access affordable energy services based on the efficient use of gaseous and liquid fuels, and on electricity, and are dependent on gathering fuelwood and fetching water. This lack of access to affordable energy constrains their opportunities for economic development and improved living standards. Women, the elderly and children suffer disproportionately because of their relative dependence on traditional fuels, and exposure to emissions from cooking is the main cause of respiratory disease in these groups. Access to electricity through transmission distribution lines is unlikely to be possible in

BOX 1.2 FLOW OF GULF OIL SUPPLIES, 2010

Almost two-thirds of the world's oil resources are in the Middle East, mostly in the Gulf region (the Islamic Republic of Iran, Iraq, Kuwait, Qatar, Saudi Arabia and the United Arab Emirates), although these six countries now account for only 26 per cent of global crude oil supplies (IEA, 2005). They are expected to double their share to 53 per cent in 2010 (UNDP, WEC and UNDESA, 2000). All OECD countries are expected to increase their dependence on oil imports over the next few years. Their imports, 56 per cent of their requirements in 2000, are expected to rise to 76 per cent in 2020. Asia-Pacific countries' crude oil imports are expected to increase to 72 per cent of their requirements in 2005 (up from 56 per cent in 1993). The Middle East is expected to account for 92 per cent of the region's imports. The Gulf region is expected to supply 18 million barrels a day to Asia-Pacific countries in 2010.

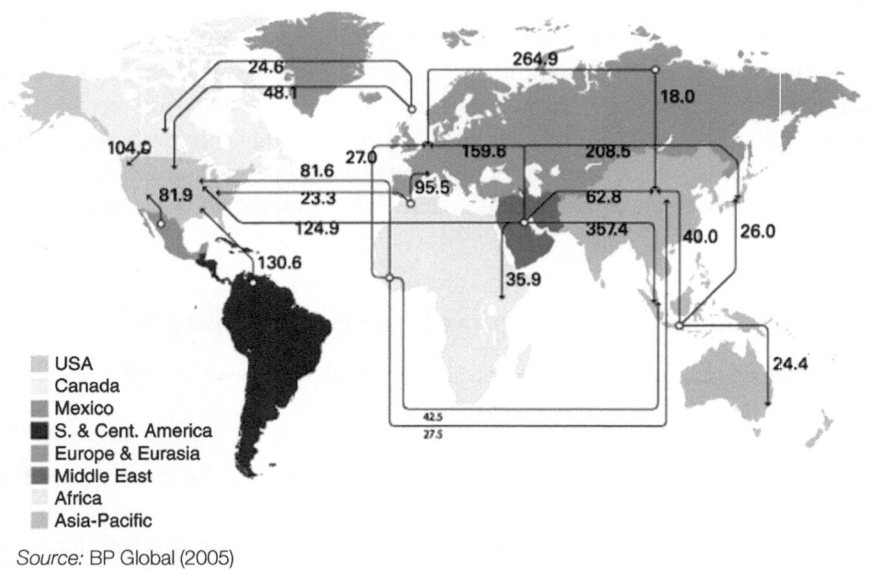

USA
Canada
Mexico
S. & Cent. America
Europe & Eurasia
Middle East
Africa
Asia-Pacific

Source: BP Global (2005)

Figure 1.1 *Major oil trade movements, 2004 (million tonnes)*

many parts of the world for a long time, so access to modern, decentralized small-scale energy technologies – particularly renewables – is an important element of successful poverty alleviation. The revenues from exported biofuels are another important element for alleviating poverty in developing countries.

Health-related impacts

The main pollutants emitted in the combustion of fossil fuels are sulphur and nitrogen oxides, carbon monoxide and suspended particulate matter. Ozone is formed in the troposphere from interaction among hydrocarbons, nitrogen oxides and sunlight. The environmental impacts of a host of energy-linked emissions – including suspended fine particles and precursors of ozone and acid deposition – contribute to local and regional air pollution and ecosystem degradation. Human health is threatened by high levels of the pollution resulting from fossil fuel combustion. At the local level, energy-related emissions from fossil fuel combustion, including in the transport sector, are major contributors to urban air pollution, which is thought to be responsible for some hundreds of thousands deaths annually around the world. At the regional level, precursors of acid deposition from fuel combustion can be precipitated thousands of kilometres from their point of origin – often crossing national boundaries. The resulting acidification is causing significant damage to natural systems, crops and human-made structures, and can, over time, alter the composition and function of entire ecosystems. Table 1.4 shows some of the consequences of the use of fossil fuels. Needless to say, renewables contribute relatively little to these emissions.

Positive spill-over effects to other sectors and further benefits

One of the problems with renewables is the fact that some of them are intermittent. This is indeed the case for photovoltaics, which requires sunshine, and for wind, which tends to be erratic in some locations. However, geothermal energy, small hydro schemes and especially biomass do not suffer from such shortcomings. In the case of photovoltaics, which is eminently suited for decentralized use in rural remote areas that cannot be reached by the electricity grid, the use of automobile or solar batteries for storage has proved to be a sensible and practical way of supplying electricity for lighting in the evenings when it is most needed, and the same applies to other uses such as for radio, television, communications and refrigeration. In the case of wind, the problem of intermittency can be solved by feeding the electricity generated into large grids, as is done in Denmark, Germany and the United Kingdom. Another

Table 1.4 *Environmental and health problems*

Insult	Human disruption index	Share of the human disruption caused by fossil fuel burning (%)
Oil added to oceans	10	44
Sulphur emissions to atmosphere	2.7	85
Nitrous oxide flow to atmosphere	0.5	12
Particulate emission to atmosphere	0.12	35
Carbon dioxide flow to atmosphere	0.05	75

Note: The human disruption index is the ratio of human-generated flow to the natural (baseline) flow.
Sources: UNDP, WEC and UNDESA (2003)

solution is the installation of hybrid systems with several sources of different energy profile.

One of the most striking successes of renewables is the ethanol programme in Brazil, where ethanol is produced from sugar cane and has replaced half the gasoline that would otherwise be consumed in the country (using roughly 4 million hectares of land). In this case, most of the energy needed for the processing of the raw material to the final product comes from the bagasse (the residue from sugar cane after the crushing process), requiring very little by way of 'external' sources of supply in the form of fossil fuels (in the agricultural part of sugar cane production). This is the reason why the energy balance for ethanol production in Brazil is 10:1 and the production of ethanol from corn in the US has an energy balance of approximately 1.5:1. The technological progress in this area, in both the agricultural and the industrial sectors, is striking and in Brazil, as well as economies of scale, has led to an impressive reduction in the costs of ethanol production. Indeed, Brazilian ethanol is presently competitive with gasoline in Rotterdam (Figure 1.2).

The Brazilian success story has led to intensive research to produce liquid fuels from cellulosic materials, a very promising area since the resource base of biomass is so widespread all over the world. Developed countries have a significant role to play in introducing biofuels into their energy systems, either directly or blended with gasoline or diesel.

Conclusions

The full potential and advantages of renewables are not clearly manifest at present because the costs of fossil fuels do not reflect their full cost. They are

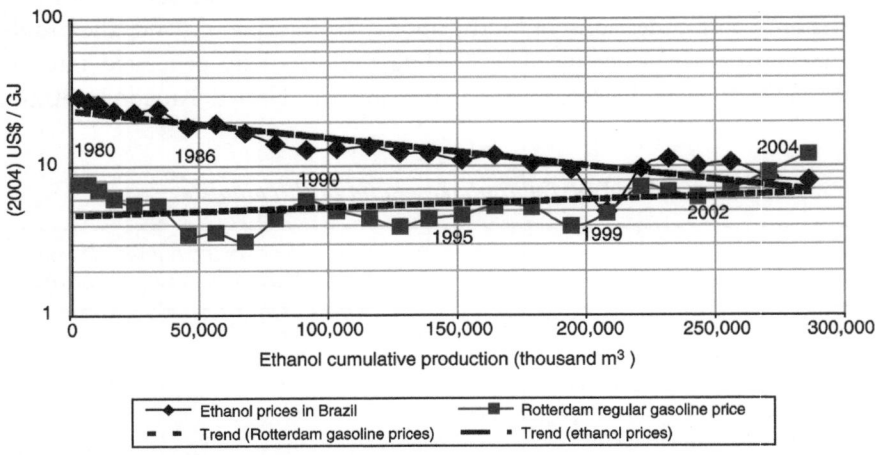

Source: Goldemberg et al (2003), updated to 2005

Figure 1.2 *Prices of gasoline in Rotterdam and ethanol in Brazil*

subsidized in several parts of the world, and the 'externalities' associated with them, such as additional health and environmental costs, are not considered. Removing subsidies from fossil fuels would make renewables competitive in many areas. Generally speaking, the use of renewables might benefit from bilateral and regional cooperation. After the Johannesburg World Summit for Sustainable Development (WSSD), a number of programmes to promote sustainable energy programmes in developing countries were presented to the United Nations Secretariat. Of these programmes, 23 had energy as a central focus and a further 16 would have a considerable impact on energy use. They included most prominently the DESA-led Clean Fuels and Transport Initiative; the UNDP/World Bank-led Global Village Energy Partnership (GVEP); the UNDP/LPG Association-led LPG Challenge; the EdF/ACCESS-led Alliance for Rural Energy in Africa (AREA); the EU Partnership on Energy for Poverty Eradication and Sustainable Development; and the UNEP-led Global Network on Energy for Sustainable Development (GNESD). Particularly important among them is the Johannesburg Coalition for Renewable Energy (JREC), to which more than 80 countries have adhered. The enhanced use of renewables, with its reliance on decentralized production, employment generation and reduction of environmental impacts, is a general characteristic of most of these programmes.

REFERENCES

BP Global (2005) 'Map of major oil trade movements', available at www.bp.com

Carbon Dioxide Information Analysis Center (CDIAC) (2005) http://cdiac.esd.ornl.gov/

Carvalho, L. C. and Szwarcz, A. (2001) 'Understanding the impact of externalities, case studies Brazil', International Development, Seminar on Fuel Ethanol, Washington DC

Goldemberg, J., Coelho, S. T., Nastari, P. M. and Lucon, O. (2003) 'Ethanol learning curve – The Brazilian Experience', *Biomass and Bioenergy*, vol. 26, no 3, pp301–304

Grassi, G. (1996) 'Potential employment impacts of bioenergy activity on employment', in Chartier, P. (ed) *Proceedings of the 9th European Bioenergy Conference*, vol 1, Elsevier, Oxford, pp419–423

Hubbert, M. K. (1971) 'The Energy Resources of the Earth', *Scientific American*, vol 60, p224

International Energy Agency (IEA) (2002a) *Energy Balances from Non-OECD Countries*, Paris, IEA

IEA (2002b) Renewable Energy . . . Into the Mainstream, IEA, The Netherlands, available at www.iea.org

IEA (2005) *World Energy Outlook*, Paris

Intergovernmental Panel on Climate Change (IPCC) (2003) 'Third Assessment Report – Climate Change 2001', available at www.ipcc.ch

Perez, E. M. (2001) *Energías renovables, sustentabilidad y creación de empleo: una economía impulsionada por el sol,* Catarata, Madrid, p270

Renewable Energy Policy Project (REPP) (2001) 'The Work That Goes into Renewable Energy', Research Report 13, Renewable Energy Policy Project, Washington, DC, available at www.repp.org

United Nations Development Programme (UNDP), United Nations Department of Economic and Social Affairs (UNDESA) and World Energy Council (WEC) (2000) *World Energy Assessment,* UNDP, New York, available at www.undp.org/seed/eap/activities/wea

UNDP, UNDESA and WEC (2003) *World Energy Assessment Overview,* update, in press

UNICA (2003) The São Paulo Sugarcane Agroindustry Union's website: www.unica.com.br

2

The Potentials of Renewable Energy

Thomas B. Johansson, Kes McCormick, Lena Neij and
Wim C. Turkenburg

INTRODUCTION

Renewable energy flows are very large in comparison with humankind's use of energy. Therefore, in principle, all our energy needs, both now and into the future, can be met by energy from renewable sources (see Box 2.1). Technologies exist that convert renewable energy flows to modern energy carriers or directly into desired energy services. Technological development during the past few decades has resulted in modern renewable energy supply becoming competitive in a number of situations. Further technological development and industrial learning will continue to bring costs down. When environmental costs and security of supply considerations are included, renewable energy has even wider markets. With decisive efforts to speed up the development and dissemination of renewable energy technologies and systems, all human energy needs could be met by rerouting a small fraction of naturally occurring renewable energy flows within a century.

Although natural flows of renewable resources are immense in comparison with global energy use – from both a theoretical and a technical perspective – the level of their future use will primarily depend on the economic performance of technologies utilizing these flows. Policies promoting the development and use of renewable energy sources and technologies can make a significant difference. Clearly, the long-term use of energy resources will probably become more an issue of the degree to which present and future societies have to balance environmental and economic trade-offs, and control greenhouse gas emissions, rather than a question of resource and technology existence. Furthermore, the growing problem of the availability of (cheap) fossil fuels will amplify energy security concerns.

Box 2.1 Definition of renewable energy

In a broad sense, the term 'renewable energy sources' refers to hydro energy, biomass energy, solar energy, wind energy, geothermal energy and ocean energy. The term 'new' renewables suggests a greater focus on modern and sustainable forms of renewable energy. In particular, these are modern biomass energy, geothermal heat and electricity, small-scale hydropower, low-temperature solar heat, wind electricity, solar photovoltaic and thermal electricity, and ocean energy (tidal, wave, current, ocean thermal, osmotic and marine biomass energy).

In 2004, renewable energy provided 17 per cent of global primary energy consumption, mostly traditional biomass, and about 19 per cent of electricity, mostly large-scale hydropower. 'New' renewables contributed only 2 per cent of the world's primary energy use. However, 'new' renewables, often based on indigenous resources, have the potential to provide energy services with low or zero emissions of both air pollutants and greenhouse gases.

Discussions on biomass as a source of energy are sometimes clouded by problems of definition. The term 'combustible renewables and waste' includes all vegetable and animal matter used directly or converted to solid fuels, as well as biomass-derived gaseous and liquid fuels, and industrial and municipal waste converted to modern energy carriers. The main biomass fuels in developing countries are firewood, charcoal, agricultural residues and dung, often referred to as traditional biomass. The major challenge facing biomass is to shift towards sustainable technologies and systems while reducing costs.

Source: WEA (2004); REN21 (2005)

A rapid expansion of energy systems based on renewable energy sources will require actions to stimulate the market in this direction. This expansion can be achieved by finding ways to drive down the relative cost of 'new' renewables in their early stages of development and commercialization, while still taking advantage of the economic efficiencies of the marketplace. Pricing based on the full costs of conventional energy sources (including phasing out subsidies and internalizing externalities) will make 'new' renewables more competitive. However, such measures remain controversial. In any case, barriers stand in the way of the accelerated development of renewable technologies, which can only be overcome by appropriate frameworks and policies.

This chapter comprises several sections. The first section discusses the theoretical and technical potentials of renewable energy resources and technologies with subsections on technology options and status, and the associated environmental and social issues. The second section explores the economic potentials of renewable energy, with a particular focus on cost reductions and technological development. The third section outlines a selection of scenarios that have been developed to illustrate future energy systems. The fourth section looks at the markets where renewable energy might compete and make a difference, particularly in the case of developing countries. The fifth section identifies the barriers that renewable energy innovations confront all along the innovation chain, and highlights the role played by industrialized countries in developing and disseminating novel technologies. The chapter concludes with policy implications and recommendations that relate to many sectors, including land use, agriculture, buildings, transportation and urban planning.

Renewable energy resources and technologies

The natural energy flows through the earth's ecosystem, and the theoretical and technical potential of what they can produce for human needs, exceed current energy use by many times, which in 2004 was approximately 470 exajoules (EJ) including traditional biomass (WEA, 2004; BP, 2005; REN21, 2005).[1] The long-term energy resource availability from the viewpoint of theoretical maximums is immense (see Table 2.1). Admittedly, it can be argued that an analysis based on recoverable resources is irrelevant because hydrocarbon occurrences or natural flows become resources only if there is demand for them and appropriate technology has been developed for their conversion and use. The appraisal of technical potential therefore takes into account engineering and technological criteria. In any case, the picture is clear: renewable energy resources will not act as a constraint on their development (WEA, 2000; Boyle, 2004; Hoogwijk, 2004; IEA, 2004; WEC, 2004).

In 2004, renewable energy sources supplied about 17 per cent of the world's primary energy use, predominantly traditional biomass, used for cooking and heating, especially in rural areas of developing countries (REN21, 2005). Large-scale hydropower supplied about 16 per cent of global electricity (BP, 2005). The scope to expand large-scale hydropower is limited in the industrialized world, where it has almost reached its economic capacity. In the developing world, potential still exists, but large hydropower projects often face financial,

Table 2.1 *Annual global renewable energy resources*

Energy resource	Energy use, 2001 (EJ/yr)	Technical potential (EJ/yr)	Theoretical potential (EJ/yr)
Hydro energy	9.4	50	150
Biomass energy	45.0	>250	2900
Solar energy	0.2	> 1600	3,900,000
Wind energy	0.2	600	6000
Geothermal energy	2.1	5000	140,000,000
Ocean energy	—	—	7400
TOTAL	**~57**	**> 7500**	**>143,000,000**

Note: The current use of secondary energy carriers (heat, electricity and fuels) is converted to primary energy carriers using conversion factors.

Source: WEA (2004)

environmental and social constraints. It is estimated that, together, 'new' renewables (modern biomass energy, geothermal heat and electricity, small-scale hydropower, low-temperature solar heat, wind electricity, solar photovoltaic and thermal electricity and ocean energy) contribute about 2 per cent of the world's energy use (WEA, 2004; REN21, 2005). However, the dissemination of several 'new' renewables displays impressive growth rates (see Figure 2.1).

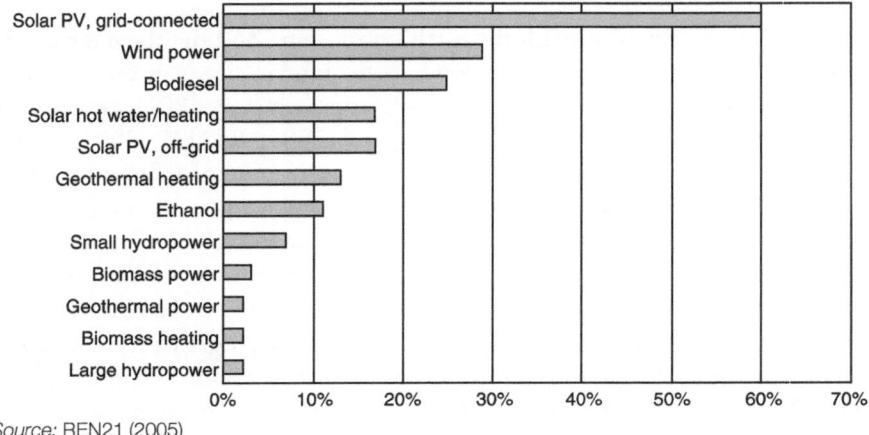

Source: REN21 (2005)

Figure 2.1 *Average annual growth rates of renewable energy capacity, 2000–2004*

Hydro energy

Hydropower is obtained by mechanical conversion of the potential energy of water in high elevations. An assessment of its energy potential requires detailed information on the local and geographical factors of runoff water. The total theoretical potential of hydro energy is estimated at 150 EJ a year and the technical potential is estimated at 50 EJ a year (WEA, 2000). Because rainfall varies by region and country, hydro energy is not evenly accessible. Rainfall may also vary in time, resulting in variable annual power output.

Technology options and status
Large-scale hydropower generation is regarded as a mature technology, unlikely to advance further. Refurbishment of plants has shown that advanced technologies can increase the energy output at essentially unchanged water flows. For small-scale hydropower, there is room for further technical development, and with the choice of very favourable sites, the use of existing administrative structures and existing civil works for flood-control purposes, the costs of small-scale projects could come down substantially. The installed capacity in 2004 can be

estimated at 720 gigawatts (GW) for large hydro and 62 GW for small hydro (REN21, 2005).

Environmental and social issues
Considering the criticism of large dams, modern construction tries to include in the system design several technologies that minimize the social and ecological impacts. Some of the negative impacts are the displacement of local communities (particularly indigenous people), changes in fish biodiversity, sedimentation, biodiversity perturbation, water quality standards, human health deterioration, and downstream impacts (WCD, 2000a). The World Commission on Dams (WCD) has done substantial work on this issue and elaborated a comprehensive set of recommendations for reconciling conflicting demands surrounding large dams (see Box 2.2).

Box 2.2 Principles for dams and development

- *Gaining public acceptance.* Wide public acceptance of key decisions is imperative for equitable and sustainable water and energy resources development.
- *Comprehensive options assessment.* Alternatives to dams do often exist. Needs for water, food and energy should be assessed and objectives clearly defined. Furthermore, assessments should involve a transparent and participatory process, applying economic, social and ecological criteria.
- *Addressing existing dams.* Opportunities exist to improve existing dams, respond to remaining social issues, and strengthen environmental and restoration measures.
- *Sustaining rivers and livelihoods.* Understanding, protecting and restoring ecosystems is important to protect the welfare of all species and to foster equitable human development.
- *Recognizing entitlements and sharing benefits.* Negotiations with adversely affected communities can result in mutually agreed and legally enforceable mitigation and development provisions. However, affected people need to be among the first to benefit from the project.
- *Ensuring compliance.* Public trust and confidence requires that the governments, developers, regulators and operators meet all commitments made for the planning, implementation and operation of dams.
- *Sharing rivers for peace, development and security.* Dams with a transboundary impact require constructive cooperation and good faith negotiation among riparian states.

Source: WCD (2000a)

It is important to note that the WCD also recognizes that hydropower projects (not including the construction phase) can produce very low greenhouse gas emissions and air pollutants. However, in some cases the greenhouse gas emissions can be high and actually surpass those resulting from thermal alternatives (WCD, 2000b). It is therefore advisable to conduct full life-cycle assessments to compare available options. This includes measuring carbon flows in the natural pre-impoundment watershed and identifying how these will be altered following the construction of a dam. Quantification of these changes is complex because it requires an understanding of the carbon cycle in the whole watershed (Fearnside, 2004).

Biomass energy

Biomass can be classified as plant, animal manure or municipal solid waste. Forestry plantations, natural forests, woodlands and forestry waste provide most woody biomass, while most non-woody biomass and processed waste comes from agricultural residues and agro-industrial activities (see Box 2.3). Sweden, Finland, Denmark and Austria are world leaders in creating working biomass markets that utilize biomass for energy purposes such as domestic heating with advanced heating systems and district heating. The growing contribution of biomass has been combined with increases in the number of companies that supply wood and wood products, as well as the number of parties that use biomass as an energy source.

Box 2.3 TYPES AND EXAMPLES OF BIOMASS

- *Woody biomass*. Trees; shrubs and scrub; bushes such as coffee and tea; sweepings from forest floor; bamboo; and palms.
- *Non-woody biomass*. Energy crops such as sugar cane; cereal straw; cotton, cassava, tobacco stems and roots; grass; bananas, plantains and the like; soft stems such as pulses and potatoes; and swamp and water plants.
- *Processed waste*. Cereal husks and cobs; bagasse; wastes from pineapple and other fruits; nut shells, flesh and the like; plant oil cake; sawmill wastes; industrial wood bark and logging wastes; black liquor from pulp mills; and municipal waste.
- *Processed fuels*. Charcoal from wood and residues; briquette and densified biomass; methanol and ethanol; plant oils from palm, rape, sunflower and the like; producer gas; and biogas.

Source: WEA (2000)

Biomass resources are abundant in most parts of the world, and various commercially available conversion technologies could transform current traditional and low-tech uses of biomass to modern energy. If dedicated energy crops and advanced conversion technologies are introduced extensively, biomass could make a substantial contribution to the global energy mix. However, the potential contribution of biomass in the long term can take a variety of estimates (see Table 2.2). Although most biomass is used in traditional ways (as fuel for households and small industries) and not necessarily in a sustainable manner, modern industrial-scale biomass applications have increasingly become commercially available. The biomass challenge is therefore not so much an issue

Table 2.2 *Annual global biomass energy resources*

Source[a]		Biomass type[b]	Biomass resources (EJ)		
			2020–2030	2050	2100
1		FR, CR, AR	31		
2[c]		FR, CR, AR, MSW	30	38	46
3		FR, MSW	90		
4					272
5		FR, CR, AR, MSW		217–245	
6[c]		FR, CR, AR, MSW	62	78	
7		FR, CR, AR	87		
8[d]	A1	EC		657	1115
	A2	EC		311	395
	B1	EC		451	699
	B2	EC		322	485

Notes:
a Data from several sources, including 1: Hall et al (1993); 2: Williams (1995), 3: Dessus et al (1992); 4: Yamamoto et al (1999); 5: Fischer and Schrattenholzer (2001); 6: Johansson et al (1993b); 7: Swisher and Wilson (1993); 8: Hoogwijk et al (2005).
b FR = forest residues, CR = crop residues, AR = animal residues, MSW = municipal solid waste, EC = energy crops.
c These studies estimated the potential contribution rather than the potential available.
d These studies are based on scenarios and therefore depict several different potentials for energy crops.

of availability as one of sustainable management, conversion, and delivery to the market in the form of modern and affordable energy services (Berndes et al, 2003).

Technology options and status
A large variety of raw materials and treatment procedures make the use of biomass a complex system that offers a lot of options. Biomass energy conversion technologies can produce heat, electricity and fuels (solid, liquid and gas). Domestic biomass-fired heating systems are widespread, especially in colder climates. In developing countries the development and introduction of improved stoves for cooking and heating has a large potential to expand. Combustion of biomass to produce electricity and heat is applied commercially in countries such as Sweden, Austria and Denmark. The globally installed capacity to produce electricity from biomass is estimated at 39 GW (REN21, 2005). Furthermore, gasification technologies can convert biomass into fuel gas with demonstration projects under way in various countries. The gas produced can be used to generate electricity but also to produce liquids such as diesel and methanol or gases such as substitute natural gas (SNG), dimethyl ether (DME) and hydrogen. Small gasifiers coupled to diesel or gasoline engines are commercially available on the market and applied in, for example, India and China.

Anaerobic digestion of biomass has been demonstrated and applied commercially with success in many situations and for a variety of feedstocks, including organic domestic waste, organic industrial waste, manure and sludge. Large advanced systems have been developed for wet industrial waste. In India there is widespread biogas production from animal and other wastes. Conversion of biomass to liquid, gaseous and solid fractions can also be achieved by pyrolysis (heating up to 500 °C in the absence of oxygen). Biodiesel can be obtained from oilseeds by using extraction and esterification techniques. In 2004 the world production was about 2.2 billion litres (REN21, 2005). Also, production of ethanol by fermenting sugars is a classic conversion route for sugar cane, maize and corn on a large scale, especially in Brazil, France and the US. In 2004 the world production of ethanol was estimated at 31 billion litres (REN21, 2005). Finally, hydrolysis of lignocellulosic biomass is an alternative option to produce ethanol, and one that is gaining more development attention in a number of countries.

Environmental and social issues
Biomass energy can be a carbon-neutral energy source, which makes it very attractive. However, assessing the sustainability of biomass production systems often requires a full life-cycle analysis of carbon flows and greenhouse gas emissions. The ratio of carbon output to carbon input needs to be calculated taking into account all components at the farm and industrial plant levels. For example, farm inputs, such as tillage operations, fertilizers, pesticides and irrigation, can result in significant greenhouse gas emissions (SRU, 2005). Furthermore, soil carbon degradation can occur through the removal of residues or decomposable matter, which in turn depletes soils and reduces productivity (Lal, 2005).

Erosion is a problem related to the cultivation of many annual crops. The best-suited energy crops are perennials with much better land cover than food crops. Increased water use caused by additional demands of new vegetation could become a concern in some regions. Furthermore, the use of pesticides can affect the quality of groundwater and surface water, which in turn impacts on plants and animals. The use of crop residues and plantation biomass will result in the removal from the soil of nutrients that have to be replenished in one way or another (Lal, 2005). Ash recycling is a possible method, but the issue of soil quality and nutrients requires further research to ensure the sustainability of biomass systems (Dornburg, 2004).

Land use competition is a significant issue for the future of bioenergy, particularly energy crops. The availability of agricultural land for energy crops will be defined by population growth rates, diets (levels and types of consumption), and the intensity of farming (Hoogwijk et al, 2005). Biomass plantations will always be limited by the available land. The efficiency of land use in terms of climate mitigation and energy security is therefore an issue of importance when it comes to biomass plantations and conversion processes. For example, utilizing biomass for biofuels (especially biodiesel) is at present not as efficient from a climate change perspective as is utilizing biomass for heat and power (CONCAWE et al, 2004; Lal, 2005).

Biomass plantations can be criticized because the range of biological species they support is much narrower than what natural forests support. However, if plantations are established on degraded land or excess agricultural lands, the restored lands are likely to support a more diverse ecology. Finally, the process of collection, transport and use of biomass increases the use of vehicles and infrastructures, and causes emissions to the atmosphere. A wide variety of social issues, some related to environmental factors, must also be addressed to expand bioenergy.

Solar energy

Solar energy has immense theoretical potential. The amount of solar radiation intercepted by the Earth is much higher than annual global energy use. Large-scale availability of solar energy depends on a region's geographic position, typical weather conditions and land availability (see Table 2.3). The assessment here is made in terms of primary energy. In other words, the energy before the conversion to secondary or final energy is estimated. The amount of final energy will depend on the efficiency of the conversion device used (such as the photovoltaic cell applied).

Table 2.3 *Annual global solar energy resources*

Region	Minimum EJ	Maximum EJ
North America	181	7410
Latin America and Caribbean	112	3385
Western Europe	25	914
Central and Eastern Europe	4	154
Former Soviet Union	199	8655
Middle East and North Africa	412	11,060
Sub-Saharan Africa	371	9528
Pacific Asia	41	994
South Asia	38	1339
Central Asia	115	4135
Pacific OECD	72	2263
TOTAL	**1575**	**49,837**

Note: The minimum and maximum reflect different assumptions on annual clear sky irradiance, annual average sky clearance and available land area.

Source: WEA (2000)

Technology options and status

Solar energy is versatile and can be used to generate electricity, heat, cold, steam, light, ventilation or hydrogen. It appears that several factors will determine the extent to which solar energy is utilized. These include the availability of efficient and low-cost technologies, effective energy storage technologies and high-efficiency end-use technologies. Solar thermal systems that produce high temperature heat can be used to generate electricity. Examples of solar thermal electricity (STE) technologies are parabolic trough systems, parabolic dish systems and solar power towers. The total installed capacity is currently about 0.4 GW (REN21, 2005). STE systems can be designed for solar-only applications, but also hybridized with fossil fuels to allow power production without sunlight. There are expectations that STE can begin to play a larger role in energy supply in the near future (Greenpeace and ESTIA, 2003).

Another technique to produce electricity is the direct conversion of solar light to electricity using photovoltaic (PV) systems. The cumulatively produced solar PV (off-grid and grid-connected) capacity up to the year 2004 is estimated at 4 GW (REN21, 2005). The average annual global growth rate for grid-connected solar PV is currently 60 per cent (see Figure 2.1). The major component of PV systems is the solar module, normally a number of cells connected in series. At present, crystalline silicon cells and modules are dominating the market. The conversion efficiency of commercially available modules is 10–17 per cent. This figure may increase to 12–20 per cent by 2010 and up to 30 per cent or more in the longer term (WEA, 2000). Very high efficiencies may be achieved by stacking cells with different optical properties. There are many types of solar cells under development or in production and it is still too early to identify winners or losers among the PV technologies. However, there is reasonable consensus that thin-film technologies generally offer the best long-term prospects for very low production costs and an energy payback time of less than a year.

The world's low- and medium-temperature heat consumption, estimated at about 100 EJ a year, can at least partially be met using solar collectors. At present, the total installed collector area is about a hundred million square metres (WEA, 2004). The solar domestic hot water (SDHW) system is the most important application. The SDHW systems in Northern and Central Europe are designed to operate on a solar fraction of 50–65 per cent. Subtropical climates generally achieve solar fractions of 80–100 per cent (WEA, 2000). Large water heating systems find widespread use in swimming pools, hotels, hospitals and homes for the elderly, among other applications. Heat pumps can generate high-temperature heat from a low-temperature (solar) heat source. Tens of millions of these appliances have been installed worldwide. Solar cooling, using absorption or adsorption cooling technologies, may become a feasible option as well.

The application of passive solar principles in building designs contributes to the reduction of (active) energy demands for heating, cooling, lighting and ventilation. Some of these principles include having good insulation, having a responsive, efficient heating system and avoiding over-shading by other buildings. Technologies involved include low-emission double-glazed windows, low-cost opaque insulation material and high insulating building elements, transparent insulation material, high-efficiency ventilation heat recovery and advanced high-efficiency lighting systems. Recent developments of building technology together with advanced, well-calculated system technology reduce the demand for heat energy by a factor of 10–15 in comparison with houses built some 30–40 years ago (WEA, 2000). In such a low-energy house, renewable energies can contribute up to 100 per cent of the energy demand.

Environmental and social issues
Solar technologies do not cause emissions during operation, but they do cause emissions during manufacturing and possibly on decommissioning (unless produced entirely by 'solar breeders'). One of the most controversial issues for PV was whether the amount of energy required to manufacture a complete system is smaller or larger than the energy produced over its lifetime. Nowadays the energy payback time of grid-connected PV systems is 2–6 years, and it is expected to decrease to 1–2 years in the longer term (Alsema, 2003). For stand-alone PV systems with battery storage the situation is less favourable. The energy payback time of modern solar home systems is now 1–2 years if compared with kerosene lamps, but 10–20 years in comparison with a diesel generator set. These results are mainly determined by the batteries, with their high energy requirements and short lifetime (Alsema, 2003). The availability of some of the elements in thin-film PV modules (like indium and tellurium) is also a subject of concern, although there are no short-term supply limitations. Of special concern is the acceptance of cadmium-containing PV modules, although the cadmium content of these modules appears to be well within limits for safe use.

Wind energy

Winds develop when solar radiation reaches the Earth, meeting clouds and uneven surfaces and creating temperature, density and pressure differences. The atmosphere circulates heat from the tropics to the poles, also creating winds (WEA, 2000). A region's mean wind speed and its frequency distribution have to be taken into account to calculate the amount of electricity that can be produced by wind turbines. Technical advances are expected to open up new areas to development. Furthermore, learning through experience will continue to improve the output of windfarms (see Table 2.4).

Table 2.4 *Annual global wind energy resources*

Region	Land surface with sufficient wind conditions		Wind energy resources without land restrictions	
	%	Thousands km²	TWh	EJ
North America	41	7876	126,000	1512
Latin America and Caribbean	18	3310	53,000	636
Western Europe	42	1968	31,000	372
Eastern Europe and former Soviet Union	29	6783	109,000	1308
Middle East and North Africa	32	2566	41,000	492
Sub-Saharan Africa	30	2209	35,000	420
Pacific Asia	20	4188	67,000	804
China	11	1056	17,000	204
Central and South Asia	6	243	4000	48
TOTAL	**27**	**30,200**	**483,000**	**5800**

Note: The assessment includes regions where the average annual wind power density exceeds 250–300 watts per square metre at 50 metres high. The energy equivalent is calculated based on the electricity generation potential of the referenced sources by dividing the electricity generation potential by a representative value for the efficiency of wind turbines, including transmission losses, resulting in a primary energy estimate. Also, the total excludes China.

Source: WEA (2000)

Technology options and status

Modern electronic components have enabled designers to control output and produce excellent power quality. These developments make wind turbines more suitable for integration with electricity infrastructure and ultimately for higher penetration. However, wind (and solar) are intermittent energy sources, so sufficient back-up remains an important issue. There has been a gradual growth in the unit size of commercial machines, from 30 kW of generating capacity in the 1970s (rotor diameter of 10 metres) to 5 MW (rotor diameter of 110–120 metres), and more improvements are likely (WEA, 2000).

Market demands have driven the trend towards larger machines through economies of scale, less visual impact on the landscape per unit of installed power, and expectations that offshore potential will soon be greatly developed. Special offshore designs are being implemented (GWEC, 2002). Modern wind turbines also have fewer components. By the end of 2004, worldwide installed capacity had topped 48 GW, with much of the development being in Europe, mainly Germany, Spain and Denmark (REN21, 2005). In fact, electricity production from grid-connected wind turbines has been growing at an impressive rate of about 30 per cent per year (see Figure 2.1).

Large penetration of wind energy and other intermittent renewable energy technologies would become very attractive if a cheap form of large-scale electricity storage were developed. At present, there is only the virtual storage of electricity on the grid. For example, if compressed air energy storage (CAES) can be developed to become economically attractive, it could transform wind-generated electricity into a baseload power supply (WEA, 2000). In addition, there is the developing high-voltage direct current (HVDC) technology, which allows the transmission of electricity over long distances by overhead lines or submarine cables. With the HVDC system, power flow can be controlled rapidly and accurately.

Environmental and social issues
Environmental and social aspects come into play in the various phases of a wind turbine project: building and manufacturing, normal operation, and decommissioning. Negative environmental aspects connected to the use of wind turbines include acoustic noise, visual impact on the landscape, impact on bird behaviour, moving shadows caused by the rotor, and electromagnetic interference with radio, television and radar signals. In practice the noise and visual impact cause the most problems for the development of windfarms.

Geothermal energy

Geothermal energy is generally defined as heat coming from the Earth. It has a large theoretical potential but only a much smaller amount can be classified as resources and reserves. Still, even the most accessible part, classified as reserves, exceeds current annual consumption of primary energy (see Table 2.5). But as with other renewable resources, geothermal energy is widely dispersed. Thus, the technological ability to use geothermal energy, not its quantity, will determine its future share. High-temperature fields used for conventional power production (with temperatures above 150°C) are largely confined to areas with young volcanism and seismic activity. But low-temperature resources suitable for direct use can be found in most countries.

Table 2.5 *Annual global geothermal energy resources*

Region	Millions EJ	%
North America	26	18.6
Latin America and Caribbean	26	18.6
Western Europe	7	5.0
Eastern Europe and Former Soviet Union	23	16.4
Middle East and North Africa	6	4.3
Sub-Saharan Africa	17	12.2
Pacific Asia	11	7.8
China	11	7.8
Central and South Asia	13	9.3
TOTAL	**140**	**100**

Source: WEA (2000)

Technology options and status

Geothermal use is commonly divided into two categories: electricity production and direct application. The technology to use geothermal energy is relatively mature. The conversion efficiency of geothermal power plants is rather low, at about 5–20 per cent. In 2004 the global installed capacity for power production was 8.9 GW (REN21, 2005). Major applications can be found in the US, the Philippines, Italy, Mexico, Indonesia, Japan and New Zealand.

Direct application of geothermal energy can involve a wide variety of end uses, such as space heating and cooling, industry, greenhouses, fish farming and health spas. It uses mostly existing technology and straightforward engineering. The technology, reliability, economics and environmental acceptability of direct use of geothermal energy have been demonstrated throughout the world. Compared with electricity production from geothermal energy, direct use has several advantages, such as much higher energy efficiency. In 2004 the installed capacity for direct use (heating and heat pumps) was 28 GW (REN21, 2005).

Geothermal energy for electricity production previously had considerable economic potential only in areas where thermal water or steam is found concentrated at certain depths. This has changed recently with developments in the application of ground source heat pumps using the Earth as a heat source for heating or as a heat sink for cooling, depending on the season. These pumps can be used basically everywhere. Important applications can be found in Switzerland and the US, among others countries.

Environmental and social issues

Geothermal fluids contain a variable quantity of gas, largely nitrogen and carbon dioxide with some hydrogen sulphide and smaller proportions of ammonia, mercury, radon and boron. Most of the chemicals are concentrated in the disposal water, routinely re-injected into drill holes, and thus not released into the environment. The concentrations of the gases are usually not harmful. The gas emissions from low-temperature geothermal resources are normally only a fraction of the emissions from the high-temperature fields used or electricity production.

Ocean energy

Tidal energy, wave energy and ocean thermal energy make up the types of ocean energy resources that appear most likely to move beyond speculative assumptions. The theoretical potential of each type of ocean energy is quite large, but dominated by ocean thermal energy (see Table 2.6). However, like other renewables, these energy resources are diffuse, which makes it difficult to use the energy. The difficulties are specific to each type of ocean energy, so technical approaches and progress differ as well. In a recent development the world's first

Table 2.6 *Annual global ocean energy resources*

Energy resource	TWh	EJ
Tidal energy	22,000	79
Wave energy	18,000	65
Ocean thermal energy	2,000,000	7200
TOTAL	**2,040,000**	**7344**

Note: The potential of ocean thermal energy is difficult to assess but is known to be much larger than for the other types of ocean energy. The estimate used here assumes that the potential for ocean thermal energy is two orders of magnitude higher than for tidal or wave energy.

Source: WEA (2000)

commercial wave farm to generate renewable electricity from ocean waves is being constructed in Portugal (Petroncini and Yemm, 2005).

Technology options and status

The energy of the oceans is stored partly as kinetic energy from motion of the waves and currents, and partly as thermal energy from the sun. The rise and fall of the tides creates, in effect, a low-head hydropower system. Tidal energy has been exploited in this way for centuries in the form of water mills. The largest modern scheme was built in France in the 1960s. A handful of smaller schemes have also been built. Wave energy remains at an experimental stage, with only a few prototype systems actually working. Total grid-connected wave power is very small, consisting of several oscillating water column (OWC) devices. A new generation of larger OWC devices is under development. Marine currents can also be used to generate electricity if the velocity of the current is high enough. The various turbine rotor options that are developed to use marine current energy generally coincide with those used for wind turbines.

Exploiting natural temperature differences in the sea by using some form of heat engine, potentially the largest source of renewable energy of all, has been considered and discussed for the best part of a century. But the laws of thermodynamics demand as large a temperature difference as possible to deliver a technically feasible and reasonably economic system. Ocean thermal energy conversion (OTEC) requires a temperature difference of about 20 °C, and this limits the application of this technology to a few tropical regions with very deep water. Offshore OTEC is technically difficult because of the need to pipe large volumes of water from the seabed to a floating system, the huge area of heat exchanger needed, and the difficulty of transmitting power from a floating device in deep water to the shore.

Environmental and social issues

Offshore environmental impacts for marine energy technologies tend to be minimal. Few produce pollution while in operation. An exception is tidal barrages, where the creation of a large human-made seawater lake behind the barrage has the potential to affect fish and bird breeding and feeding. Another exception is OTEC, which may cause the release of carbon dioxide from seawater to the atmosphere. The main issues, however, tend to be conflicts with other users of the seas, for fishing, marine traffic and leisure activities. None of the technologies discussed seems likely to cause measurable harm to fish or marine mammals.

ECONOMIC POTENTIALS OF RENEWABLE ENERGY

Substantial cost reductions in the past few decades in combination with government policies have made a number of renewable energy technologies competitive with fossil fuel technologies in certain applications. However, making these renewables competitive will require further technology development and market deployment, as well as an increase in production capacities to mass-production levels (Johansson and Goldemberg, 2002; van Sark et al, 2005). The present status of 'new' renewables shows that substantial cost reductions can be achieved for most technologies (see Table 2.7).

Experience curves for renewable energy

Because many renewable technologies are small in scale and modular, they are good candidates for continued cost cutting (Neij, 1997; Junginger, 2005). Such cost reduction can be illustrated using experience curves, which describe how cost declines with cumulative production, where cumulative production is used as an approximation for the accumulated experience in producing and employing a certain technology (see Figures 2.2 and 2.3). Cost data are hard to find, and prices are often taken as a proxy for cost, introducing uncertainties, especially in non-competitive markets. In addition, the cost reductions illustrated by the experience curves only show the cost reduction of technologies. The cost reduction of generated heat or electricity could be larger, owing to additional sources of cost reduction such as reduced installations costs and improved availability (Neij et al, 2003). For some resources, such as hydro and wind, cost reductions of generated electricity may level off when all 'good sites' are occupied. Technologies may also mature. Furthermore, the slope of experience curves may depend on the chosen timeframe and system boundaries. The experience curves depicted here represent only a limited number of experience curves developed over recent years (see Figures 2.2 and 2.3). Experience curves have been developed for additional energy technologies, and several experience curves have been developed for one and the same technology (Junginger, 2005).

Table 2.7 *Status of renewable energy resources and technologies, 2001*

Resource and technology	Expansion in energy production from 1997 to 2001 (%/yr)	Operation capacity (2001)	Capacity factor (%)	Energy production (2001)	Turnkey investment costs in 2001 (US$/kw)	Current energy cost	Potential energy cost
Biomass energy							
Electricity	~2.5	~40 GWe	25–80	~170 TWh (e)	500–6000	3–12 ¢/kWh	4–10 ¢/kWh
Heat	~2	~210 GWth	25–80	~730 TWh (th)	170–1000	1–6 ¢/kWh	1–5 ¢/kWh
Ethanol	~2	~19 bln litres		~450 PJ		(8–25 $/GJ)	(6–10 $/GJ)
Biodiesel	~1	~1.2 bln litres		~45 PJ		(15–25 $/GJ)	(10–15 $/GJ)
Wind energy							
Electricity	~30	23 GWe	20–40	43 TWh (e)	850–1700	4–12¢/kWh	3–10 ¢/kWh
Solar energy							
Photovoltaic electricity	~30	1.1 GWe	6–20	1 TWh (e)	5000–18,000	25–160 ¢/kWh	5 or 6–25 ¢/kWh
Thermal electricity	~2	0.4 GWe	20–35	0.9 TWh (e)	2500–6000	12–34 ¢/kWh	4–20 ¢/kWh
Heat	~10	57 GWth	8–20	57 TWh (th)	300–1700	2–25 ¢/kWh	2–10 ¢/kWh
Hydro energy							
Large	~2	690 GWe	35–60	2,600 TWh (e)	1000–3500	2–10 ¢/kWh	2–10 ¢/kWh
Small	~3	25 GWe	20–90	100 TWh (e)	700–8000	2–12 ¢/kWh	2–10 ¢/kWh
Geothermal energy							
Electricity	~3	8 GWe	45–90	53 TWh (e)	800–3000	2–10 ¢/kWh	1 or 2–8 ¢/kWh
Heat	~10	16 GWth	20–70	55 TWh (th)	200–2000	0.5–5 ¢/kWh	0.5–5 ¢/kWh
Ocean energy							
Tidal barrage	0	0.3 GWe	20–30	0.6 TWh (e)	1700–2500	8–15 ¢/kWh	8–15 ¢/kWh
Wave	—	—	20–35	0	2000–5000	10–30 ¢/kWh	5–10 ¢/kWh
Tidal stream	—	—	25–40	0	2000–5000	10–25 ¢/kWh	4–10 ¢/kWh
OTEC	—	—	70–80	0	8000–20,000	15–40 ¢/kWh	7–20 ¢/kWh

Source: WEA (2004)

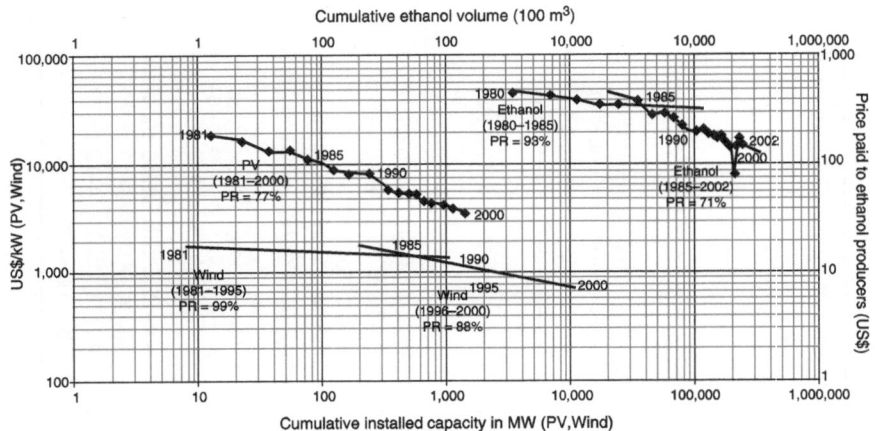

Note: Data from several sources, including wind turbines produced in Denmark (Neij et al, 2003), photovoltaics worldwide (Parente et al, 2002), and ethanol produced in Brazil (Goldemberg et al, 2004). Costs are expressed in year 2000 prices.

Source: WEA (2004)

Figure 2.2 *Experience curves for photovoltaics, wind turbines and ethanol production*

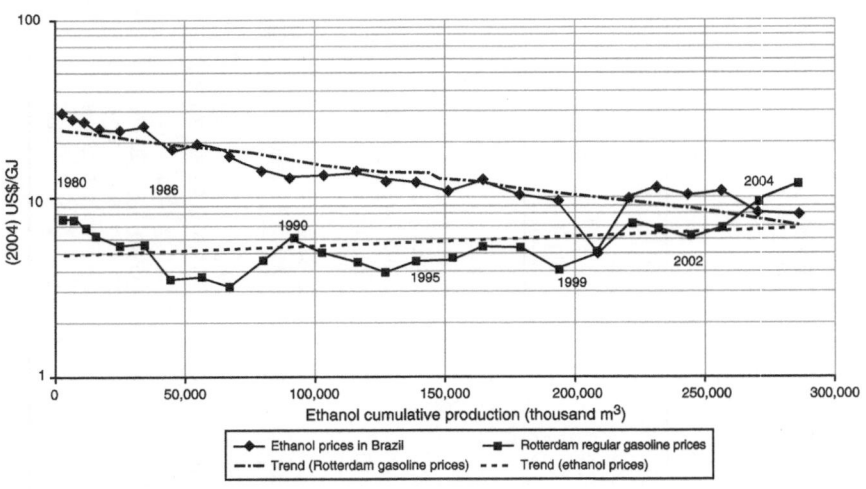

Source: Coelho (2005)

Figure 2.3 *Experience curves for ethanol and gasoline*

Present costs for renewable energy

Wind power in coastal and other windy regions is a promising energy source (see Figure 2.4). Other potentially attractive options include low-temperature solar heat production, and solar electricity production in remote applications (see

Figure 2.5). Wind and solar thermal or electric sources are intermittent, and not fully predictable. Nevertheless, they can be important in rural areas where grid extension is expensive. They can also contribute to grid-connected electricity supplies in appropriate hybrid configurations. Intermittent renewables can reliably provide electricity supplies in regions covered by a sufficiently strong transmission grid if operated in conjunction with hydropower or fuel-based power generation (Hoogwijk, 2004). Emerging storage possibilities and new strategies for operating grids offer promising indications that the role of intermittent technologies can be extended much further. Alternatively, hydrogen may become the medium for storing intermittently available energy production.

Modern, distributed forms of biomass, in particular, have the potential to provide rural areas with clean forms of energy based on biomass resources that have traditionally been used in inefficient, polluting ways (see Figure 2.6). Biomass can be economically produced with minimal or even positive environmental impacts through perennial crops. In the US, cellulosic biofuels could be cheaper than fossil fuel gasoline and diesel (NDRC, 2004). Biomass production and use currently is helping to create international bioenergy markets, stimulated by policies to reduce carbon dioxide emissions. Bioenergy is complex and may be differentiated into different subsystems including different resources, supply systems, conversions systems, and energy carriers (Hoogwijk, 2004). Each

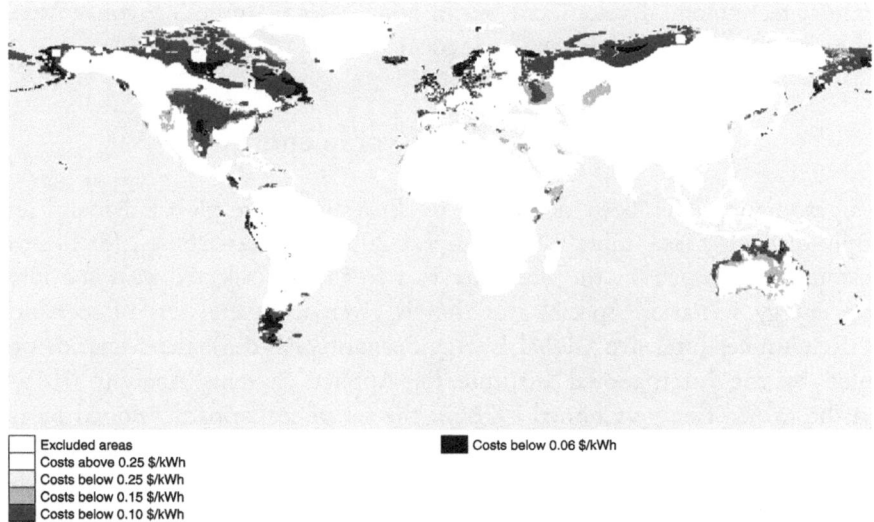

Excluded areas
Costs above 0.25 $/kWh
Costs below 0.25 $/kWh
Costs below 0.15 $/kWh
Costs below 0.10 $/kWh

Costs below 0.06 $/kWh

Note: The costs of wind electricity may come down as a result of further technological development.
Source: Hoogwijk (2004)

Figure 2.4 *Geographical distribution of present costs for wind energy*

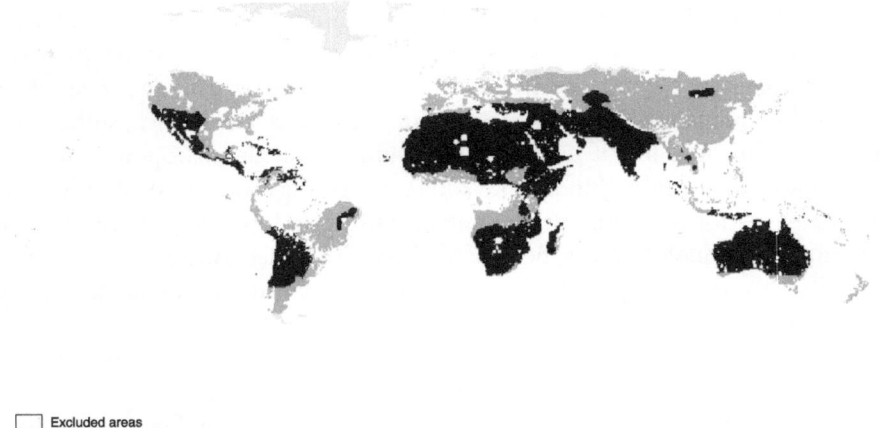

Excluded areas
Cost above 1.5 $/kWh
Cost below 1 $/kWh
Cost below 0.7 $/kWh
Cost below 0.5 $/kWh

Note: The costs of solar electricity may come down as a result of further technological development.

Source: Hoogwijk (2004)

Figure 2.5 *Geographical distribution of present costs for solar energy*

subsystem includes different technologies with individual learning processes for cost reductions. The development of bioenergy will in some cases be based on modular technology development but in other cases be more like conventional technologies for heat and power production.

Scenarios for renewable energy

Many scenarios have been developed to illustrate future global demand and supply of energy (see Table 2.8). The year 2050 has been chosen for illustrative purposes, which in the literature is a frequently selected year for long-term energy scenarios. Special attention is given to several sets of scenarios: the Renewables-Intensive Global Energy Scenario (RIGES); the scenarios presented by the International Institute for Applied Systems Analysis (IIASA) and the World Energy Council (WEC); the set of scenarios developed by the Intergovernmental Panel on Climate Change (IPCC); and scenarios defined by Shell International (SI). It is important to note that a larger renewable energy share of total energy supply does not necessarily mean a high quantity of renewable energy, because it depends on the total energy use in the scenario.

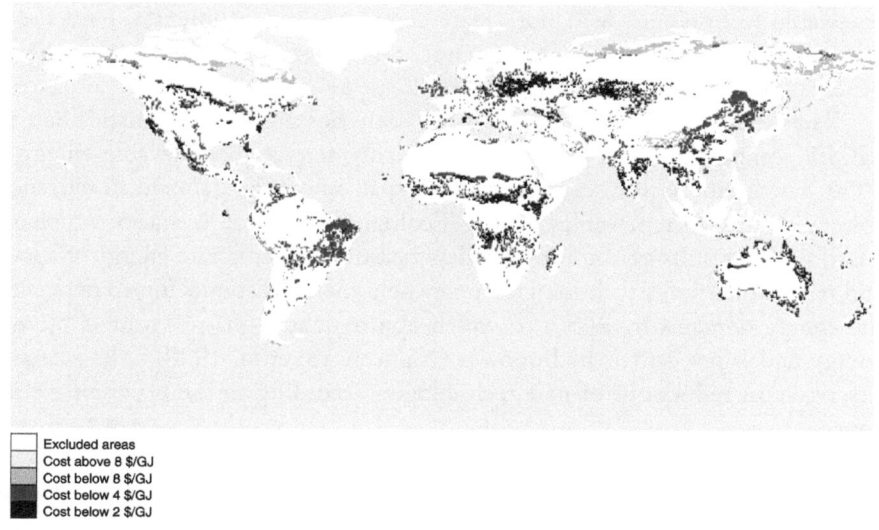

Excluded areas
Cost above 8 $/GJ
Cost below 8 $/GJ
Cost below 4 $/GJ
Cost below 2 $/GJ

Note: The costs of biomass electricity may come down as a result of further technology development.

Source: Hoogwijk (2004)

Figure 2.6 *Geographical distribution of present costs for biomass energy*

Table 2.8 *Scenarios for renewable energy, 2050*

	Total energy demand and supply (EJ)	Total renewable energy (EJ)	Total renewable energy (%)
RIGES[a]	512	237	46
IIASA and WEC[b]	479–1040	96–308	22–40
IPCC[c]	642–1611	73–444	9–35
SI[d]	852 or 1217	282 or 336	33 or 28

Notes:
a   Johansson et al (1993b)
b   Nakicenovic et al (1998)
c   IPCC (2000)
d   SI (2001)

Description of scenarios

The RIGES scenario, which involves an intensive introduction of renewables, illustrates the potential markets for renewable energy assuming that market barriers will be removed by comprehensive, and even accelerated, policy measures (Johansson et al, 1993b). In the scenario it is assumed that renewable energy technologies will capture markets whenever renewable energy is no more expensive on a life-cycle cost basis than conventional alternatives, and the use of renewable technologies will not create significant environmental, land use or other problems. The analysis does not consider the credits of any external benefits from renewable energy.

The results of the IIASA and WEC scenarios are less optimistic than the RIGES scenario but still present a significant increase in renewable energy by 2050. Furthermore, the scenarios show a span in energy demand, total renewables and share of renewables. In the 'Ecologically Driven' scenario, which also describes the results of ambitious policy measures to accelerate energy efficiency and renewable energy technologies, renewable energy accounts for 40 per cent of the energy demand by 2050, of which approximately 30 per cent is biomass energy and 8 per cent is hydropower (Nakicenovic et al, 1998). The scenarios describe cost reductions of new technologies according to the experience curve concept.

The IPCC scenarios published in 2000 have a wider span regarding the contribution of renewable energy in the future. In some scenarios the share of renewable energy is even expected to be lower than today. The 40 scenarios involve a range of driving forces, such as demographic change, social and economic development, and rate and direction of technical change (IPCC, 2000). Even with the same assumptions on driving forces, the models come up with different results. This indicates the sensitivity of the scenarios not only to assumptions made but also to the methods used. The scenarios including the highest share of renewable energy have the lowest cumulative CO_2 emissions from 1990 until 2100. These scenarios are characterized by the introduction of clean and resource efficient technologies, and for some scenarios also rapid changes in economic structures towards a service and information economy with reduction in material intensity.

The energy scenarios by SI published in 2001 show a considerable increase in the future share of renewable energy. The 'Dynamic as Usual' scenario reflects a social shift in priority to a clean, secure and sustainable energy system and an intense competition between new and old technologies. The scenario describes a gradual shift to low-carbon fuels and electricity supported by gas until 2025. The 'Spirit of Coming Age' scenario illustrates a higher demand of energy to meet the energy needs related to consumer preferences for mobility, flexibility and convenience (SI, 2001). At the same time, new energy technologies are introduced in developing countries as well as in industrialized countries, making renewable energy an important source of energy.

Comparison of scenarios

The scenarios are generally based on assumptions of economic growth, socio-economic development, future energy efficiencies and policy approaches. These assumptions, which may differ considerably for the various scenarios, result in differences in energy demand. The amount of renewable energy in the scenarios depends on assumptions regarding technologies available, the competitiveness of technologies, cost developments and resources available.

In all, the scenarios suggest that the amount of renewable energy can increase considerably until 2050, from 73 EJ to more than 400 EJ. The share of renewable energy in 2050 may range from 9 per cent to almost 50 per cent. An increasing share of renewable energy does not necessarily mean that the supply needs to reach 300–400 EJ; it can be lower, if the total energy demand is limited by restricted social and economic development, improved technology development or energy efficiency measures. Even though most scenarios show a huge increase in renewable energy until 2050, a major shift to a total renewable energy system by that date is not suggested in any of these studies, in spite of the fact that the renewable energy resources are more than sufficient. The reason for this is that energy supply systems have long lifetimes, as well as limitations in growth rates of new systems, economic acceptance and technological turnover, apart from other lock-in features of present energy systems. The scenarios illustrate that a transition to a global energy system based on renewable energy will require significant time.

It is important to keep in mind that scenarios are thought-experiments. No likelihood of the realization of any scenario can be assigned. This is due to the important fact that policies will affect the conditions in the marketplace, and ongoing research and development will provide new opportunities. The introduction of renewable energy will not just take place; it will depend on the support of technology development, market deployment and early adoption of new technologies. Such support will give rise to learning opportunities, cost reductions of new technologies, and capacity development, which will make the development and diffusion of renewable energy technologies possible. For some technologies, subsidies can be used to accelerate the learning process. This is often called 'buying down' the experience curves. Due to subsidies, investments will be made in relatively expensive technologies, and due to the increased number of sold and produced units, costs will go down and make the technology more competitive.

The scenarios not only show the possible share of renewable energy in the future, but also describe the importance of energy efficiency, and especially end-use efficiency, for the reduction of CO_2 emissions. Today the global energy efficiency of converting primary energy to useful energy is approximately one-third, which means that two-thirds of primary energy is dissipated in the conversion processes, mostly as low-temperature heat. Furthermore, significant losses occur on the demand side when final energy delivers the energy service. Numerous opportunities exist for energy efficiency improvements, especially on the demand side. When renewable energy is applied in decentralized systems, energy efficiency improvement can be an important strategy to reduce the investment costs of these systems. Clearly, scenario analysis indicates significant opportunities by 2050, and more thereafter, for using renewable energy in the world energy system.

MARKET DEVELOPMENT FOR RENEWABLE ENERGY

Looking at the markets where renewable energy carriers might compete facilitates an understanding of the demand for renewable energy. The potential markets for renewable energy and the role played by the public sector to develop these markets depend on specific conditions in each country and region. Providing efficient energy-using technologies and renewable energy is a public good in many developing countries, with a wide range of benefits for sustainable development. Thus, governments must find an effective balance between liberalization and directing markets towards wider social goals. It is within developing countries that much work is necessary to develop markets for renewable energy. This implies a change in focus, away from the historically dominating resource and technology assessments. A market perspective brings into question what underlies a market, such as social conditions, demand for products and services, and consumer knowledge (UNDP, 2005).

The use of renewable energy is either direct or indirect. Direct use is the immediate use of renewable energy flows to satisfy energy service needs. Examples include passive solar heating, day lighting and solar crop drying. There are often no energy markets involved here. However, policies related to other areas could advance the direct use of renewable energy – for example, building codes or other instruments in the buildings area to promote passive solar heating and day lighting. Energy services cannot be measured on a dollar per kilowatt-hour basis; thus, many comparisons of costs of local and integrated renewables with the costs of, for example, electricity generation by conventional power plants are incorrect and misleading.

Indirect use of renewable energy refers to the generation of an energy carrier that is then applied in end-use equipment to provide the desired energy service. Such energy carriers include electricity, biogas, mechanical (shaft) power and liquid biofuels. For some of these energy carriers there exist established markets. In other cases the use is local, such as small hydro or wind energy providing shaft power, or stand-alone electricity use that serves niche markets, such as solar photovoltaics for illumination and communication uses.

In industrialized countries and many developing countries, most renewable energy use takes place through markets for heat, electricity and fuels. Such markets increasingly exist in all developing countries, with some having nation-wide systems for electricity, and well-developed fuel markets, while others rely more heavily on local markets and direct uses of renewable energy. The development of these energy markets thus relies on the use of a battery of incentives and regulations. In developing countries it is useful to consider the direct end-uses and look at the opportunities for renewable energy to expand (see Box 2.4). Many of these applications encourage increased decision-making and participation from a variety of stakeholders, including the end-users (UNDP, 2005).

BOX 2.4 RENEWABLE ENERGY MARKETS IN DEVELOPING COUNTRIES

- *Rural residential and community lighting, television, radio and telephony.* Over 50 million households are served by small hydro village-scale mini-grids; 10 million households get lighting from biogas; over 1 million households have solar PV home systems or solar lanterns; and 10,000 households are served by solar, wind and diesel hybrid mini-grids.
- *Rural small industry, agriculture, and other productive uses.* Up to 1 million water pumps are driven by wind turbines, and over 20,000 water pumps are powered by solar PV; up to 60,000 small enterprises are powered by small hydro village-scale mini-grids; and thousands of communities receive drinking water from solar PV-powered purifiers and pumps.
- *Residential and commercial cooking and hot water.* A total of 220 million households have more efficient biomass stoves; 10 million households have solar hot water systems; and 800,000 households have solar cookers.
- *Transport fuels.* A total of 14 billion litres per year of ethanol vehicle fuel is produced from biomass; and 180 million people live in countries mandating the mixing of ethanol with gasoline.

Source: Martinot et al (2002)

Lighting, television, radio and telephony

Access to electricity opens up opportunities that are taken for granted by those who enjoy continuous access. Yet 350–400 million households in developing countries lack access to electricity. This largely means that television and radio are not available, lighting comes from candles and fires, and telephone services are absent (Martinot et al, 2002). A number of options to use renewable energy for electrification exist and the markets are growing. Solar home systems usually consist of a PV solar panel, battery, charging controller and end-uses such as lighting or heating. Lanterns powered by solar energy provide lighting only. In recent years, large markets have developed, particularly in rural areas of developing countries. Installations may service single households or public buildings, such as schools and health centres.

A biogas digester can convert wastes (animal and plant) into fuels for lighting, heating, cooking and electricity generation. Digesters can be small and serve a household, or larger and provide fuels for many households. Unfortunately, market development is hampered by community and political issues, as well as some technical challenges. Small-scale grids can provide electricity for communities with a high density. Traditionally, mini-grids have been powered by diesel generators or small hydro. However, solar PV, wind turbines or biomass digesters, often in hybrid combinations, can replace or supplement diesel power. Wind power–battery systems for a single household have been piloted in a few countries. Performance of these systems has been good, except sometimes during the summer when winds drop. Many households are therefore upgrading their systems with solar PV to complement the wind resource.

Industry, water pumping and drinking water

The emerging uses of renewable energy are for agriculture, small industry, water pumping and cottage applications (sawmills and mechanical power). Furthermore, social services, such as education and health care can be supported by renewable energy. Water pumps driven by wind have historically played a role in rural areas. More recently, interest is growing in solar PV-powered water pumps, along with biogas for water pumping in engines run on diesel and biogas. Stand-alone energy systems can power small industries, thereby creating local jobs and opportunities. In fact, the development of mini-grids and industry goes hand in hand. As small businesses grow, the economic viability of mini-grids increases. With the availability of energy, new possibilities open up. Renewable energy can also power mechanical pumping and filtering (as well as ultraviolet disinfection) to provide clean drinking water. This is emerging as a potential major market in developing countries.

Cooking and heating water

Direct combustion of biomass supports residential and commercial cooking as well as hot water in rural areas of developing countries. However, the decline in forest resources in some countries has encouraged governments to look at more efficient technologies for biomass use, as well as solar cookers. Research and development for these technologies is still urgently needed. Markets are primarily found where resource constraints are appearing. Solar hot water heaters for residential and commercial uses are cost-effective in many regions. A large market exists for domestic solar hot water collectors worldwide (Martinot et al, 2002).

Transport

Biomass-derived liquid fuels can power motor vehicles in several ways. First, ethanol can power specially designed vehicles that run on pure ethanol. Second, ethanol is mixed, in for example Brazil and the US, with gasoline or diesel fuel to produce gasohol for use in ordinary vehicles. Furthermore, the commercial viability of converting sugar cane to ethanol for motor vehicles has also been demonstrated. The competitiveness of ethanol and gasohol relative to conventional gasoline has continued to improve, although the global energy and automotive industries greatly affect the prospects of biomass-derived fuels, in the absence of accounting for external costs and benefits.

Renewable energy innovations

Technological innovation is critical to the reshaping of energy systems in ways that encourage sustainable development (Johansson et al, 1993a). However, the development and diffusion of sustainable and affordable renewable energy technologies is not occurring fast enough or widely enough (Turkenburg, 2002). The challenge of stimulating novel technologies is primarily one for industrialized countries, which have the technical and economic resources for sustained research and development, and for the dissemination of renewable energy technologies. Without effective policy it is unlikely that new technologies can overcome barriers and penetrate the market to any significant extent (see Table 2.9).

Key barriers

Innovations face barriers all along the innovation chain (from research and development over demonstration projects and cost buy-down to widespread diffusion). Some of these barriers reflect market imperfections; some inadequacies in the public sector domain; and some differences of views about needs, corporate priorities, relevant time horizons and reasonable costs. The amount of public support needed to overcome such barriers will vary from one technology to the next, depending on its maturity and market potential (PCAST, 1999). Direct government support is more likely to be needed for entirely new technologies than for incremental advances, where the private sector functions relatively effectively.

Major criteria for deciding whether government should finance a particular field of energy research can be the contribution of that area to achieving a transition to a sustainable energy future and to strengthening the competitiveness of (national) industries. It is also important that the research infrastructure in the field of interest is good enough to achieve these goals. Interventions should aim at helping the most promising energy innovations surmount bottlenecks wherever they occur in the innovation chain. Increasingly, however, this chain is viewed as a complex, interactive system requiring networks of innovation, knowledge sharing and demand 'pull' as well as supply 'push' (Turkenburg, 2002). Over the past couple of decades, industrialized countries have experimented with a growing number of policy instruments from target setting and procurement policies to green labelling and fiscal incentives.

Policy options

Cost buy-down and widespread dissemination can be advanced through a number of policy measures. A very effective policy option appears to be use of temporary subsidies, as tried in Germany and Spain with very good results in terms of

Table 2.9 *Key barriers and policy options for renewable energy in industrialized countries*

	Research and development	Demonstration	Diffusion	
			Early deployment	Widespread dissemination
Key barriers	• Governments consider R&D funding problematic • Private firms cannot appropriate full benefits of their R&D investments	• Governments consider allocating funds for demonstration projects difficult • Difficult for private sector to capture benefits • Technological risks • High capital costs	• Financing for incremental cost reduction (which can be substantial) • Uncertainties relating to potential for cost reduction • Environmental and other social costs not fully internalized	• Weaknesses in investment, savings and legal institutions and processes • Subsidies to conventional technologies and lack of competition • Prices for competing technologies exclude externalities • Weaknesses in retail supply, financing and service • Lack of information for consumers and inertia • Environmental and other social costs not fully internalized
Policy options	• Formulating research priorities • Direct public funding • Tax incentives • Technology forcing standards • Stimulating networks and collaborative R&D partnerships	• Direct support for demonstration projects • Tax incentives • Low-cost or guaranteed loans • Temporary price guarantees for energy products of demonstration projects	• Temporary subsidies • Tax incentives • Government procurement • Voluntary agreements • Favourable payback tariffs • Competitive market transformation initiatives	• Phasing out subsidies to established energy technologies • Measures to promote competition • Full costing of externalities in energy prices • Green labelling and marketing • Concessions and other market-aggregating mechanisms • Innovative retail financing and consumer credit schemes

Source: PCAST (1999)

expanding electricity generation from renewable energy. Green certificate markets are another option now in use in many countries. From an investor's point of view, the temporary subsidy would provide a better-known and predictable economic situation, as no assumption on the price of the green certificates has to be made. Carbon taxes have also proven effective in expanding the use of renewables. For example, a CO_2 tax in Sweden supported the shift from coal to biomass in district heating systems, contributing to the fact that biomass now provides 25 per cent of Sweden's primary energy. In 2002, Brazil adopted a law to promote adoption of wind energy, PV, small-scale hydro and biomass. The law was designed to protect the national interest where the market alone cannot.

Dealing with imperfections in innovation systems may require government actions on a number of fronts, including making sure that there is enough funding for new knowledge creation; improving linkages in the system; helping actors to find one another; shaping strong user–supplier links; being patient in the process of adjusting the institutional set-up; taking note of the need for variety and consistency in the applied policies; and stimulating prime movers (PCAST, 1999; Turkenburg, 2002). It is indeed a challenging task to identify the most important barriers and what policy options are capable of negating or overcoming the barriers.

POLICY IMPLICATIONS AND RECOMMENDATIONS

The degree to which there will be demand for renewable energy depends on many factors. At present, only 2 per cent of the world's primary energy is 'new' renewables (WEA, 2004; REN21, 2005). One fundamental issue is that the environmental and social benefits of using renewable energy appear at the societal level, while costs have to be borne by households and investors, typically without seeing the benefits reflected in market conditions. Therefore, the demand for renewable energy is strongly linked to the market situation, and can be dramatically affected by changes in market conditions. Policies for renewable energy relate to many sectors, including land use, agriculture, buildings, transportation and urban planning. Some specific areas that have to be addressed include the following:

- Understanding local renewable energy flows and their potential use. There is a pressing need to disseminate methodologies to estimate the local renewable energy flows and to create integrated (holistic) and sustainable solutions. Furthermore, continued research and development is required on local renewable energy technologies, such as heat pumps, building-integrated PV, passive solar and demand-side systems (integration of efficiency improvements and renewable energy).
- Supporting all steps in the innovation chain for renewable energy technologies and systems, including allocating a larger share of public sector funding for energy research and development to renewable energy; supporting demonstration projects (especially for modern biomass in developing countries), perhaps as public–private partnerships; and buying down the

relative cost of 'new' renewables in their early stages of development and commercialization, while still taking advantage of the economic efficiency of the marketplace.

- Setting ambitious but realistic targets and timetables in combination with effective policies, such as the use of green certificates that can be traded at a national or international market combined with agreements to reduce emissions; favourable uptake prices for renewable electricity delivered to the grid; tax credits for investments in renewables; subsidies with 'sunset' clauses; and concessions for the development of renewable energy resources.

- Methods and procedures for calculating the value of distributed generation need to be improved and disseminated, especially in situations with liberalized markets without vertical integration, where benefits may not be captured by investors in generation but by distributors. Energy services cannot be measured on a dollar per kilowatt-hour basis; thus, many comparisons of costs of local and integrated renewables with the costs of electricity generation are incorrect and misleading.

NOTES

1   Global energy consumption in 2004 was 470 EJ, based on statistics from BP (2005) and calculations by REN21 (2005). The primary energy value of hydroelectricity generation (and nuclear) was calculated by defining the equivalent amount of fossil fuel required to generate the same volume of electricity in a thermal power station, assuming a conversion efficiency of 38 per cent. Hydropower was therefore 634 Mtoe in 2004, or 6.2 per cent of global primary commercial energy. It is common to see hydropower as 2.4 per cent of global primary commercial energy in other studies. From the work by BP (2005) and REN21 (2005), the total world primary energy in 2004 was 10,224 Mtoe (commercial) + 1010 Mtoe (traditional) = 11,234 Mtoe. If 1 Mtoe = 41.9 PJ then the result is 470 EJ.

REFERENCES

Alsema, E. (2003) 'Energy Pay-back Time and CO_2 Emissions of PV Systems', in Markvart, T. and Castañer, L. (eds) *Practical Handbook of Photovoltaics: Fundamentals and Applications*, Oxford, Elsevier

Berndes, G., Hoogwijk, M. and van den Broek, R. (2003) 'The Contribution of Biomass in the Future Global Energy Supply: A Review of 17 Studies', *Biomass and Bioenergy*, vol 25, pp1–28

Boyle, G. (2004) *Renewable Energy*, Oxford, Oxford University Press

British Petroleum (BP) (2005) *Statistical Review of World Energy*, www.bp.com/

Coelho, S. (2005) 'Brazilian Sugarcane Ethanol: Lessons Learned', *Scientific and Advisory Panel of the Global Environmental Facility Workshop on Biofuels*, 29 August to 1 September 2005, New Delhi

Dessus, B., Devin, B. and Pharabod, F. (1992) *World potential of renewable energy*, Paris, CNRS-PIRSEM

Dornburg, V. (2004) *Multi-functional Biomass Systems*, Utrecht, Utrecht University

Fearnside, P. (2004) 'Greenhouse Gas Emissions from Hydroelectric Dams: Controversies Provide a Springboard for Rethinking a Supposedly Clean Energy Source', *Climatic Change*, vol 66, pp1–8

Fischer, G. and Schrattenholzer, L. (2001) 'Global bioenergy potentials through 2050', *Biomass and Bioenergy*, vol 20, pp151–159

German Advisory Council on the Environment (SRU) (2005) *Reducing CO$_2$ Emissions from Cars*, www.umweltrat.de/

Global Wind Energy Council (GWEC) (2002) *Wind Force 12: A Blueprint to Achieve 12% of the World's Electricity from Wind Power by 2020*, Brussels, GWEC

Goldemberg, J., Coelho, S. T., Nastari, P. M. and Lucon, O. (2004) 'Ethanol Learning Curve: The Brazilian Experience', *Biomass and Bioenergy*, vol 26, pp301–304

Greenpeace and European Thermal Power Industry Association (ESTIA) (2003) *Solar Thermal Power: Exploiting the Heat from the Sun to Combat Climate Change*, www.greenpeace.org/

Hall, D. O., Rosilo-Calle, F., Williams, R. H. and Woods, J. (1993) 'Biomass for energy: supply prospects', in Johannson, T. B., Kelly, H., Reddy, A. K. N. and Williams, R. H. (eds) *Renewable Energy: Souces for Fuels and Electricity*, Washington DC, Island Press

Hoogwijk, M. (2004) *On the Global and Regional Potential of Renewable Energy Sources*, Utrecht, Utrecht University

Hoogwijk, M., Faaij, A., Erickhout, B., de Vries, B. and Turkenburg, W. (2005) 'Potential Biomass Energy out to 2100 for four IPCC SRES Land-use Scenarios', *Biomass and Bioenergy*, vol 29, pp225–257

Intergovernmental Panel on Climate Change (IPCC) (2000) *Emission Scenarios*, Cambridge, Cambridge University Press

International Energy Agency (IEA) (2004) *World Energy Outlook*, Paris, IEA

Johansson, T. B. and Goldemberg, J. (2002) *Energy for Sustainable Development: A Policy Agenda*, New York, United Nations Development Programme

Johansson, T. B., Kelly, H., Reddy, A. K. N. and Williams, R. H. (1993a) *Renewable Energy: Sources of Fuels and Electricity*, Washington, DC, Island Press

Johansson, T. B., Kelly, H., Reddy, A. K. N. and Williams, R. H. (1993b) 'Renewable Fuels and Electricity for a Growing World Economy', in Johansson, T. B., Kelly, H., Reddy, A. K. N. and Williams, R. H. (eds) *Renewable Energy: Sources of Fuels and Electricity*, Washington, DC, Island Press

Junginger, M. (2005) *Learning in Renewable Energy Technology Development*, Utrecht, Utrecht University

Junginger, M., Faaij, A. and Turkenburg, W. (2005) 'Global Experience Curves for Wind Farms', *Energy Policy*, vol 33, pp133–150

Lal, R. (2005) 'Land Area for Establishing Biofuel Plantations', *Scientific and Advisory Panel of the Global Environmental Facility Workshop on Biofuels*, 29 August to 1 September 2005, New Delhi, India

Martinot, E., Chaurey, A., Lew, D., Moreira, J. B. and Wamukonya, N. (2002) 'Renewable Energy Markets in Developing Countries', *Annual Review of Energy and the Environment*, vol 27, pp309–348

Nakicenovic, N., Grübler, A. and McDonald, A. (1998) *Global Energy Perspectives*, Cambridge, Cambridge University Press

Natural Resources Defense Council (NRDC) (2004) *Growing Energy – How Biofuels Can Help End America's Oil Dependence*, Washington, DC, NRDC

Neij, L. (1997) 'Use of Experience Curves to Analyse the Prospects for Diffusion and Adoption of Renewable Energy Technology', *Energy Policy*, vol 23, pp1099–1107

Neij, L., Andersen, P. D., Durstewitz, M., Helby, P., Hoppe-Klipper, M. and Morthorst, P. E. (2003) *Experience Curves: A Tool for Energy Policy Assessment*, Lund, Lund University

Oil Companies European Organisation for Environment, Health and Safety (CONCAWE), European Council for Automotive Research and Development (EUCAR) and Joint Research Centre of the European Commission (JRC) (2004) *Well to Wheels Analysis of Future Automotive Fuels and Powertrains in the European Context*, Ispra, Italy

Parente, V., Goldemberg, J. and Zilles, R. (2002) 'Comments on experience curves for PV modules progress', *Photovoltaics: Research and Applications*, vol 10, pp571–574

Petroncini, S. and Yemm, R.W. (2005) *Introducing Wave Energy into the Renewable Energy Marketplace*, www.oceanpd.com/

President's Council of Advisors on Science and Technology (PCAST) (1999) *Powerful Partnerships: The Federal Role in International Cooperation on Energy Innovation*, Washington, DC, PCAST

Renewable Energy Policy Network for the 21st Century (REN21) (2005) *Draft Renewables 2005 Global Status Report*, Washington, DC, Worldwatch Institute

Shell International (SI) (2001) *Energy Needs, Choices and Possibilities: Scenarios to 2050*, London, SI

Swisher, J. and Wilson, D. (1993) 'Renewable energy potentials', *Energy*, vol 18, pp437–459

Turkenburg, W. C. (2002) 'The Innovation Chain: Policies to Promote Energy Innovations', in Johansson, T. B. and Goldemberg, J. (eds) *Energy for Sustainable Development: A Policy Agenda*, New York, United Nations Development Programme

United Nations Development Programme (UNDP) (2005) *Achieving the Millennium Development Goals: The Role of Energy Services*, New York, UNDP

van Sark, W., Patel, M., Faaij, A. and Hoogwijk, M. (2005) 'The Potential of Renewables as a Feedstock for Chemistry and Energy', in Dewulf, J. and van Langenhove, H. (eds) *Draft Renewable-Based Technology: Sustainability Assessment*, Hoboken, Wiley

Williams, R. H. (1995) *Variants for a low CO_2-emitting energy supply system (LESS) for the world*, Richland, Pacific Northwest Laboratories

World Commission on Dams (WCD) (2000a) *Dams and Development: A New Framework for Decision-Making*, London, Earthscan

WCD (2000b) *Does Hydropower Reduce Greenhouse Gas Emissions?* www.dams.org/

World Energy Assessment (WEA) (2000) *Energy and the Challenge of Sustainability*, New York, United Nations Development Programme

WEA (2004) *Overview Update*, New York, United Nations Development Programme

World Energy Council (WEC) (2004) *Survey of Energy Resources 2004*, Oxford, Elsevier

Yamamoto, H., Yamaji, K. and Fujino, J. (1999) 'Evaluation of bioenergy resources with a global land use and energy model formulated with SD technique', *Applied Energy*, vol 63, pp101–113

3

Renewable Energy for Transport

Dirk Aßmann, Niklas Sieber and Ricardo Külheim

INTRODUCTION

Worldwide, the transport sector consumes about one-fifth of primary energy (IEA, 1997) and is responsible for almost 60 per cent of oil consumption in Organisation for Economic Co-operation and Development (OECD) countries, and consumption in the developing world is increasing. After the energy sector, transport is the largest emitter of carbon dioxide. Unlike the energy sector, energy consumption and greenhouse gas emissions of the transport sector are still increasing in many industrialized countries. The problem is even more severe in developing countries, where enormous growth is required to catch up with the industrialized world.

Figure 3.1 demonstrates the rapid increase of the worldwide motorization since 1930. Between 1980 and 1995 the global fleet of cars, trucks and buses increased by 60 per cent, with a third of the increase taking place in developing countries. World Bank research (Ingram and Liu, 1999) has revealed that vehicle fleets grow at the same pace as economic growth. These authors estimate that the vehicle fleet will increase sevenfold by 2050.

Transport has positive development effects linked to the economic growth process, but its negative impacts, such as the emissions of greenhouse gases and pollutants, are well known. Development strategists face a dilemma, since economic growth is highly desired, but not its negative side effects. Thus, a sustainable transport strategy has to take into account the growing transport demands in developing countries and reduce the emissions at the same time. The key question is: what are the main pillars for a sustainable transport system?

Transport emissions are dependent on transport volumes, specific fuel consumption and emissions per unit of fuel. Transport planners have various

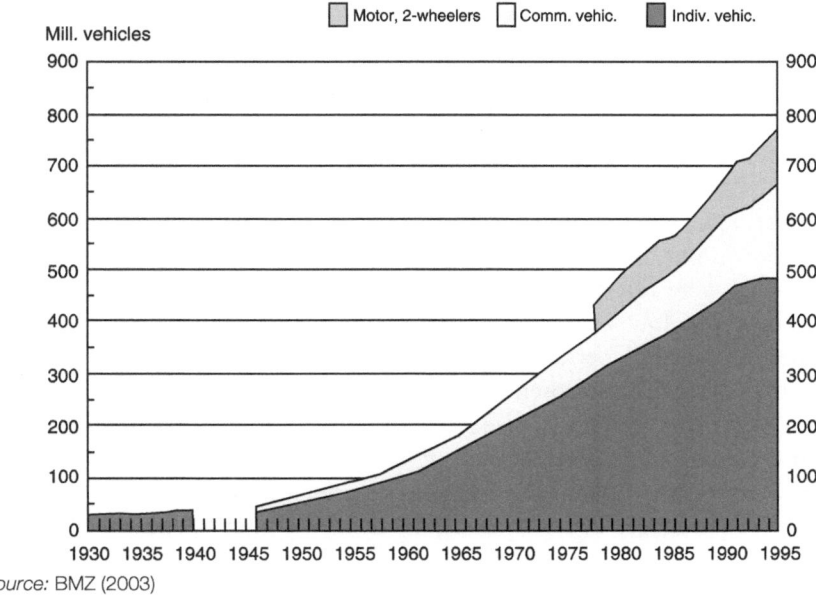

Source: BMZ (2003)

Figure 3.1 *Motorization trend worldwide, 1930–1995*

leverage points in order to change emission volumes and energy consumption. This chapter will give an overview on the most relevant measures for developing countries:

- technical options for conventional engines, such as engine technologies, vehicle characteristics, inspection and maintenance and fuel quality;
- options for renewable energies in transport, such as ethanol and biogas;
- planning approaches for efficient transport systems, such as transport demand management, land use planning and fuel pricing.

These options do not represent a comprehensive set, but represent the most prominent measures for improving transport systems in the developing world. They will be discussed with regard to their relevance for developing countries, their technical feasibility and competitiveness compared to existing solutions.

CONVENTIONAL ENGINES

In industrialized countries, public pressure has resulted in an increase of the energy efficiency of engines in transport, accompanied by improved emission regulations entailing air pollution reduction technologies, such as three-way catalytic converters and filters. While during the past four decades emissions of pollutants have generally decreased, fuel consumption and thus greenhouse gas (GHG) emissions resulting from transport have increased, owing to growing transport volumes and more powerful engines. The example of Germany (Figure 3.2) shows that during the past 40 years average engine power has more than doubled. Since the 1960s this process has been accompanied by an increase in specific fuel consumption. From the 1980s onwards a slight decline (16 per cent) can be observed, which is due to improvements in technical efficiency and a larger proportion of diesel-engined vehicles in the fleet. However, the efficiency improvements were not able to compensate entirely for the increasing engine power. Thus, the low consumption values from the 1960s have not been maintained. In North America a similar trend can be observed.

However, a significant potential for a further decrease of average fuel consumption exists in the industrialized countries. Volkswagen has demonstrated that a car driven by a one-piston diesel engine can reduce energy consumption to less than 1 litre per 100 km. However, this vehicle, a slow two-seater with very little space for baggage, is not available for everyday use. Commercially, low-consumption cars that consume 3 litres of diesel per 100 km are available; an example is the VW Lupo 3L TDI. The disadvantages of diesel engines are the higher CO_2 emissions per litre of fuel and the particulates when not using a filter. The best commercial petrol cars, such as the Daihatsu Cuore or the Opel Corsa Eco, consume slightly less than 5 litres per 100 km and are competitively priced. Hybrid engines combining petrol and electric engines have even higher energy efficiency, but are presently much more costly.[1] However, in the past the market for low-consumption cars was very small. In Germany, sales of the Lupo, which consumed only 3 litres per 100 km, were only 15,000 and thus Volkswagen decided to cease production. By contrast, 4.3 million high-powered vehicles with a maximum speed of more than 210 km/h are registered and the use of SUVs is

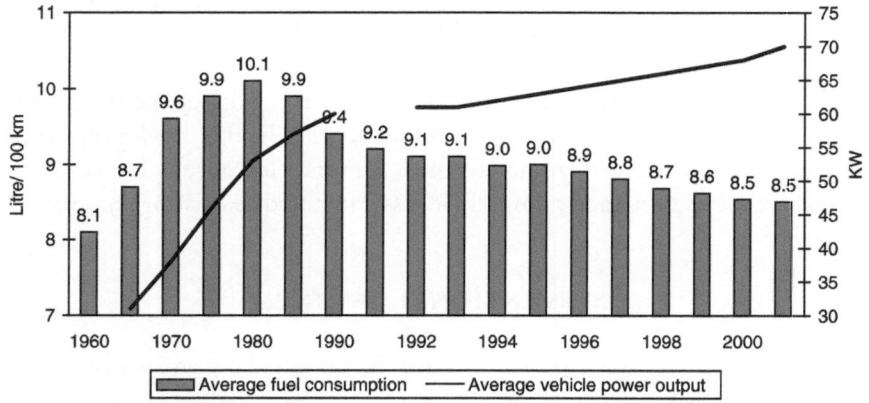

Source: Verkehr in Zahlen 2002/2003, own graph

Figure 3.2 *Fuel consumption and vehicle power of private cars in Germany*

increasing rapidly. The question is, how much the fuel price has to rise until this attitude changes.

In most developing countries a large share of the vehicle fleet consists of second-hand vehicles purchased from industrialized countries. The general rule is, the richer the country, the smaller the share of second-hand vehicles. In low-income countries, especially in Africa, used cars are very common, while in emerging economies the share of new cars is considerably higher. Therefore, in poor countries the emission levels can be compared to those of used cars in the developed world, and any efficiency improvements implemented in the North will also have long-run effects in the South. Thus, without additional measures the energy efficiency of vehicles in developing countries will be no better than it used to be five to ten years ago in the developed world. In developing countries, fuel prices are often fairly low (see later in the chapter), and low-consumption vehicles make up only a small market share, which is balanced by the large number of off-road vehicles, which are badly needed on the poor roads.

The rapid motorization in the successful Asian economies entails growing environmental and congestion problems. China may serve as an example, since it has the world's fastest-growing automobile market. The country's vehicle population increased from 6.6 million in 1990 to 45.3 million in 2003, which implies an average annual increase of 21 per cent.[2] During the first half of 2005, 2.8 million vehicles were produced in the country.[3] China's increasing energy demand for transport not only had an impact on the international oil market, but led to national shortages. In order to save the country's scarce energy resources, the National Standardization Administration has set up a fuel consumption cap for new passenger cars. For example, a light-duty vehicle (between 1.2 and 1.3 tonnes) should be designed to consume no more than 9.5 litres per 100 km until 2008. From 2008 on, the limit will be lowered to 8.6 litres.[4] At present it is inconceivable that Western countries would enact similar measures in order to reduce GHG emissions.

A significant decrease in pollutant emissions could be achieved in developing countries if European emission standards such as Euro II to IV were implemented. Many countries have already introduced the Euro I standard, some Euro II, and a few even Euro III.[5] Preconditions are the introduction of legal binding requirements for mandatory roadworthiness tests, combined with a set of regulations dealing with fines and other sanctions. However, this requires a dedicated government willing to undertake a considerable effort in order to implement this comprehensive set of measures. A first step towards the necessary higher environmental standards is an improvement in the fuel quality, for example by reducing the sulphur content of petrol and diesel.

Anybody travelling in developing countries can easily observe that many vehicles produce black smoke. Emission standards are simply not met, because engines are badly adjusted and vehicles poorly maintained. The reason is the lack of a system of inspection and maintenance, or the inadequacy of the system. A good inspection and maintenance system can reduce fuel consumption by 3–7 per cent and hence leads to a similar decrease in CO_2 emissions.[6] This holds especially true for two- and three-wheelers, which emerge as a major problem in many Asian towns. In Delhi, two- and three-wheelers produce three-quarters of the particulates and roughly one-third of the hydrocarbon emissions.[7] Especially for two-stroke engines, the pollution problem is often visible. Since the vehicles are relatively cheap, reliable and easy to repair, they are a very popular mode for the poor. Because a replacement by cleaner technologies is not feasible in the short run, inspection and maintenance and tighter emission standards are the only measures available to limit the negative effects.

RENEWABLE ENERGIES IN TRANSPORT

Alternatives to fossil-based fuels

Renewable energies can be used in various forms and in different sectors for different purposes. For transport purposes there are in general three options: the transfer of solar radiation (or other RE) through, for example, solar thermal power plants into hydrogen, the transfer of RE into electricity which is directly used as 'fuel', or the conversion of biomass into liquid or gaseous fluids. As the hydrogen path is still too expensive and various technical problems still need to be solved, and the electricity path does not fulfil transport requirements, biofuels are highlighted in all discussion on alternative fuels.

If biomass is produced sustainably (for example, within organic land use systems, or the use of industrial waste as biomass), it is classified as a renewable energy. However, pure biomass production in the absence of an awareness of other environmental aspects may lead to deforestation or erosion.

Biomass can be used in two ways:

1   Traditional or non-commercial biomass, such as crop residues, fuel wood and animal dung, is frequently used with very low efficiencies for cooking and heating in many developing countries. This is a major cause of serious indoor pollution, particularly affecting women, small children and the elderly, and leads to diseases and overuse of natural resources. Therefore, programmes to develop and disseminate improved biomass stoves in many African and Latin American countries, China and India are of major importance.

2   Modern biomass is biomass produced in a sustainable manner and used for electricity generation, heat production and transportation (liquid fuels). It includes wood and forest residues from reforestation or sustainable forest management, energy crops from organic production, and rural and urban residues such as solid waste and liquid effluents.

Today an enormous variety of technologies are available to convert the energy provided by plants into fuel energy for transportation. The use of pure plant oil as a fuel for transportation has a long history, leading us back to Rudolf Diesel's earliest compression engines working on peanut and hemp oil. Pure plant oils (PPO) can be extracted from almost any oil seed crop such as rapeseed, sunflower or soybean. Because of their different combustion properties as compared with fossil diesel, PPO cannot be used in normal engines but requires special refitting, or dedicated engines such as the Elsbett technology. As yet, no car manufacturers have offered plant oil motors; there are just small companies offering engine conversions. Apart from some niche sectors (e.g. sensitive areas, fleet owners), the use of pure plant oil does not seem to be appropriate for broad markets.

The most widely used biofuel is *ethanol*, accounting for more than 90 per cent of total biofuel production. Ethanol is produced by fermenting sugar, such as from sugar cane or sugar beet, or by converting corn starches (and other starchy feedstocks) into sugar before fermentation. Ethanol can either be blended with gasoline in low concentrations without any engine modifications or, with modified engines, it can be burned in higher concentrations up to 100 per cent.

The special characteristics of ethanol lead to higher octane numbers and thus increase engine efficiency. New developments such as flex-fuels (flexible-fuel vehicles), which are able to switch between various mixtures of ethanol and gasoline, are gaining market relevance since the automobile industry successfully introduced them on a large scale in Brazil. It was expected that 60 per cent of the new cars sold in Brazil in 2005 would be flex-fuels. Furthermore, ethanol can also be used as an additive. Ethyl tertiary-butyl ether (ETBE) can replace the fossil-based methyl tertiary-butyl ether (MTBE). Recent data on the energy balances of ethanol show that the ethanol production chain is highly efficient. Brazil, with its production based on sugar cane, has by far the highest energy output/input relationship of all biofuels, varying from a factor of 8.3 to 10.2.

Biodiesel is the second important biofuel. It is also known as fatty acid methyl ester (FAME), because of the transesterification process of plant oils. Biodiesel has similar characteristics to those of conventional diesel and can thus be used in conventional vehicles without major changes. Minor refitting, mostly concerning seals, is required when biodiesel is used in high proportions. Blending biodiesel to conventional diesel in a proportion of 5 per cent is technically no problem. Research on the biodiesel life cycle concluded that owing to its closed carbon cycle it has a strong impact on greenhouse gas mitigation, reducing carbon dioxide emission by more than 75 per cent compared to fossil-based diesel. Most other emissions will also be reduced. However, like diesel engines, biodiesel causes particulate emissions that need to be reduced by particulate traps. As NO_x emissions also rise by 10 per cent, using biodiesel does not mean that catalytic converters can be dispensed with.

In parallel to the biofuels options already mentioned, there are other alternatives, such as gaseous or liquid biofuels. Some of them are not yet marketable; others are at a preliminary development stage and still need further research before becoming a technically and economically interesting option.

Biogas is produced by means of a process known as anaerobic digestion. It is a process whereby organic matter is broken down by microbiological activity in the absence of air. It occurs naturally at the bottom of ponds and marshes, and gives rise to marsh gas or methane, which is similar to natural gas. There are two common technologies for obtaining biogas. The first, more widespread, is the fermentation of human and/or animal waste in specially designed digesters. The second is a more recently developed technology for capturing methane from municipal waste landfill sites.[8]

At the moment, the biogas sold is usually collected from municipal waste landfills. Also, cattle farmers produce it in small fermenters, and China and India have had much success with family-sized biodigesters. The gas is compressed and used in co-generation plants, in vehicles equipped for natural gas or for cooking. However, compared to other forms of energy, biogas has a very small market share, but there is still a huge potential for both stationary and mobile use.

Although *hydrogen* is often highlighted as 'the fuel of the future', it is not expected to be available before 2020. Since there is still no mature technology for hydrogen production based on renewables, a large-scale launch does not seem realistic within the next 10 or 15 years.

Along with hydrogen, fuel cells are the technology that is most often mentioned. However, modified engines would also be able to utilize hydrogen; it has not yet been decided which technology will be at the forefront. The use of hydrogen could lead to a big reduction in GHG emissions if renewable energies were used to produce the fuel, and if the efficiency of the entire production and consumption chain were high. At present, most fuel cells run on natural gas. This may lead to better efficiencies but the improvements are fairly small. On the other hand, technologies based on biomass are not sufficiently developed to be used on a large scale.

Research into so-called designer fuels, such as 'Synfuel', developed by Choren Industries, is under way. The *biomass-to-liquids* (BtL) process is mostly based on the Fischer–Tropsch reaction, but other research is being done as well. In general, solid biomass will be gasified and afterwards liquefied. It is still in the development phase but the first demonstration plant will probably begin to produce in 2006. Therefore, BtL is rather more developed than hydrogen technology, and most scientists expect it to be on the market within the next five to ten years. However, many technical problems still need to be solved. As the characteristics of 'designer fuels' will be similar to those of fossil-based fuels, there will be little need to modify conventional engines – an important aspect for their marketing and for their acceptance by car manufacturers.

Natural gas vehicles (using compressed natural gas, *CNG*) do not use a renewable energy form, but emit 15–20 per cent less GHG than do petrol vehicles. As regards its environmental impacts, liquefied petroleum gas (*LPG*) has similar features to those of natural gas, but GHG emissions are slightly higher. Since LPG is collected from oil wells, where it is usually burned, its supply is limited.

Biofuels – today's market

Until industrialization in the 18th and 19th centuries, energy supply was mainly based on renewable energies, especially biomass and hydro. Rising demand and new technologies led to better living and working conditions, but also to an ecologically harmful change in the energy supply. Today there are important differences in the energy systems of OECD countries and developing countries respectively, as indicated in Table 3.1.

Table 3.1 *Characteristics of global primary energy demand, 2000*

	Fossil fuels (%)	Renewable energy (%)	Nuclear (%)	Growth rate, all sources (%/yr, 1971–2000)
OECD	82.7	6.2	11.0	1.6
Developing countries	71.7	27.6	0.7	3.9

Source: IEA (2004)

In OECD countries, which have reached a very high level of development, gains in energy efficiency have been the main strategy used to decrease energy consumption. In developing countries, where renewables (mainly biomass) are already very important (see Table 3.1), albeit used in inefficient ways, modernization of their usage seems the better strategy to follow.

As Figure 3.3 shows, around 14 per cent of total world energy (around 425 EJ in 2002) is provided by renewable energies. Apart from biomass (meaning here combustible renewables and waste), hydro has the highest share. The 'new' renewables such as solar, wind and geothermal energy contribute less than 1 per cent of global production.

In developing countries the non-commercial use of solid biomass ranks bioenergy before hydropower and other renewable sources of energy. According to the IEA Renewable Information, 48.4 per cent of global biomass consumption takes place in Africa. This is mainly accounted for by domestic cooking and household heating in stoves, fireplaces and domestic and district heating boilers.

Today, about 33 billion litres per year of biofuels is used commercially in North and South America, Europe and South Africa. As a whole, Europe has the largest biodiesel production capacity, estimated at 2.3 million litres per year in

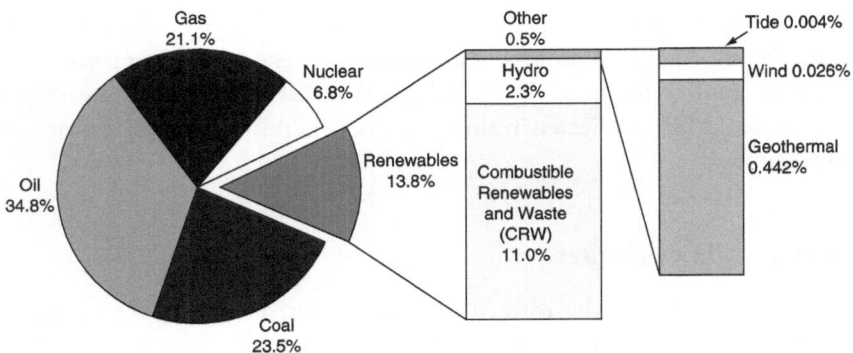

Figure 3.3 *Global primary energy demand, 2000*

more than 40 dedicated installations, mostly in Germany, France, Italy and Austria. Germany is currently the largest producer, accounting for almost half of the global biodiesel production.

World ethanol production rose to nearly 41 billion litres in 2004, of which 73 per cent was for fuel. The countries with the highest current use of fuel ethanol are Brazil (14.6 billion litres, corresponding to roughly 30 per cent of the country's non-diesel motor fuel use in 2004) and the US (14.3 billion litres). The two countries together are responsible for almost 70 per cent of total production, followed by China (9 per cent), India (4 per cent) and France.[9]

Production and costs of renewables

For the assessment of the future competitiveness of renewable energies, the price of crude oil is a crucial topic. The rapidly increasing energy demand from the growing economies in Asia has resulted in a tremendous increase in crude oil prices in recent years. First peaks were reached in November 2004, when a barrel of oil reached US$55, which expressed in real terms compares to the second oil price shock at the beginning of the 1980s. However, in September 2005, after Hurricane Katrina, the price temporarily passed the US$70 threshold. On 12 September the price per barrel was down to US$64, and petrol cost around US$0.52 per litre.[10] Long-term forecasts to the end of the decade range between US$42 (Merryl Lynch) and US$60 (Goldman Sachs).[11]

Given these price levels, renewable energies are becoming increasingly competitive. The most successful example is ethanol production in Brazil. The Proálcool Programme is, worldwide, the largest application, and initially was promoted by massive government support. Recent experience shows that even though government subsidies have been cut, ethanol prices are close to being competitive with petrol. The break-even was reached at the beginning of the decade, when oil prices began to rise while ethanol continually became cheaper

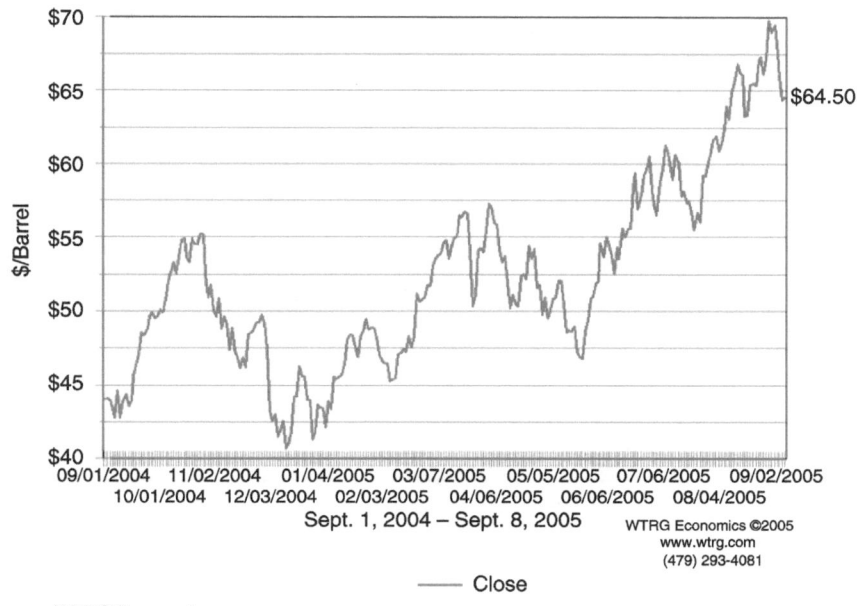

Figure 3.4 *Crude oil prices, 2004–2005*

(see Figure 1.2 in Chapter 1) due to economies of scale entailed by increased production. Table 3.2 shows that in 2003 the cost of biofuel production based on sugar cane in Brazil was competitive with that of petrol. Given the expected oil price level of at least US$42/bbl, ethanol will be economical in the long run.

In the future, further cost reduction can be expected, since the new conversion technologies described above, that make use of ligno-cellulosic feedstock, either from waste materials or grown as dedicated energy crops, will reduce costs further. The International Energy Agency (IEA) estimates that production cost will initially be $0.53/petrol equivalent and decrease to $0.43 in the near term and to $0.27 after 2010 (IEA, 2004, p78). This has to be compared to the petrol price of around US$0.5 in September 2005.

The biodiesel case is different, as Table 3.3 shows. With oil price levels at their 2004 levels, biodiesel has not yet achieved competitiveness. In Germany it is used only because of tax exemptions. Increased oil prices, new technologies and reduced production costs in developing countries might increase its competitiveness in the future.

According to Berndes et al (2005), advanced biofuels such as bio-dimethyl ether (bio-DME), biomethanol and compressed substitute natural gas (SNG) might be competitive with an oil price of US$60–100/bbl, and further cost reduction might decrease the threshold to $40. However, the availability of the new production technologies is a determining factor. Furthermore, other

Table 3.2 *Ethanol and petrol production costs before taxes, 2003*

Fuel type	Source	Country	Production costs (US$/litre petrol equivalent)
Ethanol	Corn	US	0.43
	Sugar beet	Europe	0.63–0.90
	Wheat	Europe	0.53–0.93
	Sugar cane	Brazil	0.20–0.30
Petrol	Crude oil		0.26–0.28

Note: The volumetric energy content of ethanol is one-third less than in the case of petrol. To compare, the costs per litre are therefore transferred into petrol equivalents.

Table 3.3 *Biodiesel and diesel production costs, 2003*

Fuel type	Source	Country	Production costs (US$/ litre diesel equivalent)
Biodiesel	Rape seed	Europe	0.35–0.80
	Soy beans	US	0.48–0.73
Diesel	Crude oil		0.20–0.24

Note: The volumetric energy content of biodiesel is a few per cent less than in the case of diesel. To compare, the costs per litre are therefore converted to diesel equivalents.

processes, such as the use of hydrogen or the liquefaction of coal, might also become economic.

In Europe a replacement of 5 per cent of the fuel demand (petrol and diesel) would require about 20 per cent of available cropland for production (IEA, 2004, pp17–18). The reduced food crop production would increase food prices and thus would reduce the need for agricultural subsidies in industrialized countries. Thus, in industrialized countries, without major importation only a small proportion of the transport energy demand can be satisfied. This gives developing countries the opportunity to improve their trade balance and generate additional income for their rural populations, where poverty is prevalent. Assuming adequate productivity, the low labour costs of developing countries might reduce production cost and make biofuels even more competitive. The IEA estimates that in the near term (2020) biofuels can replace 20 per cent of world gasoline consumption and in the long run (2050–2100) a third or more of road transportation fuels.

Social and ecological aspects

The Brazilian Proálcool Programme was started in the 1970s with research and development activities in the field of ethanol for reasons of energy security and self-sufficiency. The programme is known to have created major social and environmental problems on the agricultural side (Weber, 2004, pp69–70). Proálcool has been criticized for its negative environmental impacts, the result of large-scale monoculture. All ecological investigations show that the evaluation of biofuels greatly depends on the type of agriculture used to produce the plants. The results are quite simple: the lower the use of mineral fertilizers and chemical pesticides, the more appropriate the crop rotation; and the more efficient the use of by-products (such as bagasse or straw in co-generation), the better the ecological evaluation. Therefore, in order to achieve environmentally sound production, environmental standards and criteria through the whole production and consumption chain need to be developed and implemented.

Another important aspect of the production of biofuels is the future land use. If a significant share of worldwide road fuels is to be replaced, large-scale agricultural production will be necessary. The IEA assumes that low-value grazing lands will be put under cultivation. Proálcool had the advantage of using old farmland formerly cultivated for sugar production, but past experience shows that agricultural expansion has often entailed a reduction or degradation of tropical forests. Furthermore, the production of biofuels should never affect food production.

In the past five to seven years a social component has been added to the discussion. The discussion focused on the question of whether farmers could be the 'oil sheikhs of tomorrow'. As with ecological evaluation, social impact assessment did not lead to a clear and distinct result in Brazil. The argument about farmers becoming oil sheikhs is, in general, a myth. The overwhelming majority of ethanol producers are not small-scale farmers but agro-industrials whose farm workers are paid extremely low wages, whereas producers of bio-diesel in Europe are 'normal' farmers. However, here the market is not yet developed enough to undertake a social impact assessment. Some experience in the field of solid biomass, for example in power plants, shows that big players such as energy utilities try to use their power to make extra profit by paying the farmers extremely low prices.

In this regard the Malaysian case is quite interesting, as the production of plant oil does not automatically lead to poverty reduction. Malaysia is the world's biggest producer of palm oil, with agriculture having a share of 11 per cent in the economy, and palm oil contributes 70 per cent to it. However, in order to become such a big player, Malaysia destroyed large parts of its primary rainforest and thus minimized the habitat for certain species, especially tigers and orang-utans. At the same time, the average salary (2003: US$92) of the farm workers is 80 per cent below the poverty level, and has increased by only $1 within the past 45 years.

The question of whether production of biofuels will be of general benefit in developing countries remains open. A number of factors influence the sustainability, such as existing land tenure systems, land use conflicts, trade protectionism from industrialized countries, technology transfers, wage levels in agriculture and industries, availability of adequate finances for small-scale farmers, and access to transport infrastructures. A good example of a sustainable approach is that of German Technical Cooperation (GTZ), which has started a public–private partnership project on biofuels in India that might lead to a three- to fourfold increase in farmers' incomes within a few years – while protecting the environment and generating new business opportunities.

Planning approaches for efficient transport systems

Since at present technical improvements and the use of renewable energies in transport are limited in developing countries, planning approaches are needed to reduce energy in order to meet both the increasing transport demand and greenhouse gas reduction targets. Like planners in industrialized countries, transport planners in developing countries have a large variety of options to manage transportation demand. Since transportation demand management has already been discussed in depth in the transport literature (McClintock, 2001; Newman and Kenworthy, 1999), this section will highlight how transport planning has to be adapted to the needs of developing countries. But what makes the difference between planning in the North and planning in the South? The following factors are involved:

- a high demand for low-cost means of public transport;
- the abundance of non-motorized transport;
- high-density urban settlements with expected future urban sprawl;
- severe financial constraints on public budgets.

Planning low-cost public transport systems

Individual means of transport have a higher energy consumption per person transported compared to public and non-motorized modes of transport. In general, passengers driving in cars in industrialized countries consume roughly three to five times more energy than passengers transported by public modes of transport.[12] Thus, energy efficiency in transport can be increased tremendously by shifting passengers from individual to public and non-motorized means of transport. Transport demand management encompasses many measures, ranging from traffic calming, alternative work schedules, encouragement of walking and cycling, road pricing to the improvement of mass transport systems, to mention just a few.

Most important for transportation demand management is the choice of the system for mass transportation, since large-scale investments are related to its implementation. Experience in industrialized countries shows that mass rapid transit systems, such as metros, commuter rails or light rail transit, are often not economically viable. Practically no European mass transit system operates without large-scale subsidies. Investment costs are extremely high if subterranean metro lines are to be constructed.

In developing countries the problems are more acute, since public funds are even scarcer than in industrialized economies. Appropriately, low-cost solutions have been developed in Latin America, using high-quality buses for public urban transport. A very efficient bus rapid transit system has been developed with the following features (Wright, 2004, p31):

- segregated busways with transit prioritization at intersections;
- rapid boarding and alighting using pre-paid tickets;
- clean, secure and comfortable stations with real-time information displays;
- modal integration of railways and non-motorized means of transport at bus stations and terminals;
- 'soft' characteristics, such as marketing identity (brand) or advanced customer service.

A comparison with conventional mass rapid transit systems shows that bus rapid transit can achieve results comparable to those given by rail systems as regards carrying capacity and service, but average speed is slower, mainly because of the lack of grade separation and the lower maximum speed achieved on the busways. The main advantage is a significantly lower capital cost, especially when compared to subterranean metro systems. The main disadvantage is the large surface areas required by the buses, especially in the case of four-lane operations.[13]

Planning for non-motorized transport

Walking is the dominant mode of transport for the poor. The poorer the country, the remoter the region, the smaller the integration into markets, then the more non-motorized means of transport dominate the picture. Even though walking consumes less energy, it is not the most efficient means of transport, as Figure 3.5 demonstrates. The graph lists typical costs in Africa for goods transported a medium distance (50 kilometres) on a good road and a short distance (5 kilometres) on a poor road. The most expensive mode of transport is walking.[14] But the good news is that tremendous efficiency improvements can be made by using other low-cost means of transport, such as bicycles and animal carts.

The efficiency improvement in transport which non-motorized means can generate is a lesson to be learned by transport planners, especially in rural areas, where export of agricultural goods is essential in the struggle against poverty. The solution is intermodal logistic transport systems, with the low-cost means taking care of the 'first mile's' journey from the field to the collection point, where the trucks pick up their loads.

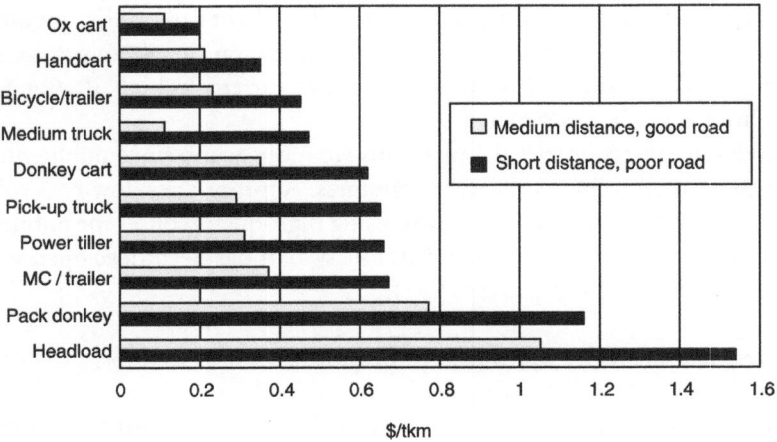

Figure 3.5 *Typical transport costs in Africa*

In urban areas, especially in Asian cities, non-motorized means of transport have an important role in transporting goods and passengers. More often than not, transport planners educated in the North ignore these modes and simply forget to implement adequate facilities, such as walkways, footbridges, bicycle lanes or loading facilities at markets. The lack of these facilities makes walking and cycling even more dangerous, cumbersome, time-consuming and costly. As a matter of course, this lack will have negative impacts on the use of public transport systems.

Improved land use planning

The spatial development in industrialized countries may be characterized as showing a concentration in the cities, which took place during the industrialization phase, followed by a process of massive suburbanization after the establishment of the automobile as the main mode of transport. During this period, urban sprawl took place not only along the railway lines, but in the entire suburban area around conurbations. Cars allowed for a decentralization of urban functions, making it possible to live in the countryside in a detached house and commute to the urban centre for work. Urban sprawl produced spatial structures that result in long commuting distances and developments that are less easy to accommodate to public transport systems. Given the lower energy requirements of public transport, suburbia is a settlement structure that entails large-scale energy inefficiencies in transport. An international comparison of cities (Kenworthy and Laube, 1999) shows that a strong relationship exists between urban density and car use (mileage). Higher-density locations result in lower specific energy consumption in transport.

CAIRO   LOS ANGELES

Figure 3.6 *Similar population but different size*

In many developing countries the process of suburbanization is still in its early stages. Even so-called megacities, such as Cairo, cover a surface area that is only a small fraction of that occupied by cities in industrialized countries, such as Los Angeles, even though the two conurbations have about the same number of inhabitants. However, the tremendous urban growth, reaching levels up to 10 per cent annually, will lead to a rapid expansion of urban areas in most developing countries. Additionally, it can be expected that in the future, income in developing cities will increase and similar processes of suburbanization can be expected in the South.

Since spatial developments are very difficult to reverse, it is advisable to influence urban sprawl before car-dependent settlement structures have developed. The target is to create urban regions amenable to public transport, including a city of short travel distances. A number of urban and transport planning measures can be taken in order to achieve this goal. The city of short distances is liveable, needs less transport infrastructure, produces fewer external effects and consumes less energy in transport. If early measures are taken, the mistakes made in industrialized countries during past decades can be avoided in developing cities.

Financial planning under severe budget constraints

Tight budget constraints and constant lack of financial resources are salient features of developing countries. The fuel sector plays an important role in the economy of these countries. In the transport sector more than 90 per cent of transport modes use conventional fuel, much of which has to be imported. In low-income countries, fuel imports often represent up to one-third of all foreign imports, causing large imbalances of trade. Thus, measures to increase energy efficiency in transport have strong impacts on the whole economy of developing countries.

In addition, the fuel sector can be an important source of revenue for the state. Taxes on fuel and energy may comprise up to 36 per cent of all state taxes – or in the reverse case fuel subsidies may form a burden of up to 8 per cent of the state budget. Worldwide, revenues from fuel taxation finance roads. In Africa a fuel tax amounting to only US$0.10 is sufficient to finance the adequate maintenance of the whole road network. However, an overview of fuel taxation worldwide (GTZ, 2003a) shows that many developing countries do not even generate this minimum amount and thus the road networks continue to deteriorate. This not only implies a depreciation of the state's largest assets, but also forces the individual user to pay for higher transport costs. Every euro saved on maintenance costs the user three additional euros, mainly through increased vehicle operating costs. Therefore, pricing and taxation issues for conventional fuels are of prime importance in developing countries.

Next to the issue of revenue collection, fuel taxes may serve as a means of environmental taxation. Economic instruments are most important to achieve improvements of energy efficiency in transport. The instruments may serve to reinforce or support the above-mentioned measures regarding the introduction of renewable energy sources, transportation demand management and transport-avoiding land use planning. Conventional taxes such as fuel levies and vehicle licence fees may focus on the reduction of environmental effects of transport, such as by favouring more energy-efficient engines. Transport-specific instruments, environmentally oriented vehicle taxation, road tolls and spatially or temporally differentiated road pricing systems may serve as efficient tools with which to move regional transport patterns towards higher energy efficiency.

The potential for changing behaviour should not be underestimated. Training drivers to achieve an economically and ecologically sound driving style is a cheap method to save fuel and money, as well as contributing to road safety and the reduction of greenhouse gas emissions. The average reduction of fuel consumption amounts to 10 per cent. Payback periods of the training lessons are reasonably short. Nevertheless, development of the market for eco-driving is slow, at least in Europe.

CONCLUSIONS FOR INTERNATIONAL DEVELOPMENT COOPERATION

The cheapest and most promising means to reduce energy use in transport is through the improvement of energy efficiency, either by technical improvements or by modern urban planning approaches. This fact should be in the mind of each decision-maker who is interested in reducing environmental damage, human disease and the societal costs caused by the transport sector. But if serious predictions come true (for example, the price of oil remains at high levels of between US$40 and US$60/bbl) or further political targets are implemented (such as the European Directive on biofuels), renewable energies such as ethanol and biodiesel will come to play a more important role in transport. However, in each case energy efficiency should be paid special attention.

Currently, many developing countries are fascinated by biofuels, as they might have positive impacts on their economy and social situation. In industrialized countries the production capacities of fossil fuels as well as the potential of renewables can only satisfy a small share of the energy demand in transport. Therefore, developing countries expect to have the option of producing biofuels for export. As, ultimately, farmers have to produce and sell plants, this might have positive effects on rural development, where poverty prevails. However, social and economic conditions are so complex that no general predictions on sustainability can be made. Here, the key question is: what are the market conditions for the farmers? Can they take part in the creation of value or do they only produce for a very small income?

Furthermore, the ecological question is also not easy to answer. Environmental advantages are related to reduced GHG emissions; this is clear so far. But negative effects may occur through eutrophication, acidification and ozone depletion caused by excessive use of nitrogen fertilizers. Additionally, a serious risk is related to the further reduction and degradation of tropical forests. Appropriate farming methods, such as organic farming, are absolutely necessary to avoid substituting one problem by another.

The above deliberations show that the use of renewables in transport is related to a number of uncertainties regarding their sustainability, and the usage will be limited in the next decade. Therefore, a number of strategies to reduce energy consumption and thus improve energy efficiency in transport are recommended. Improved and more frequent technical inspection and maintenance of vehicles can improve energy efficiency significantly. Other strategies comprise measures from the 'traditional' repertoire of transport planners. Especially in the rapidly growing cities in developing countries, large-scale impacts can be expected. A modal shift towards more energy-efficient means of transport – or the retention of the existing high proportion of people walking, cycling or using transit systems – and the planning of cities with short distances are two issues that will be of crucial importance in order to secure liveable cities and reduce GHG emissions.

An important aspect is the expected urban sprawl of developing cities. If planning mistakes from the North, such as the establishment of car-dependent settlement structures, are to be avoided in the South, a planning approach that integrates transport and regional planning has to be taken. So far, neither donors nor decision-makers in developing cities seem to be aware of the future problems and they have yet to take appropriate measures.

To date, emission reductions from the transport sector have not played a significant role in the Kyoto mechanisms, such as the Clean Development Mechanism or Joint Implementation. Only a very few transport-related projects have been registered under these mechanisms. Action is urgently needed.

Notes

1 The Toyota Prius, which can be used as a family vehicle, consumes 4.3 litres and, especially in urban areas, is without competitors with respect to noise and emissions (owing to the high efficiency of electric engines in urban transport).
2 *China Statistical Yearbook 2004.*
3 *Stuttgarter Zeitung*, 10 September 2005, p11.
4 *China Daily*, 31 October 2004.
5 GTZ (2003b, Module 5a: 'Air Quality Management', p28).
6 GTZ (2003b, Module 4b: 'Inspection & Maintenance and Roadworthiness').
7 GTZ (2003b, Module 4c: 'Two- and Three-Wheelers').
8 ITDG Practical Action Technical Brief on biogas and liquid fuels, www.itdg.org/docs/technical_information_service/biogas_liquid_fuels.pdf (accessed 1 November 2005).
9 Brazilian ethanol production from BEN 2005; US ethanol production from Renewable Fuels Association, 'Industry Statistics', at www.ethanolrfa.org (viewed 1 November 2005); world ethanol production from F.O. Licht World Ethanol and Biofuels Report, cited in Renewable Fuels Association, ibid; European production from Observ'ER, Biofuels Barometer', *Systèmes Solaire* (European Commission Directorate General for Energy and Transport), 13 June 2005.
10 On 9 October 2005 the price per gallon was $1.9597 (New York Harbor).
11 CNN, 19 August 2005.
12 Data used are from *Umweltbundesamt und Verkehr in Zahlen*, 2002/2003.
13 More information on bus rapid transit can be retrieved from GTZ (2003b, Module 3b).
14 For walking, travel time valuations are essential in order to estimate transport costs. Assuming an average load of 25 kg/person and a speed of 3 km/h, the transport of 1 tonne-kilometre requires 27 hours of travel time. Forty return trips covering a distance of 80 kilometres have to be walked.

References and Further Reading

Aßmann, Dirk (2003) *Akteure, Strukturen und Technologien für ein zukunftsfähiges Energiesystem*, Frankfurt am Main, Peter Lang Verlag

Berndes, Göran, Wakker, A., Egging, R., van Thujil, E., et al (2005) *Biofuel and Bio-energy Implementation Scenarios*, final report of VIEWLS WP5 modelling studies, Energy Research Centre of the Netherlands, Unit Policy Studies, Petten, the Netherlands, and Chalmers University of Technology, Energy and Environment, Department of Physical Resource Theory, Göteborg, Sweden

Bundesministerium für Wirtschaftliche Zusammenarbeit und Entwicklung (BMZ) (2003) *Transport und Verkehr – Zukunftsfähige Mobilität für eine gerechte und nachhaltige Entwicklung*, Bonn

Crossley, P. and Ellis, S. (1997) *A Handbook of Rural Transport Vehicles in Developing Countries*, Bedford, Silsoe College, TRL

EIA (1999) *Annual Energy Outlook 2000 – with Projections to 2020*, US Department of Energy

EIA (2004) *Annual Energy Outlook 2005 – with Projections to 2025*, US Department of Energy

Gesellschaft für Technische Zusammenarbeit (GTZ) (2003a) *International Fuel Prices*, 3rd edn, May, Eschborn. Download at worldbank.org/transport or http://zietlow.com/docs/engdocs.htm

GTZ (2003b) *Sustainable Transport: A Sourcebook for Policy Makers in Developing Cities*, Eschborn. See www.sutp.org/docs/sourcebook/sourcebook.aspx

GTZ (2004) *Transport in Developing Countries: Renewable Energy or Energy Reduction?*, a reader prepared for the Side Event at the Renewables 2004 Conference, edited by Niklas Sieber, Eschborn

Goldemberg, J., Coelho, S. T., Nastari, P. M. and Lucon, O. (2003) 'Ethanol Learning Curve – the Brazilian Experience', *Biomass and Bioenergy*, vol 26, no 3, pp301–304

Ingram, G. K. and Liu, Z. (1999) *Determinants of Motorization and Road Provision*, World Bank, Washington, DC

International Energy Agency (IEA) (2004) *Biofuels for Transport – An International Perspective*, Paris, IEA

Johansson, T., McCormic, K., Neij, L. and Turkenburg, W. (2004) *The Potentials of Renewable Energy*, Thematic Background Paper, International Conference for Renewable Energies, Bonn

Kartha, S., Leach, G. and Sudhir Chella Rajan (2005) *Advancing Bioenergy for Sustainable Development*, Guideline for Policymakers and Investors, three vols, Stockholm, Stockholm Environment Institute, Energy Sector Management Assistance Program (ESMAP)

Kenworthy, J. and Laube, F. (1999) *An International Sourcebook of Automobile Dependence in Cities 1960–1990*, Updated Edition, Boulder.

McClintock, H. (2001) *Comprehensive Transportation Planning Bibliography*, Institute of Urban Planning, University of Nottingham, www.nottingham.ac.uk/sbe/planbiblios/bibs

Newman, P. and Kenworthy, J. (1999) *Sustainability and Cities: Overcoming Automobile Dependency*, Island Press, Boulder

Quirin, M. et al (2004) *CO$_2$ Mitigation through Biofuels in the Transport Sector*, Heidelberg, IFEU Institut für Energie und Umweltforschung (IFEU)

Schmitz, N. (ed) (2005) *Innovationen bei der Bioethanolerzeugung und ihre Auswirkungen auf Energie und Treibausbilanzen*, Schriftenreihe 'Nachwachsende Rohstoffe', Band 26, Fachagentur Nachwachsende Rohstoffe, Gülzow

Sieverts, T. et al (1998) *Zwischenstadt zwischen Ort und Welt, Raum und Zeit, Stadt und Land*, Braunschweig and Wiesbaden, Bauwelt-Fundamente, Verlag Vieweg

Weber, M. (2004) *Bioethanol – Herstellung in Deutschland und Brasilien. Ein technischer, ökonomischer und ökologischer Vergleich*, Offenburg, University of Applied Sciences

World Bank (1994) *World Development Report 1994 – Infrastructure for Development*, Oxford University Press, New York
World Business Council for Sustainable Development (WBCSD) (2001) *Mobility 2001 – World Mobility at the End of the Twentieth Century and Its Sustainability*, WBCSD, Geneva
Wright, L. (2004) 'The Limits of Technology: Achieving Transport Efficiency in Developing Nations', GTZ, *Transport in Developing Countries: Renewable Energy or Energy Reduction?*, a reader prepared for the Side Event at the Renewables 2004 Conference, edited by Niklas Sieber, Eschborn
WTRG (2005) *Brent Spot Prices*, 25 January 2005, www.wtrg.com, London, WTRG Economics

PART TWO

POLICIES

4

National Policy Instruments: Policy Lessons for the Advancement and Diffusion of Renewable Energy Technologies Around the World

Janet L. Sawin

EXECUTIVE SUMMARY

For renewable energy to make a significant contribution to economic development, job creation, reduced oil dependence and lower greenhouse gas emissions, it will be essential to improve the efficiency of technologies, reduce their costs and develop mature, self-sustaining industries to manufacture, install and maintain renewable energy systems. The goal must not be simply to install capacity, but to provide the conditions for creation of a sustained and profitable industry, which, in turn, will result in increased renewable energy capacity and generation, and will drive down costs. To achieve this end, a viable, clear and long-term government commitment is critical. Also essential are policies that create markets and ensure a fair rate of return for investors.

During the past decade the world has witnessed double-digit growth in the wind and photovoltaic (PV) industries, significant advances in these technologies, and dramatic cost reductions. Today half a dozen countries represent roughly 80 per cent of the world market for these technologies. Those countries have demonstrated that it is possible to create vibrant markets for renewable energy and to do so very rapidly; but the record also shows that the renewable energy policies of most countries have been unsuccessful to date.

Most of the renewable energy development experienced thus far has been driven by countries with feed-in, or pricing, systems. At the same time, a combination of policies is required, including standards, education, stakeholder involvement, and incentives to bring down the initial costs of investment and reduce risk, whether real or perceived. Ultimately, the effectiveness of policies in promoting renewable energy will depend on their design, enforcement, how well they address national circumstances, and the extent to which they are consistent and sustained.

INTRODUCTION

For renewable energy to make as large as possible a contribution to economic development, job creation, lower oil dependence and reduced greenhouse gas emissions, it is essential to improve the efficiency of technologies, reduce their costs and develop mature, self-sustaining industries to manufacture, install and maintain those systems. Today's energy markets include several obstacles that frustrate efforts to achieve these goals. Among them are lack of access to the electric grid at reasonable prices, high initial cost compared to conventional energy sources, and a widespread lack of awareness about the scale of resources available, the pace of development of renewable technologies (RETs), or the potential economic advantages of renewable energy.

These barriers have been largely overcome in several countries, allowing a period of sustained double-digit growth in the solar and wind markets over the past decade, and providing policy models for other countries to adopt. These successful models show that a sustained renewable energy market can be developed quickly and efficiently if the right combination of policies is in place.

This chapter examines which policies have been most effective in promoting RETs. It focuses primarily on grid-connected electricity and vehicle fuels, and briefly on remote uses of photovoltaics (PV) and heat systems, with the assumption that policy recommendations can be carried over to other types and uses of RETs. 'Success' of policies is defined to cover positive impacts on a range of factors, including installed capacity and energy generation from renewable energy technologies; technological advances; reductions in cost and price; domestic manufacturing capacity and related jobs; and public acceptance.

Major barriers to renewable energy

While most renewable fuels are free, renewable energy projects have high up-front costs, and a number of factors combine to make many renewable energies more expensive than conventional energy. Distortions resulting from unequal tax burdens and existing subsidies, and the failure to internalize all costs and benefits of energy production and use, erect high barriers to RETs. Additional cost barriers range from the cost of technologies themselves (and the need to achieve economies of scale in production) to the lack of access to affordable credit, and the costs of connecting with the grid and transmission charges, which often penalize intermittent energy sources. Import duties on RETs and components also act to make renewable energy more costly. As a result, many of the cost barriers to renewables are perceived rather than real.

In many countries, electric utilities maintain monopoly rights to produce, transmit and distribute electricity. High costs or a lack of standards for connection and transmission also discourage renewable energy projects. And everywhere, renewable energy must compete with financial and regulatory systems that have evolved to promote the development and use of fossil fuels and nuclear power, and that often discriminate against the use of RETs.

In addition, lack of information about available renewable energy resources and about the current state of RETs, misperceptions, lack of experience or training, negative past experiences with old technologies, and a lack of understanding about the benefits associated with renewable energy all act as barriers to their use. Each of these factors works to increase the perceived risks – technical and financial – of investing in renewable energy.

For the most part, barriers in developing countries are similar to those in the industrial world. However, specific national characteristics, particularly within the developing world, can play an important role in determining barriers from one country to the next. Additional obstacles in many developing countries include long travel distances to remote areas, poor transport and communication infrastructure, lack of trained personnel, and low literacy rates (Martinot et al, 2002). In addition, the perceived risk of investing in renewable energy projects in developing countries is high, owing to uncertainties about political, regulatory and market stability (Frost, 2003). In the past, donor aid has inhibited commercial markets, often reducing perceived value of RETs (if they are free), while focusing little on models of development that can be viable, sustainable and replicable. And if people expect to be connected to the grid soon, as a result of unrealistic plans for grid extension or political promises, there is no incentive to invest in alternatives (Martinot, 2003).

Finally, even government policies that are enacted to promote RETs can have negative impacts if they are inappropriate, inconsistent, or are too short-term.

Policy mechanisms to be discussed

Governments have a number of options that they can use to promote renewables. The first is to support the use of voluntary measures, particularly through education and information dissemination. This option has varying and limited effects. The second option involves environmental standards or energy taxes. The third option is to promote RETs through direct support, which is the focus of this chapter. Generally, a mix of instruments is essential for success. The combination of policies needed depends on the costs of the technology used, location and conditions.

There are five major categories of relevant policy mechanisms:

1   regulations that govern capacity access to the market/electric grid and production or purchase obligations;
2   financial incentives;
3   industry standards, permitting and building codes;
4   education and information dissemination;
5   stakeholder involvement.

Each of these policy mechanisms will be discussed. There is not necessarily a direct link between these policy mechanisms and specific obstacles to greater use of renewable energy, as some of the policy options tackle a combination of barriers. An additional critical element is the need for a general change in government perspective on and approach to energy policy.

Government investments in research and development (R&D) are also important. Ultimately, however, it is only by creating a market for renewables that the technological development, learning and economies of scale in production can come about to further advance RETs and reduce their costs. And as markets expand and industries grow, more private money is drawn into private research and development, which is often more successful than public R&D (Sawin, 2001).

Some key points to consider

The key points to frame the discussion are as follows:

• Experience to date demonstrates that considerable intervention in energy markets is required to introduce significant amounts of renewable energy into the mix. Every country that has succeeded thus far in developing renewable energy on a substantial scale has been committed over the long term to this goal, with consistent policies that include a package of policy mechanisms.
• The effectiveness of government policies depends on how well they are designed and whether or not they are enforced. The use of a particular policy type does not guarantee success. In addition, policy-makers must be cognizant of the projects and technologies they are trying to promote, as such decisions determine the necessary policy framework. For example, to promote technologies such as PV, solar thermal, heat pumps and wind turbines on a small-scale, distributed basis, support should be granted to the end customer; to promote large wind, biomass, geothermal or marine technologies, the investment is more likely to be channelled through a large entity

or company (comments on draft text by Robert Kleiburg, Vice President Strategy and Planning, Shell International Renewables, provided 28 November 2003). Further, each country has unique circumstances and must design its own system and enact a combination of policies, based on needs, circumstances and available resources.

- The experiences of countries such as Denmark, Germany, Japan, Spain and Brazil have demonstrated that the key to steady and significant cost reductions is the development of consistent and reliable markets. Such conditions allow for the entry and maturation of small- and medium-scale enterprises, which have provided the bulk of the technological innovation that has driven down renewable energy costs. In addition to the 'global learning curve' that exists for technologies such as wind turbines and PV cells, there is a 'national learning curve' as individual countries develop domestic industries to manufacture, install and maintain renewable energy systems using local equipment and labour. Countries that do not yet have sizeable industries in place can expect dramatic price reductions in the first few years after effective new policies are introduced.

- Most of the policies to be discussed involve some sort of subsidy, direct or indirect. Energy markets are not now and never have been fully competitive and open, and today's markets include substantial institutional barriers, as well as long-term subsidies for conventional energy, that act as obstacles to renewable energy. Even market-oriented countries such as the US and United Kingdom now agree that subsidizing renewable energy makes sense. Support for renewables is important not only to incorporate the external costs (environmental, social and security) of energy production and use, and make up for decades of past support for conventional energy. It is also essential to account for the environmental, social and security benefits associated with renewables – including the reduced risk of fuel price volatility, a more diversified portfolio of energy options, a cleaner environment and better health, and job creation and economic development. Well-designed, modest production-based subsidies provided up front can work rapidly to close the cost gap between renewables and conventional energy systems.

- To date, feed-in – or pricing – systems have been responsible for most of the additions in renewable electricity capacity and generation, while driving down costs through technology advancement and economies of scale, and developing domestic industries. The record of quota systems is more uneven thus far, with a tendency towards stop and go, and boom and bust markets. It is important to recognize that both quota and pricing systems involve subsidies. But pricing systems have provided increased predictability and consistency in markets, which in turn have encouraged banks and other financial institutions to provide the capital required for investment.

- In developing countries, markets are apt to be particularly sensitive to the need for relatively uncomplicated access to the electric grid and low transaction costs. Pricing laws allow for ease of entry into the marketplace and tend to favour smaller companies and incremental investment, making them particularly suited to developing countries, where power markets are often small and dispersed. As in the industrial world, it is critical to focus on development models that are viable, sustainable and replicable, and that emphasize local participation and ownership; to date, donor aid projects have tended to reduce the perceived value of renewable energy while inhibiting commercial markets.

REGULATIONS GOVERNING MARKET ACCESS

Access to the market, such as the electric grid, is imperative for renewables to gain a foothold. It is also important that systems not discriminate against, or give preference to, any particular technology or generator with regard to access and charges for grid connections and transmission. However, many countries have transmission-pricing penalties for intermittent generation. And in many cases, where power markets have been opened to competition, investment in renewables (which are capital-intensive) has been hindered in the absence of strong policies to promote renewable energy. In general, the impacts of privatization have depended on the specific policies and regulations in place.

The regulatory framework is at least as important as subsidies for renewable energy. Two main types of regulatory policies have been used to open electricity grids to RETs. One guarantees price, another ensures market share through government-mandated targets or quotas. The first is the fair access and standard pricing law, also called the renewable energy feed-in law.

Feed-in laws: pricing systems

Under the feed-in law – or pricing system, as it will be called from here on – electric utilities are obligated to enable renewable energy plants to connect to the electric grid, and they must purchase any electricity generated with renewable resources at fixed, minimum prices. These prices are generally set higher than the regular market price, and payments are usually guaranteed over a specified period of time. Tariffs may have a direct relationship with cost or price, or may be chosen instead to spur investment in renewables.

The precursor to the pricing law was enacted in California during the 1980s. The US Public Utilities Regulatory Act of 1978 (PURPA) required utilities to interconnect with and buy energy from 'qualifying facilities', including renewable energy plants, at incremental or avoided costs of production. In California the implementation of PURPA involved the use of standardized long-term contracts with fixed (and, in some cases, increasing) payments for all or part of the contract term. The costs of the contracts were covered through higher electric rates for consumers. While these contracts proved costly, it is widely believed that the alternative (nuclear power) would have been even more expensive. The duration of the contracts (15–30 years for wind projects), combined with fixed energy prices for much of that time, assured producers of a market for their product and finally gave them something they could take to the bank to obtain financing. While most other US states saw little development during the 1980s, California for a time became the world's leader in renewable energy use (Sawin, 2001).

The early pricing laws in Europe, in Denmark and Germany, also required that utilities give small wind and other private generators access to the electric grid, and they guaranteed producers a minimum share of the retail rate – at least 85 per cent in Denmark, and 90 per cent in Germany. The German system was revised in 2000, and today most pricing laws provide a fixed payment for renewable electricity that varies by technology type, plant size and occasionally by location, and is generally based on the costs of generation. Payments guaranteed to new projects in Germany decline annually, and are adjusted every two years. The tariffs last for 20 years from date of project installation. German electric utilities now qualify for these payments as well (discussion with G. Gerdes of the Deutsches Windenergie Institut (German Wind Energy Institute, DEWI), 7 December 2000).

The costs of higher payments to renewables are covered by an additional per kilowatt-hour (kWh) charge on all consumers according to their level of use (e.g. Spain, Germany as of 2000), a charge on those customers of utilities required to purchase green electricity (e.g. Germany until 2000), or by taxpayers, or a combination of both (Denmark through feed-in rates and reimbursement of the carbon tax). Laws similar to Germany's pricing law had been enacted in more than 35 states, provinces and countries as of mid-2005 (Martinot, 2005).

While pricing laws have not succeeded in every country that has enacted them, to date those countries that have experienced the most significant market growth and have created the strongest domestic industries have had pricing laws. In order to succeed, tariffs must be high enough to cover costs and encourage development of particular technologies; they also must be guaranteed for a time period long enough to assure investors of a sufficient rate of return. The success of pricing laws is also determined by factors such as charges for access to the electric grid, limits set on qualifying capacity, and the ease of permitting and siting (influenced by the existence and specifics of national or regional standards).

A variation on pricing laws, 'net metering', permits consumers to install small renewable systems at homes or businesses and then to sell excess electricity into the grid. This excess electricity must be purchased at wholesale market prices by the utility. In some cases, producers are paid for every kilowatt-hour they feed into the grid; in other cases they receive credit only to the point where their production equals their consumption. This option is available in at least 35

US states and at least 32 countries, including Japan, Thailand and Canada (DSIRE, 2005; Martinot, 2005). It is of benefit to electricity providers as well as system owners, particularly in the case of PV, because excess power generated during peaking times can improve system load factors and offset the need for new peak load generating plants.

Net metering differs from the access and pricing laws in Europe primarily in scale and implementation. Success in attracting new renewable energy investments and capacity depends on limits set on participation (capacity caps, number of customers, or share of peak demand); on the price paid, if any, for net excess generation; on the existence of grid connection standards; and on enforcement mechanisms. Without other financial incentives, net metering is not enough to advance market penetration. Neither California nor Texas saw much benefit from net metering for wind power, let alone for more costly renewables like solar PVs, until other incentives were added to the mix. However, net metering would most likely have a greater impact if private generators received time-of-use rates for the electricity they fed into the grid – particularly in the case of PVs, which generate electricity at peak demand times when the value of their power is highest. Mandated targets or quotas, to be discussed in the next subsection, and net metering can be used simultaneously.

Quotas: mandating capacity or generation

While pricing laws establish the price and let the market determine capacity and generation, mandated targets work in reverse: the government sets a target and lets the market determine the price. Typically, governments mandate a minimum share of capacity or electricity generation (generally grid-connected only), or a share of fuel, to come from renewable sources. The share required often increases gradually over time, with a specific final target and end-date. The mandate can be placed on producers, distributors or consumers.

The simplest form of quota system is one in which the government imposes a mandate on one producer/supplier. For example, during the 1990s the Minnesota Public Utilities Commission ordered the electric utility Northern States Power to install successive amounts of wind energy capacity, thereby helping to open up the wind market in that US state (Sawin, 2001). Quotas have also been used to promote the use of renewables off the grid, including alternative fuels. The European Commission aims for biodiesel to represent 5.75 per cent of diesel fuel by 2010, and Brazil has become the world leader in ethanol by requiring since 1975 that it make up a specific share (generally in the 20–24 per cent range) of all gasoline sold in Brazil.

The use of quotas for renewable electricity is a relatively new type of policy, first introduced in the late 1990s, so there is relatively little experience with quota systems to date. There are two main types of quota systems used today for electricity generation: obligation/certificate and tendering systems.

The Renewables Portfolio Standard (RPS), widely used in the US, is in the former category. Under an RPS, a political target is established for the minimum amount of capacity or generation that must come from renewables, with the amount generally increasing over time. Investors and generators then determine how they will comply – the technology to be used (unless specific targets are established by technology type), the developers to do business with, and the price and contract terms they will accept. At the end of the target period, electricity generators (or suppliers, depending on policy design) must demonstrate, through the ownership of credits, that they are in compliance in order to avoid paying a penalty. Producers receive credit – in the form of 'Green Certificates', 'Green Labels' or 'Renewable Energy Credits' – for the renewable electricity they generate. Such credits can be tradable or sellable, to serve as proof of meeting the legal obligation and to earn additional income (some countries have set floors and/or ceilings for the value that these certificates can achieve). Those with too many certificates can trade or sell them; those with too few can build their own renewable capacity, buy electricity from other plants (generally through a bidding process) or buy credits from others. Once the system is established, government involvement includes certifying of credits, and compliance monitoring and enforcement.

Under tendering systems, regulators specify an amount of capacity or share of total electricity to be achieved, and the maximum price per kilowatt-hour. Project developers then submit price bids for contracts. The United Kingdom's Non-Fossil Fuel Obligation (NFFO) was an early example of this policy type. Governments set the desired level of generation from each resource, and the growth rates required over time. The criteria for evaluation are established before each bidding round. In some cases, governments require separate bids for different technologies, so that solar PV is not competing against wind energy projects, for example. Generally, proposals from potential developers are accepted starting with the lowest bid and working upwards, until the level of capacity or generation required is achieved. Those who win the bid are guaranteed their price for a specified period of time; on the flip side, electricity providers are obligated to purchase a certain amount of renewable electricity from winning producers at a premium price. The government covers the difference between the market reference price and the winning bid price. Each bidding round is a one-time competition for funds and contracts. In contrast, under the RPS, companies and projects must constantly compete in the marketplace, with existing and new projects, unless they have signed long-term contracts.

As with the pricing law, the additional costs of renewable energy under quota systems are paid through a special tax on electricity or by a higher rate charged to all electricity consumers.

More than one-third of US states have mandated quotas through RPS laws, representing a significant share of the nation's total electricity load. Outside the

US, quota systems had been enacted in more than 15 states, provinces and countries by mid-2005 (Martinot, 2005).

Pricing systems compared with quota systems

Quota systems can be used for a range of technologies and fuels, while the pricing laws and RPS and tendering systems can be used only for electricity. The discussion that follows covers electricity only. It looks at several issues that have been raised and debated regarding pricing laws and quotas, from their impacts on renewable capacity and generation, to costs and innovation.

Renewable energy capacity and generation

Because quota systems establish specific targets for renewable capacity or generation there is, theoretically,[1] certainty regarding the future share of the market, and quotas can be tied directly to other government policies such as emissions reductions. Quotas can provide producers and manufacturers with a predictable, steadily growing market for renewable energy (Lauber, 2004). With pricing laws it is not possible to know in advance how much generation or capacity will result or whether the share of renewable energy generation will increase over the long term. However, tariffs can be adjusted up or down to encourage more or less investment in renewable energy in order to bring installations in line with desired targets. In addition, under a quota system, the speed with which technologies are introduced is based on a political decision that might be largely unrelated to technical progress and the efficiency of using renewable energy (Krohn, 2000b). Those countries with pricing laws have regularly surpassed national renewables targets (Menanteau et al, 2003; Lauber, 2004; Meyer, 2003).

There are some concerns that mandated targets or quotas can set the upper limit for development. At least to date, Texas is evidence that this is not necessarily the case, as current renewables capacity in Texas is well above the required level. An RPS law is primarily responsible for the rapid growth of wind energy in Texas since 1999, when the state required that 2000 MW of additional renewable capacity be installed within a decade. Texas was more than halfway there with wind alone by the end of 2002, and the target was nearly met by early 2006. In 2001, Texas installed more wind capacity than the entire US had ever had before during the course of one year. At the same time, the combination of excellent winds in Texas and the US federal Production Tax Credit (to be discussed later) make wind energy cost-competitive or better, so it pays to invest in wind energy. Where circumstances are different, there might be no incentive to install more than the mandated amount of renewable capacity. As this type of system is still relatively new, it is not possible to know whether this will be the case.

Other US states with RPS laws have had far less success to date than Texas, at least in part because of the fact that some RPS requirements are not well designed. For example, they apply only to a small segment of the market, they have uncertain purchase obligations and/or end-dates, penalties for non-compliance are too low – or they are not enforced. Any one of these factors can limit the potential for a quota system to advance renewable energies (Wiser et al, 2000). But again, most of these laws have not been in place long enough for it to be possible to determine what their ultimate impact will be.

As with pricing laws, many of these problems associated with quota systems can be overcome with careful system design. Regardless, some analysts believe that the lower purchase prices common under bidding or quota systems result in lower levels of installed capacity (Menanteau et al, 2003). In fact, pricing laws have consistently proved most successful at promoting the growth of renewable electricity capacity and generation. While more than 46 countries have installed wind capacity, just three – Germany, Spain and Denmark, all of which have used pricing laws – accounted for 59 per cent of total installed capacity by the end of 2004.

When Spain passed a pricing law in 1994, relatively few wind turbines were spinning in the Spanish plains or mountains; by the end of 2002 the country ranked second in the world for wind installations, surpassed only by Germany. These advances in renewable capacity and electricity generation have translated into successes in other areas as well, from job creation and economic development to reductions in greenhouse gas emissions. For example, Denmark's CO_2 emissions dropped 11 per cent between 1990 and 2002, despite a 28 per cent increase in gross national product, due primarily to increased use of renewable energy and fuel-switching from coal to natural gas (DME, 2003).

Spain has had less success with solar PV. In 1998, Spain set PV tariffs that were similar to those provided in Germany. While Germany's market took off, Spain installed little PV capacity over the next few years, despite having better solar resources, because major barriers remained. For example, owing to a lack of grid connection regulations, utilities could set their own, often exorbitant, charges to cover safety and other factors. Once this issue was resolved in 2001, with national technical standards for grid connection, another barrier remained. PV producers who sold electricity into the grid, even households, had to register as businesses and pay income tax on their sales, a cumbersome and costly process that discouraged potential projects (Bravo, undated; Miquel Muñoz, affiliated with the Institut de Ciència i Tecnologia Ambientals, Universitat Autònoma de Barcelona, numerous emails and conversations, November and December 2003). Significant growth in Spanish PV manufacturing during these years was attributable primarily to the nearby German market (Ristau, 2003).

In France, creation of a pricing law was soon followed by a wave of applications for grid-connected windfarms – about 16,000 MW by early 2003 (Forster, 2003). But just over 180 MW had been installed nationally by then (BTM Consult ApS, 2003), and only 386 MW were installed by the end of 2004 (EWEA, 2005). High hurdles, including onerous building approval procedures and turbine spacing and capacity limits, have slowed development (Forster, 2003; Choy, 2003). Italy's pricing law had little success for several reasons, including a lack of confidence in the continuity of the policy, lack of standards, difficulty in obtaining financing, and problems in accessing the electric grid (Dieter Uh, Thematic Advisor, Secretariat, International Conference for Renewable Energies, comments on draft text, provided 2 December 2003, and email to author dated 8 January 2004).

Clearly, success with the pricing law is dependent upon the specifics of the law, and other policies enacted in parallel, particularly connections standards and charges.

Technological innovation, domestic industries and economic benefits

Many analysts have argued that pricing laws do not encourage innovation (Martinot, 2002). Indeed, generous tariffs alone are no guarantee that a domestic industry will develop. For instance, for most of the 1990s, renewable energy producers in Italy received more generous payments than did those in Germany, yet there was little impact on Italy's manufacturing industries, despite significant wind resources, largely because of obstacles already discussed (Lauber, 2004). Others argue that once producers achieve a certain level of profit, they invest in private R&D to reduce costs and increase profit – a situation that is more favourable to 'radical innovations' that require long payback periods than the circumstances created under quota systems (Lauber, 2004; Menanteau et al, 2003). With pricing systems, technological improvements increase profits, thereby encouraging innovation.

Under quota systems, the surplus may go entirely to consumers and, as a result, producers do not receive enough profit (or reliable long-term profits) to invest in R&D and thereby reduce their costs. At the same time, pressure to minimize costs under quota systems often encourages producers to turn to overseas manufacturers of technology (Lauber, 2004; Menanteau et al, 2003). In the United Kingdom, under the NFFO, developers turned to foreign technology to keep costs down, and it became unprofitable for domestic manufacturers to remain in the market (Martinot, 2002). Further, bidding rounds can be time-consuming and costly, and can create cycles of stop and go.

Because quotas often create on–off cycles, they do not allow for continuous market development, they discourage innovation, and they make it difficult to establish a strong domestic industry because investment in production facilities will take place only with a short-term perspective. This in turn limits potential job growth and economic development benefits associated with renewable energy (Wagner, 2000; Martinot and Reiche, 2000).

No matter what type of policy is used, companies will try to maximize their profits. But in order to drive down system costs, it is essential to have sustained and growing markets. To date, payment systems have most consistently provided such markets. Success of the wind industries in Denmark, Germany and Spain seems to bear this out. Turbine manufacturers in these three countries account for the majority of the world's turbine market, supplying about 82 per cent of the market in 2004, and have driven most of the technological development in the wind industry (BTM Consult ApS, 2005). By 2003, about 100,000 people worldwide were employed in the wind industry; of these, three-quarters live in the European Union, with nearly half in Germany (Millais, 2003; Sawin, 2003; Cox, 2003). Approximately 130,000 people work in the renewable energy industries in Germany (Höhn, 2003; Martinot, 2005). In Spain, about 350 companies were involved in renewables industries by late 2003; in Navarre and Castilla-La Mancha alone, a new high-tech industry and more than 3600 new jobs have been created by the growth in renewable energy markets (IDAE in Hernández Gonzálvez, 2003; Iturriagagoitia, 2002).

Geographic and ownership distribution
Quota-type schemes tend to promote the least-cost projects, thus restricting them geographically to the areas with the best resources, and encouraging larger-scale, centralized projects (Lauber, 2004; Meyer, 2003). In Texas, for example, most wind capacity has been installed in the windiest (western) part of the state.

Such was the case with the early pricing laws as well. In the early 1990s most wind development in Germany occurred in the northern coastal states, leading regional utilities and consumers to bear the greatest financial burden of renewables development, and creating strong local opposition to wind energy (J. Twele, Bundesverband WindEnergie eV (BWE), Berlin, Germany, discussions with author, April 1999 and 5 December 2000; Sawin, 2003). This problem was overcome by adjusting payments to reflect differing costs of production in different regions – for example, wind turbines erected in areas (onshore) with better wind resources have received lower payments than others.

Thus, some consider the pricing system to be a more flexible means for exploiting available resources because it allows for development in areas with varying levels of resource potential, assuming that tariffs vary by location (Meyer, 2003). Such adjustments would be necessary under a regional system as well, to ensure that development is more evenly dispersed. This does make pricing systems more complex, however. Similar adjustments are not possible under quota-type systems, although specific targets could be set by region, state or country.

This situation raises concerns among some analysts that quota systems could have negative impacts on public acceptance of RETs (due to heavy development in particular regions) and political support. Moreover, countries with relatively few resources (and thus experiencing less development, little job creation, etc.) would be less willing to support the more ambitious promotion of renewables in the future (Lauber, 2004). The Netherlands provides an example of this case: the government established a voluntary quota system with tradable credits that resulted in increased use of renewable energy. But about three-quarters of the credits and accompanying subsidies went to foreign producers, leading the government to abandon this system (Lauber, 2003).

In the longer term there is concern that relying only on the windiest or sunniest regions will be insufficient to meet the growing need for renewable energy. In a system with one common price for green certificates, for example, there will be no incentive to develop in less resource-intensive areas if prices are too low; but if prices are too high, developers of good sites will make windfall profits (Meyer, 2003). A system will then be required that promotes the use of a variety of renewable resources in areas with vastly different potential (Meyer, 2003).

Some argue that pricing laws offer no inherent incentive for utilities to reduce institutional barriers to development of renewable energy. In fact, utilities can be driven to raise them, requiring the implementation of grid connection and charging standards (Lauber, 2004). On the other hand, the lack of need for negotiated contracts, combined with the fact that anyone has the right to install RETs on their property and sell electricity to the grid, tends to ease entry into the marketplace. Pricing laws tend to favour smaller companies (even individuals or cooperatives) and incremental investment, leading to varying sizes of companies and projects. This aspect of pricing laws makes them particularly suited to developing countries, where power markets are often small and dispersed (Christopher Flavin, President of the Worldwatch Institute, 2003, and member of the International Steering Committee for Renewables, 2004, multiple conversations in November and December 2003).

Quota systems are more likely to fully integrate renewables into existing electricity supply infrastructure as they put utilities in charge. At the same time, they could also result in serving primarily the interests of major suppliers or utilities (Lauber, 2004). Because they rely on competitive bidding, quotas can limit participation to large players, concentrating renewable energy development in the hands of a few, often the major power generators. For example, one company in the US (a subsidiary of the utility Florida Power & Light) is the nation's largest producer of wind and solar power (FPL Energy, 2005). Local or smaller projects are often unable to compete with larger ones on the basis of cost alone (Wagner, 2000). Local investors are rarely wealthy, particularly in rural areas, and can seldom assume the risks and uncertainties associated with development under quota systems.

Pricing systems also enable the average citizen to benefit from investments in renewable energy projects, and encourage installation of the most optimally sized project for a location, rather than capacities that meet only individual household or business needs. Also, because development is more geographically dispersed under pricing systems, there is generally less opposition to projects at the local level. As a result, the German pricing law has created a constituency in favour of renewable energy, such that farmers, lawyers, union workers, landowners, construction companies, renewable energy companies and others have lobbied alongside banks in favour of the law. This broad support has helped to overcome powerful political opposition that favours support for existing fossil fuel and nuclear power over that for renewable energies.

Technological diversity

Because quota systems tend to encourage the least-cost technologies, they are best at promoting technologies that are closest to market competitiveness (Espey, 2000). For example, in Texas, where wind energy has advanced so rapidly, the RPS has done little to encourage the use of more expensive technologies such as solar PVs, despite the vast solar resources in Texas. Higher-cost renewables, such as PV, offshore wind and marine energy, cannot compete against the lowest-cost technologies, meaning that quota systems on their own will not create markets for them and thus will not drive them down their 'learning curves' (comments by Robert Kleiburg, 2003, as before).

Pricing laws, on the other hand, encourage a diversity of technologies, assuming that payments vary according to technology type. Because they can create a market for all renewables, they can more easily support technologies from early development to market competitiveness. Quota systems might get around this shortcoming with specific standards for different technologies, as with the RPS enacted in the US state of Nevada (requiring solar to constitute a specific share of generation). However, this tends to be more complicated, and to date there is not enough experience with this policy to know how effective it will be.

Costs, prices and competition

Quota-based systems are not inherently cheaper, nor are pricing systems inherently more costly; the costs per unit of electricity produced depend on the details of those systems.

It had been argued that it is difficult to control the costs of pricing laws over the short term, whereas subsidies can be controlled under bidding or quota systems. If tariffs under pricing laws are set too high, they can encourage significant development and dramatically increase electric rates; if they are not high enough, little development will occur (Wiser et al, 2000). The pricing law could be more expensive than tendering programmes or an RPS per kilowatt-hour of electricity produced. According to one estimate, in 1998 the Danish government paid out more than €100 million in subsidies to renewable energy, and additional costs in Germany totalled €200 million in 2000 (Menanteau et al, 2003). However, several studies have concluded that the average additional cost per German household has been minimal.[2]

It is also argued that there is less competition and cost minimization under pricing laws than with quota systems, in which developers must compete to win bids or gain contracts. Historically, it has been assumed that pricing laws do not inherently encourage cost or price reductions, and do not ensure least-cost development. The pricing law can drive down costs by driving economies of scale and innovation, and manufacturers and developers will compete for the lowest possible costs in order to achieve higher profit margins, which promotes cost reductions. Yet developers have little incentive to pass these cost savings on to consumers as long as tariffs remain unchanged. Furthermore, under pricing systems, utilities and customers in resource-rich areas might bear the main burden of costs associated with renewable energy development.

However, pricing policies can address such concerns with regular adjustments to tariffs that respond to changes in technologies and the marketplace. This is now the case in several countries, including Germany, where the law was changed in 2000 from being based on a percentage of the retail rate to being based on fixed tariffs. Furthermore, tariffs can be established with help from research institutes (neutral consulting) and the renewables industry (with insight into production costs), as in Germany. The introduction of declining tariffs has brought the costs of the pricing and quota systems much closer together (Menanteau et al, 2003). And at least one analyst believes that pricing laws have delivered renewable electricity more cheaply than have quota or green-certificate policies (Environment Daily, 2003).

There is also evidence, according to Nitsch et al, that it may be cheaper to provide significant national investment over a period of perhaps 15–20 years – in order to bring RETs rapidly down their learning curves and reduce costs very quickly – rather than to introduce renewable energy relatively slowly, with an associated slower reduction in costs (Nitsch et al, 2001/2002; Dieter Uh, email to author dated 8 January 2004).

Further, as already discussed, pricing systems encourage development of local manufacturing industries, which leads to a large number of companies and in itself creates competition. Even where pricing laws are more expensive per unit of energy produced, they drive technological development and strengthen or establish new businesses, thereby supporting industry and agriculture, leading to job creation and furthering economic growth (Dieter Uh, comments on draft text, provided 2 December 2003).

Pricing laws, if well designed, can avoid the need for a host of additional subsidies. They also help to 'internalize external costs' of conventional energy and compensate for the benefits of renewable energy (Andreas Wagner, GE Wind and European Wind Energy Association, personal email communications, 10 and 18 September 2002; personal discussion, 10 September 2002). In early years, pricing laws encourage higher growth rates than quota systems generally do, and they encourage long-term innovation. Finally, concerns about heavy burdens in resource-rich areas can be addressed, as was the case in Germany, by spreading the costs around the entire country so that regions pay according to their total electricity consumption, rather than according to resource base.

Quota systems are generally credited with encouraging competition and dramatically driving down the cost and price of renewable energy. This appears to be true in a number of cases. One example often cited is the decline in wind energy prices under the United Kingdom's NFFO during the 1990s. Wind bids declined dramatically, from US$0.189/kWh in the first round to US$0.043/ kWh in the last (Wiser et al, 2000). At the same time, it is unclear whether these reductions resulted from the quota system (Espey, 2000). There is evidence that at least part of the reductions was due to the pricing policies of other countries, which drove technological improvements and brought down costs (Moore and Ihle, 1999). In addition, some of the later cost reductions under the NFFO were due to changing terms and conditions, including a longer contract period (comments by Robert Kleiburg, 2003).

There is also speculation that the low costs and prices driven by the RPS in parts of the US and Australia are due, at least in part, to the availability of wide open spaces with good resources. This could explain the difference in wind energy costs between those countries and Germany and Spain (Lauber, 2004). Taking into account the relationship between wind speeds and the resultant power output (wind power is proportional to the cube of wind speed), costs under quota systems will come more in line with those of pricing laws once the best resources are no longer available (Dieter Uh, comments on draft text, provided 2 December 2003).[3]

Particularly early on, when a country has few domestic manufacturers or developers, only a small number of companies might respond to bidding rounds, limiting choice and competition (Martinot and Reiche, 2000). According to some sources, if the pool of participants is very small, cartels and the abuse of market power can result (Espey, 2000). And if the price of credits or certificates is high, this can increase the electricity price paid by consumers, as is the case with pricing laws. However, this would most likely be a short-term situation, as higher certificate prices would encourage more development and thereby reduce certificate prices. Finally, if purchase obligations are large enough, quotas can lead to economies of scale, thereby reducing both costs and prices.

There is some evidence that in quota systems that lack differentiation among technologies, such as the current Renewable Obligation Certificate (ROC) programme in the United Kingdom, there is a tendency to oversubsidize lower-cost renewables such as onshore wind and biomass waste-to-power, a factor that will lead to higher costs (comments by Robert Kleiburg, 2003). Under the ROC system the price paid for renewable electricity (most of which is wind power, in a country with Europe's best winds) is similar to payments for wind energy in Germany (Mitchell, 2003). As a result, a great deal of planning is under way, with a surge in market growth projected for the coming years. But this makes it clear that the costs of renewably generated power are at least as dependent on how a particular policy is structured as they are on the system that is chosen.

Financial security

Many believe that the higher risks and lower profits associated with quota systems make them less attractive for investors than pricing laws (Menanteau et al, 2003). Ultimately, what is most important is political stability and long-term, credible, consistent policies.

Under a pricing system, the long-term certainty resulting from guaranteed prices over perhaps 20 years ensures that companies are willing to invest in technology, train staff and establish other services and resources with a longer-term perspective. This certainty also makes it easier to obtain financing, as banks and other investors are assured a guaranteed rate of return over a specified time period. In fact, even banks in Germany lobbied the Bundestag for a continuation of pricing laws in 2000.

With quota systems there are potential uncertainties through many steps in the process from project planning to operation. For example, there can be substantial preparation costs for projects submitted for bids, adding an element of risk and uncertainty that many potential developers cannot afford (Menanteau et al, 2003). Without long-term contracts, under quota systems existing developers could be undersold by future projects and will always be competing against new developments. While some see this as a disadvantage, others view it as an incentive to reduce costs (Espey, 2000). This challenge has been resolved in Texas with 10- to 25-year contract requirements (Lauber, 2004). Unless such contracts are standardized, renewable energy developers must negotiate contracts with utilities or suppliers on an individual basis. While this could be a problem, to date in several US states and elsewhere it does not seem to be a major drawback.

Further, under quota systems potential investors must assess future supply and demand balance during the lifetime of the project (often 20 years or more) by developing a forward price curve. Yet demand is created by political targets, which could change, thereby resulting in a degree of uncertainty. In addition, estimating supply is a complex process that requires an understanding of a broad range of factors. These include the current competitiveness of all eligible energy technologies; future costs (determined by learning curve effects); and cost–resource curves, or the impact on costs when the best resources are no longer available and projects must be sited where wind speeds are lower, or rely on more expensive biomass feedstock, for example. All these factors add to the level of uncertainty. Finally, if RETs enjoy subsidies or other types of support (e.g. grid connection costs, tax credits, accelerated depreciation), whose continuation over the project lifetime is also uncertain, the risks to investors will be higher, requiring a higher projected rate of return. Under these circumstances, banks will also be less willing to provide financing for renewables projects (comments by Robert Kleiburg, 2003).

Sources of income are twofold under a certificate-based quota system: first is payment for the sale of renewably generated electricity, and second is income from the sale or trade of renewable energy certificates. The price of credits or certificates can fluctuate significantly with changes in the marketplace or meteorological variability, rising when there is a shortage of renewable electricity and falling when there is a surplus. Diversifying sources and location of projects can also reduce fluctuations due to meteorological variability. Establishing minimum and maximum certificate prices can help, but does not eliminate investor uncertainty (Meyer, 2003). Trading in international markets can also work to stabilize prices, and risks can be limited through long-term contracts, or borrowing or banking of credits (Menanteau et al, 2003; Meyer, 2003). Some of these solutions, however, can increase the complexity of the system. Seasonal variations in output lead to variations in income from fixed tariffs as well, and market fixes are not built into the pricing system as they are with certificate models. Over time, however, these variations will also be smoothed out (Meyer, 2003).

Financial security is also reduced under quota systems if there is uncertainty around rules relating to green certificate trading. For instance, as system designs are altered – including changes in penalties, borrowing or banking provisions, and the status of imports – prices can be affected dramatically (comments by Robert Kleiburg, 2003).

Some analysts believe that quota systems provide more regulatory and financial stability and security than do pricing systems, which could change with the political winds (Lauber, 2004). For example, long-term purchasing contracts with private entities are enforceable under the law, which might be safer than relying on consistency of government policy (Krohn, 2000b). Others believe that pricing systems provide a greater sense of security than quota systems, particularly in developing countries, because there is not the same assurance that a market for renewable energy credits will exist and that they will be of value (Frost, 2003). Targets established under quota systems are also policy dependent and can change over time, affecting the value of certificates and creating uncertainty (Dieter Uh, comments on draft text, provided 2 December 2003). In addition, payment systems and levels are known at the outset under a pricing system; this is not necessarily the case under a quota system with certificate trading (Krohn, 2000b).

Ease of implementation
In general, pricing laws are easy to administer and enforce, and they are highly transparent. As with quota systems, policy-makers are required to establish targets and timetables, and to determine which technologies are qualified (type and scale). Where applicable, pricing laws also require the setting of tariffs for each technology type. Once the system is established, the only government follow-up required is regular adjustments of tariffs (assuming this is done).

Under quota systems many of the requirements are far more challenging. Picking optimal target levels is critical (if they are set too high, they can push prices up dramatically; if they are too low, they will not produce the economies of scale needed to reduce costs), as is the choice of timetables. Although the same can be said for the setting of tariffs under pricing laws, they can be established with input from research institutes and industries (as in Germany), and pricing laws can be created to allow for adjustments as necessary.

In addition, policy-makers must decide which technologies are eligible, and if there should be technology-specific targets – this will depend on the readiness of technologies, their costs, available resources and other factors. In order to make successful choices, it is also important to understand the cost and learning curves for the relevant renewable technologies (Berry and Jaccard, 2001). Policy-makers also need to determine which category of parties must meet the obligation (e.g. retail suppliers, grid companies or distribution companies) and whether all or just a few of those parties are required to meet the targets. The penalty for non-compliance must be established and the tradability, life span and price (e.g. floor or ceiling-prices?) of certificates or credits chosen. These decisions will all determine the impact of the quota system. Once these matters are resolved, government agencies (or other bodies) must certify renewable energy producers, issue and control certificates, monitor compliance, and collect penalties, all of which increases administrative requirements, complexities and costs (Menanteau et al, 2003).

Some argue that quota/certificate systems tend, by their very nature, to be more complex than pricing systems, difficult to administer, and open to utility manipulation, and that such problems could be even more significant in developing countries (Frost, 2003). Others have noted that the system for cost equalization under the German Renewable Energy Law 2000 is neither simple nor transparent (J. Saghir, Director, Energy and Water, World Bank, comments on draft text, provided 2 December 2003). Finally, it has been argued that bidding processes are bureaucratic, have significant transaction costs and are time-consuming for authorities and renewable energy developers (Wagner, 2000; Goldstein et al, 1999).

Flexibility
Historically, pricing laws have been criticized for being inflexible. For example, once tariffs are established, it could be difficult to reduce them (Wiser et al, 2000). However, it is possible to design systems such that payments can be adjusted on a regular basis to reflect changes in technologies and market conditions. This flexibility was incorporated into the German system in 2000, and is now featured in other national pricing systems as well.

With quota systems, once targets and timetables are established, they can be difficult to adjust – or at least made more ambitious – without lead times of several years.[4]

Summary of pricing and quota systems analysis

The following arguments for and against pricing and quota systems are based on the above analysis.

Pricing systems
The following are arguments in favour of pricing systems:

- Pricing systems are most successful to date at developing renewables markets and domestic industries, and achieving the associated social, economic, environmental and security benefits.
- They can be designed to take into account changes in technology and the marketplace.
- They have the flexibility to encourage steady growth of small and medium-scale producers.
- Transaction costs are low.
- They are easy to finance.
- They favour ease of entry.

There are also some arguments against:

- If tariffs are not adjusted over time, consumers may pay unnecessarily high prices for renewable power.
- They can involve restraints on renewable energy trade, owing to domestic production requirements.

Quota systems
The arguments in favour of quota systems are the following:

- Quota systems promote least-cost projects: cheapest resources are used first, which brings down costs early on.
- They provide certainty regarding future market share for renewables (although this is often not true in practice).
- They are perceived as being more compatible with open or traditional power markets.
- They are more likely to fully integrate renewables into the electricity supply infrastructure.

Those against are:

- They involve high risks and low rewards for equipment industry and project developers, which slows innovation.
- Prices fluctuate in 'thin' markets, creating instability and gaming.

- They favour large, centralized merchant plants and are ill suited for small investors.
- They concentrate development in areas with the best resources, causing possible opposition to projects and missing many benefits associated with renewable energy (e.g. jobs, economic development in rural areas, reduction in local pollution).
- Targets can set upper limits for development – there are no high profits to serve as incentives to install more than the mandated level because profitability exists only within the quota (without other incentives).
- Quota systems tend to create cycles of stop-and-go development.
- They are complex to design, administer and enforce.
- They involve high transaction costs.
- They are difficult to fine-tune or adjust in the short term if situations change.

It is still too early to know how successful quota systems can be.

Requirements for successful policy
Pricing law:

- Ensure regular adjustments of tariffs; incorporate incremental adjustments into law.
- Establish tariffs according to technology (and location) with input from research institutes and renewables industries.
- Provide tariffs for all potential developers, including utilities.
- Ensure that tariffs are high enough to cover costs and encourage development.
- Guarantee tariffs for a long enough time period to ensure an adequate rate of return.
- Ensure that costs are shared equally across the country or region.
- Eliminate barriers to grid connection.

Quota system:

- Apply quotas to a large segment of the market.
- Include specific purchase obligations and end-dates.
- Establish adequate penalties for non-compliance, and adequate enforcement.
- Set different bands by technology type.
- Require long-term contracts to reduce uncertainty for project developers.
- Establish minimum and maximum certificate prices.
- Do not allow a time gap between one quota and next.

What is most important for both systems is political stability and long-term, credible, enforceable and consistent policies. Additional requirements for successful implementation – such as standards, education and stakeholder involvement – will be discussed later.

Financial incentives for renewable energy

Financial incentives reduce the costs of renewable energy by lowering the price paid for RETs or energy, increasing the payment received, or reducing the cost of production. They include market compensation in the form of tax credits, rebates and payments to subsidize investment in a technology or in energy production. Such incentives have been used extensively in Europe, Japan, the US and India – one of the few developing countries with tax credits to date.[5] Long-term, low-interest loans and loan guarantees work to reduce the cost of capital. And the reduction or elimination of subsidies for conventional energy, while not a subsidy for renewable energy, helps to level the playing field so that renewables are better able to compete on a cost basis.

Tax relief

Investment and production tax credits
Investment tax credits can cover just the cost of a system or the full costs of installation. They can be helpful early in the diffusion of a technology, when costs are still high, and/or to encourage their installation in off-grid, remote locations. They directly reduce the cost of investing in renewable energy systems and reduce the level of risk. Production tax credits provide tax benefits against the amount of energy actually produced and fed into the electric grid, or the amount of biofuels produced, for example. They increase the rate of return and reduce the payback period, while rewarding producers for actual generation of energy, but must be designed carefully to avoid abuse.

To encourage investment in renewables in the early 1980s the US government and the state of California offered investors credit against their income tax. In combination with standard, long-term contracts, the credits helped to create a wind boom that many people called California's second gold rush. The lessons learned and economies of scale gained through this experience advanced wind technology and reduced its costs. But the combination of enormous tax breaks and a lack of technology standards encouraged fraud and the use of substandard equipment (in India, too, while investment subsidies for wind energy led to large investments in the 1990s, there was limited concern about maintenance and long-term performance due to a lack of standards). Inexperienced financial companies and former shopping centre developers flocked to the wind business in California, and untested designs were rushed into production – all to take advantage of credits that enabled wealthy investors to recoup anywhere from 66 to 95 per cent of their investment over the first few years, in some cases without ever generating a kilowatt-hour of power. While these early tax credits helped to jump-start the wind industry, once the credits and fixed prices expired, the industry collapsed, with ripple effects felt as far away as Denmark, and numerous wind energy firms going bankrupt (Sawin, 2001; Lauber, 2004).

One of the largest production tax credit programmes is in the US. On and off since 1994, the US government has offered a production tax credit (PTC) that reduces the income tax liability for people who supply wind-generated electricity to the grid. The PTC has encouraged wind energy development and has been credited with driving significant capacity increases in the late 1990s and early 2000s. At the same time, the PTC encouraged significant development only in those states with additional incentives (Sawin, 2001).

In general, production incentives are preferable to investment incentives because they promote the desired outcome: generation of energy. Although investment subsidies encourage installation at the optimal level for individuals or businesses, they do not necessarily result in installation at the optimal level for the society or community as a whole. Further, they do not necessarily encourage investors to purchase the most reliable systems available or to maintain them and produce as much energy with them as possible. In addition, once the credits expire, investment in the technologies they are designed to support generally declines unless costs have fallen sufficiently or other support mechanisms are in place. Production incentives, on the other hand, are most likely to encourage optimum performance and a sustained industry.

However, policies must be tailored to particular technologies and stages of maturation. Investment subsidies can be helpful when a technology is still maturing and relatively expensive, as has been the case with PVs in Japan – although rebates (to be discussed later) are a preferable means for subsidizing investment (Sawin, 2001). Further, investment support is often more appropriate for small-scale RETs such as heat pumps or small-scale PV because administrative costs are lower – with a one-time payment rather than annual payments based on metered data (comments by Robert Kleiburg, 2003), or because it may be impossible to measure output. Performance problems associated with investment subsidies can be overcome by tying investment incentives to equipment and performance standards, as long as these requirements are enforced. And if investment credits are adjustable and/or decline gradually over time and are phased out as technology costs fall, it might be possible to avoid the sudden decline in investment that often occurs when these subsidies expire.

It is also important to note that tax incentives tend to favour one type of entity over another, and they provide greater benefit to people or corporations with higher income levels and tax loads. In addition, they are often used as tax loopholes. Investment tax credits, in particular, can affect the timing of installation by encouraging investments toward the end of a tax cycle, which can negatively affect renewables industries. As with investment credits, production tax credits should decline over time and eventually be phased out.

Other tax relief

Other tax-related incentives can promote renewable energy development by reducing the costs of investment, or by accounting for the external benefits of renewable energy. The latter include eco- or carbon-tax exemptions. The former include accelerated depreciation, relief from taxes on sales and property, value added tax (VAT) exemptions, and reduction or elimination of import duties on RETs or components. It is important to note that import duties increase the upfront costs of renewable energy projects, and should be reduced, if not eliminated, until a strong domestic manufacturing industry can be established.

Rebates and payments

As an alternative to production and investment credits against taxes, some countries and US states have subsidized renewable energy through production payments or rebates. Rebates are refunds of a specific share of the cost of a technology, or share of total installation costs, or refunds of a certain amount of money per unit of capacity installed. As with investment credits, rebates are most effective when linked to technology and performance standards. Japan has provided investment subsidies through rebates for PVs; these rebates, in combination with low-interest loans, public education and net metering, have led to dramatic success with PVs in Japan.

Production payments reward energy generation through a certain payment per unit of output. For example, California enacted a production incentive that awards a per kilowatt-hour payment for some existing and new renewable energy projects. It is financed through a small per kilowatt-hour charge on electricity use, meaning that Californians share the cost of the programme according to the amount of power they consume. Provided that such payments are high enough to cover the costs of renewable generation and are guaranteed over a long enough time period, this policy integrates some key elements of a pricing law – similar in effect and perhaps more politically feasible in some countries (Sawin, 2003).

Experiences to date demonstrate that payments and rebates are preferable to tax credits for a number of reasons. Unlike tax credits, the benefits of payments and rebates are equal for people of all income levels. They also result in more even capacity growth over time (Sawin, 2001). Finally, at least one analyst believes that there is no evidence that either investment or production tax credits anywhere have led to a substantial increase in market penetration of PV (Haas, 2002). Clearly, the effects of one single instrument are limited.

With regard to rebates, it has been argued that they must cover a fixed amount per unit of capacity rather than a percentage of investment costs, because a fixed rebate encourages investors to seek out the most efficient or cheapest option (Haas, 2002). And all investment subsidies should be accompanied by standards or monitoring programmes to ensure good performance, and should be adjustable up or down as necessary.

Low-interest loans and loan guarantees

Worldwide, one of the major barriers to RETs is the high initial capital costs of renewable energy projects. Thus, the cost of borrowing plays a major role in the viability of renewable energy markets. Financing assistance in the form of low-interest, long-term loans and loan guarantees can play an important role by lowering the cost of capital, effectively reducing the average cost of energy per unit and the risk of investment. Germany addressed this through long-term, low-interest loans offered by major banks and refinanced by the federal government (J. Twele, discussions with author, 5 December 2000). Japan and some US states have also established low-interest loan programmes for solar PV and other renewables (Eckhart et al, 2003).

Even in the developing world, all but the very poorest people are able and willing to pay for reliable energy services, and the rate of on-time payment is extremely high. But the poor also need access to low-cost capital and the opportunity to lease systems (Goldemberg, 2000). The very poorest people will probably need targeted subsidies as well (Martinot, 2003). If monthly costs of solar energy systems are comparable to those for candles and kerosene lighting, households should be able to afford to substitute them (Martinot, 2003).[6]

According to PV companies in South Africa, Indonesia, India and the Dominican Republic, up to 50 per cent of prospective purchasers can afford to buy systems if reasonable third-party financing is available; otherwise, only 2–5 per cent can buy them (Eckhart et al, 2003). So, the availability of financing could increase the use of PV in some countries tenfold or more. The impacts could be similar with other RETs as well. To date, vendor-supplied credit, micro-credit and leasing/rental of renewable systems are still mostly untested systems, and the effectiveness of various consumer loan models are probably country-specific, depending on cultural, financial and legal factors. But a number of developing countries have had successful experiences with lending programmes that could be transferable to other countries.

One of the most successful means for disseminating household-scale renewable technologies in rural China has been through local public–private bodies that offer such services as technical support, materials sale, subsidies and government loans for locally manufactured technology. These bodies frequently provide revolving credit, with repayment linked to the timing of a household's income stream – for example, payments come due after crops have been harvested. As a result of this programme, more than 140,000 small wind turbines, producing power for more than half a million people, were installed in Inner Mongolia by 2002 – the greatest number of household-scale wind plants operating anywhere in the world (Martinot et al, 2002; Wu, 1995).[7] In India the terms of long-term, low-interest loans vary by technology, with the most favourable ones being for PVs. Through small-scale lending programmes, even low-income people are able to purchase small systems. In addition, the national government has worked to obtain bilateral and multi-lateral funding for large-scale projects, particularly wind (Indian MNES, 2000; CSE, 2002).

Addressing subsidies and pricing for conventional energy

Perhaps the most important step governments can take to advance renewables and reduce cost disparities is to make a comprehensive change in their perspective and approach to energy policy. Governments must eliminate inappropriate, inconsistent and inadequate policies that favour conventional fuels and technologies and that fail to recognize the social, environmental and economic advantages of renewable energy.

In the mid-1990s, governments worldwide were handing out $250–300 billion annually to subsidize fossil fuels and nuclear power (UNDP, 2000). Since then, several transitioning and developing countries have reduced energy subsidies significantly, but global subsidies for conventional energy remain many magnitudes higher than those for renewable energy (Geller, 2003). Most of these subsidies (80–90 per cent by some estimates) are found in the developing world, where the price for energy is often set well below the true costs of production and delivery (Eric Martinot, personal email communication, 9 October 2002). In the late 1990s the International Energy Agency (IEA) found that in just eight developing countries (together accounting for one-quarter of the world's energy use), energy subsidies cost $257 billion in lost GDP, or the equivalent of about 11 per cent of the nations' combined annual economic output (OECD and IEA, 1999). Even relatively small subsidies in developing countries – for kerosene and diesel, for example – can discourage the use of RETs. These subsidies should either be gradually eliminated or shifted to renewable technologies and fuels. In industrial and developing countries alike, it is important to ensure that the removal of subsidies for conventional energy does not negatively affect the poor; subsidies must be targeted to enable them to make the transition to renewable energy alternatives.

In most cases it is less a matter of finding new money to invest in renewable energy and more a matter of transferring money flows from conventional energy to renewables. It is estimated that public and private interests invest $250–300 billion annually in new energy infrastructure; $40–60 billion of this is for rural electrification. And more than $1 trillion was spent annually on direct energy purchases by about 2002 (Goldemberg, 2002); expenditure will only increase as oil prices rise. According to another estimate, each year hundreds of millions of people in the developing world spend about $20 billion on makeshift solutions such as candles, kerosene lamps and batteries for lighting and to power small appliances. And transporting diesel fuel to remote regions in the Amazon, for example, can consume two to three times as much fuel for every gallon delivered (Perlin, 1999).

The IEA projects that $16 trillion will be invested worldwide in energy supply infrastructure between 2001 and 2030. Nearly 60 per cent of this is expected to go to the electricity sector (for power generation, transmission and distribution) (IEA, 2003). According to the United Nations, investments for new power sector projects in the developing world alone are expected to be in the range of $50–60 billion annually (UNEP, 2000). Even small shifts in these expenditures and in energy subsidies would have a tremendous impact on renewable energy markets and industries, although more than a small shift is needed.

It is also important to recognize that even if all subsidies for fossil fuels and nuclear power were removed immediately, existing infrastructure in most countries – much of which was funded with public money – would continue to favour conventional energy. In addition, there exist conventional energy benefits from hidden or indirect subsidies, including obligations to purchase a specific form of energy, government investments in grid extensions, exemptions from risk or liability, and government energy purchases.

As the single largest consumers of energy in most, if not all, countries, governments should purchase ever-larger shares of energy from renewable sources and install RETs on public buildings. This would have a significant impact on renewable energy markets. In the process, governments would set an example, increase public awareness, reduce perceived risks associated with RETs, and reduce costs through economies of scale.

In addition, pricing structures must account for the significant external costs of conventional energy and the advantages of renewable energy, as Germany has begun to do through an eco-tax on fossil fuels and the renewable electricity tariffs under the Renewable Energy Law (feed-in/pricing) of 2000, and as other countries are doing with energy or carbon taxes.

OTHER ESSENTIAL POLICY MECHANISMS

Standards

Standards can prevent inferior technologies from entering the marketplace and generate greater confidence in a product, thereby reducing risks, which is important for financing. Technology standards for wind turbines, for example, can include everything from turbine blades, electronics and safety systems to performance, and compatibility with the transmission system.

Denmark adopted wind turbine standards in 1979, largely as a result of pressure from the wind industry itself. The Danish technology standards programme, combined with the sharing of performance and other relevant information among turbine owners and manufacturers, has enabled manufacturers to recognize and address problems with their technologies and to create pride in Danish machines. Standards are credited with playing a major role in Denmark's rise to become the world's leading turbine manufacturer (Krohn, 2000a; Birger Madsen, BTM Consult ApS, discussion with author, 8 December 2000, and email to author, 14 September 2002).

Germany established an investment tax credit for wind energy in 1991, and while it too has been abused as a tax loophole for wealthy people, Germany has avoided the quality control problems experienced in California and India by enacting turbine standards and certification requirements. Eventually, technology standards for all RETs should be established at the international level (Sawin, 2003).

In addition, siting and planning requirements can reduce opposition to renewables if they address other potential issues of concern, such as noise and visual or environmental impacts. Such laws can be used to set aside specific locations for development or to restrict areas at higher risk of environmental damage or injury to birds, for example. Both Germany and Denmark have required municipalities to reserve specific areas for wind turbines and have set restrictions on proximity to buildings and lakes, among other things. These policies have been extremely successful, reducing uncertainty about whether and where turbines can be sited, and expediting the planning process.

The United Kingdom offers the best example of how the lack of planning regulations can paralyse an industry. Despite having the best wind resources in Europe, the United Kingdom added little wind capacity under its renewables obligation regulations (NFFO), in part because a lack of planning regulations virtually halted the process for obtaining planning and environmental permits (Sawin, 2001; Birger Madsen, BTM consult ApS email to author, 14 September 2002). Under the current quota system only about 300 MW were added over two years (BTM Consult ApS, 2004; EWEA, 2005).

Renewables face two challenges when it comes to interconnection with the grid. First, unlike conventional plants, which have flexibility in siting, renewable plants must be sited where the resources are located. Second, some renewables, such as solar and wind, are intermittent. Both challenges necessitate the creation of fair connection standards and charges, and guaranteed access to the grid. Interconnection requirements are often overly burdensome and inconsistent, which can lead to high transaction costs for renewable energy project developers, particularly if they must hire technical and legal experts. Safety requirements are essential, but many utilities go beyond that to discourage interconnection with relatively small residential or commercial systems. For example, utilities can block transmission access or charge high prices for access to the grid and use of transmission lines (Beck and Martinot, 2004).

In addition, transmission charges on a per capacity basis put intermittent renewables at a disadvantage because they must pay for grid access even when they are not using it. This increases average transmission costs per kilowatt-hour. Thus, it is important that governments establish interconnection standards under which renewable energy developers pay only for the direct costs of connection with the grid, not for necessary upgrades to carry the additional capacity, and only for the transmission service that they actually use. Further, because most renewable electricity (with the exception of biomass) has zero marginal costs (the fuel is free), it should always have priority access to the transmission system. This makes economic sense, and all other sources (that are dispatchable) can easily be ramped down as necessary.

Building codes and standards can also be designed to promote energy efficiency and renewables such as passive solar (transparent and opaque insulation), solar thermal energy, modern biomass, geothermal and PV by requiring that these be incorporated into designs and planning processes for residential and commercial buildings. In November 2003 the London borough of Merton adopted a requirement that new non-residential buildings larger than 1000 metres square meet at least 10 per cent of their energy needs with on-site renewables such as solar thermal and PV (SolarAccess.com, 2003c). Following Barcelona's lead, several major Spanish cities soon also required that new or renovated buildings meet a significant share of their energy needs for hot water with solar thermal systems; a national law was enacted in early 2006.

Alternatively, a portion of new buildings could be required to include wiring and other hardware to make them PV- or thermal systems-ready, an addition that would add little to construction costs in many instances, while making it easier and far less costly to install such systems later. Depending on climate conditions where buildings are sited, the existence of such standards can have a significant impact on energy requirements, particularly for heating and cooling needs.

Furthermore, improving energy efficiency facilitates the use of renewable energy for two reasons: first, because the scale becomes more manageable, renewables can more easily satisfy energy needs; second, as the load is reduced, it is easier to bear higher costs per unit of output. The combination of new materials and technologies, natural cooling techniques, and passive solar heating and lighting can significantly increase the efficiency of buildings.

Education and information dissemination

Even if a government offers generous incentives and low-cost capital, people will not invest in renewable energy if they lack information regarding resource availability, technology development, the numerous advantages and potential applications of renewables, the fuel mix of the energy they use, and the incentives themselves. During the 1980s, several US states offered substantial subsidies for wind energy – including a 100 per cent tax credit in Arkansas, a state with enough wind resources to generate half of its electricity (Righter, 1996; Battelle/ PNL, 1991). But these subsidies evoked little interest, owing to a lack of knowledge about wind resources. By contrast, it was wind resource studies in California, Hawaii and Minnesota that led to interest in wind energy in these states. And cloudy Germany has more solar water heaters than the sunnier countries of Spain and France, largely because public awareness of the technology is much higher in Germany (Hua, 2002).

Lack of experience or past experiences – from failed Californian wind projects in the 1980s to early development projects in Africa – have left people in much of the world with a perception that renewables do not work, are inadequate to meet their needs, are too expensive or are too risky as investments. Above all, it is essential that government leaders recognize the inherent value of renewable energy. Then, governments, non-governmental organizations and industry must work together to educate labour organizations about employment benefits, architects and city planners about ways to incorporate renewables into building projects and their value to local communities, agricultural communities about their potential to increase farming incomes, and so on.

In India the government's Solar Finance Capacity Building Initiative educates Indian bank officials about solar technologies and encourages them to invest in projects. The Indian government has also used print media, radio, songs and theatre to educate the public about the benefits of renewable energy and government incentives, and has established training programmes (Indian MNES, 2000). In Austria, students learn about renewable energy in schools and universities, and in Germany many vocational training programmes cover renewable energy issues (Goldstein et al, 1999; German BMU, 1994).

It is often assumed that barriers and solutions to renewables are unique to particular countries or settings, but this is not necessarily the case (Kammen, 1999). Thus, at the local, national and international levels it is essential to share information regarding technology performance and cost, capacity and generation statistics, and policy successes and failures in order to increase awareness and to avoid reinventing the wheel each time. While several countries now do this on a national level, a centralized global clearinghouse for such information is clearly needed. The Renewable Energy Network for the 21st Century (REN21) – a product of the International Conference for Renewable Energies in 2004 – may play this role.

Public ownership and stakeholder involvement

Public ownership and/or participation are also essential for the success of specific projects, as well as the development of effective renewable energy policies.

Germany's renewable energy law of 2000 was designed in cooperation with research institutes and the renewables industries, and the US state of California has designed recent renewable energy programmes with stakeholder input provided at public workshops.

In Germany and Denmark, where individuals singly or as members of cooperatives have historically owned most of the turbines installed, there is strong and broad public support for wind energy. Farmers, doctors and many others own turbines or shares of windfarms, and stand beside labour and environmental groups in backing policies that support wind power. As of 2002 about 85 per cent of the installed wind capacity in Denmark was established through local initiatives and owned by farmers or cooperatives, and at least 340,000 Germans had invested about €12 billion in renewable energy projects (European Actions for Renewable Energy, 2002/2003). The 40 MW Middelgrunden project off the coast of Copenhagen is co-owned by a utility and several thousand Danes who have purchased shares in the project (EWEA and Greenpeace, 2002). Construction costs of one of the world's largest roof-mounted PV plants, which opened in Munich, Germany, in November 2002, were financed by interested citizens eager to invest in RETs (Maycock, 2003a).[8]

Through cooperatives, people share in the risks and benefits of renewable projects; often avoid the problems associated with obtaining financing and paying interest; play a direct role in the siting, planning and operation of machines; and gain a sense of pride and community (Sawin, 2001). Several surveys have demonstrated that those who own shares of projects and those living closest to wind turbines view wind power more positively than those who have no economic interest or experience with it (Damborg and Krohn, 1998). Local investment also provides an opportunity to strengthen and diversify local economies, particularly in rural areas, and can lead to new projects through the sharing of information and relevant experiences (European Actions for Renewable Energy, 2002/2003).

Public participation and a sense of ownership are as important in the South as in the North. When technologies are 'forced' on people without consultation regarding their needs or desires, or are donated as part of an aid package, people often place little value on them and do not feel they have a stake in maintaining them. On the flip side, when individuals and communities play a role in decision-making and ownership, they are literally empowered and become invested in the success of the technologies. The key to success of some projects in developing countries has been a sense of ownership among local people. For example, local participation and ownership of solar mini-grid projects in Nepal and the Indian islands of Sundarbans, have played a crucial role in the success of projects and have eliminated electricity theft (BBC News, 2003).

IMPORTANCE OF CONSISTENT, LONG-TERM POLICIES

It is important to note that policies enacted to advance renewable energy can slow the transition if they are not well formulated or are inconsistent, piecemeal or unsustained. For example, because early investment credits in the US state of California were short-lived and extensions were often uncertain, many equipment manufacturers could not begin mass production for fear that credits would end too soon (CEC, 1982). When incentives expired, interest waned and the industries and markets died with them. In the case of wind power the impact was felt as far away as Denmark, which relied on the California market for sales of Danish turbines. The US Production Tax Credit for wind energy has been allowed to expire several times, only to be extended months later. As a result, the credit has stimulated wind capacity growth but has created cycles of boom and bust in the market. Such cycles lead to suspension of projects, worker lay-offs and loss of momentum in the industry.

This on–off approach to renewables has caused significant uncertainties, bankruptcies and other problems, and has made the development of a strong industry in the US a challenge at best. Indeed, the US is the only country to have seen a decline in total wind generating capacity over the past decade (discussion with, and fax from, Paul Gipe, 1 October 1998). In India, uncoordinated, inconsistent state policies and bottlenecks imposed by state electricity boards have acted as barriers to renewables development (CSE, 2002). Even in Denmark, years of successful wind energy growth ended in 1999 when the government changed course, and uncertainty overtook years of investor confidence. By late 2003 the future of some planned offshore windfarms was uncertain, as was Denmark's target to produce half its electricity with wind by 2030 (Møller, 2002), and the number of jobs in the domestic industry had begun to decline (BWE, 2003b).

Consistent policy environments are necessary for the health of all industries. Consistency is critical for ensuring continuous growth and stability in the market, enabling the development of a domestic manufacturing industry, reducing the risk of investing in a technology and making it easier to obtain financing. It is also cheaper (Sawin, 2001). With stop-and-go policies, each time around the funds must be appropriated, a new programme must be administered, information must be distributed to stakeholders, and so on. As a result, costs of administering the programme could approach those of the incentives themselves (Dieter Uh, comments on draft text, provided 2 December 2003).

Clearly, government commitment to develop renewable energy markets and industries must be strong and long-term, with a clear intent to advance these technologies, just as it has been with fossil fuels and nuclear power.

CONCLUSIONS AND RECOMMENDATIONS

For renewable energy to reach its full potential – to make as large as possible a contribution to economic development and job creation, improve domestic energy security and reduce oil dependence, and reduce the health and environmental impacts of energy production and consumption – it is essential to create the conditions that allow for development of sustained markets and industries, which in turn will result in increased renewable capacity and generation, and will drive down costs.

To date, feed-in/pricing systems have been responsible for most of the additions in renewable energy capacity and generation, while also driving down costs through technology advancement and economies of scale, and developing domestic industries and jobs. Pricing systems, where well implemented, have provided increased predictability and consistency in markets, which in turn have encouraged banks and other financial institutions to provide the capital required for investment, and have attracted private investment for R&D.

The record of quota systems is more uneven thus far. Quota systems, if designed well, have the potential to work effectively. But they are harder to get right and have a tendency to lead to stop and go, and boom and bust markets. Once renewable electricity markets and industries are well established, and renewables can compete favourably with conventional energy, quota systems are likely to be the most appropriate means for furthering the development and use of RETs.

A combination of policies is required, whether for grid-connected electricity or other uses, including production-based incentives and financing support to lower initial investment costs and reduce risk, whether real or perceived. Ultimately, the effectiveness of policies in promoting renewable energy will depend on their design, enforcement, how well they address needs and national circumstances, and the extent to which they are reliable and sustained. Long-term, consistent and clear government commitment to renewable energy is imperative.

The following are recommendations for policies to advance a range of renewable energy technologies:

- *Establish regulatory frameworks* needed to provide access to relevant markets and favourable climates for investment, such as well-designed quota or pricing systems for grid-connected renewable electricity.
- *Provide net metering* for small-scale renewable systems where pricing laws are not in use. Time-of-use metering should be implemented for PVs.
- *Provide financing assistance* to reduce costs through production payments (rather than tax credits) for more advanced technologies, and long-term, low-interest loans with investment rebates (rather than tax credits) for more expensive technologies and/or those that are off-grid such as solar PV, water

heating and space heating. Investment incentives should always be tied to technology and/or production standards. It can also be useful to require reporting of operational data for subsidized projects on a regular basis. Regarding rebates, a fixed amount per unit of capacity might be preferable to a percentage of investment costs for encouraging the most efficient and least-cost options, and thus for reducing costs. All subsidies must be 'smart subsidies' that gradually phase out over time; subsidies are not likely to lead to sustainable markets unless they help create the conditions in which they eventually are not needed.

- *Be careful if setting capacity limits* on large-scale projects in the context of financing assistance and quota or pricing laws. At a minimum, capacity restrictions should allow for project sizes that can achieve cost reductions through scale economies; otherwise, they can discourage development of some renewable resources. When capacity restrictions are necessary, they might be addressed better under siting regulations, with variations by region and technology as appropriate.
- *Research, educate and disseminate information* regarding resource availability, the benefits and potential of renewable energy, capacity and generation statistics, government incentives (for renewable and conventional energy), and policy successes and failures on local, national and international levels. Establish a centralized global clearinghouse for such information or ensure that REN21 plays this role. And establish national training programmes in vocational schools, universities and other appropriate institutions. Provide information at all levels of education (from primary schools to universities) about the potential and benefits of renewable energy, the state of the technologies, and other relevant issues.
- *Encourage stakeholder/public ownership and participation* in renewables policy formulation and in project planning and decision-making. Citizen involvement can be critical to project development and viability, while introducing the concepts of democracy and self-determination into the process of energy procurement.
- *Establish standards* for technology performance, safety and siting, and create or strengthen building codes to improve energy efficiency and encourage the integration of RETs into buildings. Standards must also be set for grid connection to eliminate burdensome utility interconnection requirements and charges.
- *Incorporate external costs and benefits* of energy technologies and fuels in pricing structures and consider them when developing policy objectives. In addition to environmental, health and employment benefits, considerations should include factors such as reliability and security benefits associated with modular, distributed and smaller-scale RETs, the reduced risk of fuel price volatility, and benefits of a more diversified (and domestic) portfolio of energy options.

- *Shift government subsidies* from conventional to renewable energies, in line with the previous point.
- *Increase government procurement* of renewable energy. Governments should purchase ever-larger shares of energy from renewables and install RETs on public buildings and in public spaces (where appropriate) to set an example, increase public awareness, reduce perceived risks and reduce costs through learning and economies of scale.

The devil is in the details, and certain types of policies can be successful or not depending on how they are implemented. For any of the above policies to be effective, they must be:

- *Predictable, long-term and consistent, with clear government intent.* These characteristics are critical to provide certainty in the market to draw investors into the industry, and to provide enough lead time to allow industries and markets to adjust to change.
- *Appropriate.* The right types of support are needed: policies must match objectives and might vary by resource potentials, location, technology type, and timing. It is also important that the level of support not be too high or too low.
- *Flexible.* It is essential to design policies such that adjustments (fine-tuning, but not wholesale changes or elimination of policies) can be made on a regular, predetermined time schedule if circumstances change. Governments must be able to address existing barriers as they become apparent, and new barriers as they arise. Policies also must be designed to allow developers/generators flexibility for meeting government mandates.
- *Credible and enforceable.* If policies are not credible, or are not enforceable (or enforced), there will be little incentive to abide by them.
- *Clear and simple.* Policies must be easy to implement, understand and comply with. Procedures of permission and administration, where necessary, must be as clear and simple as possible.
- *Transparent.* Transparency is important for suppliers and consumers of energy and is necessary to avoid abuse. It facilitates enforcement, maximizes confidence in policies and helps ensure that mechanisms are open and fair.

NOTES

1   Note that some US states, including Massachusetts, have had trouble meeting their RPS targets.
2   A BET study estimates that the price increase for electricity consumers caused by the pricing law was only 0.11 €cent/kWh in 2000, and will be 0.19 €cent/kWh in 10 years assuming a doubling of renewables' share of total generation (Lackmann, 2002). A German government article estimates the extra cost at an average of €8 per

German household each year (German Embassy, United Kingdom, 2003). A third estimate puts additional costs at 0.25 €cents/kWh in 2001, a number that has been accepted by authorities of the German federal states (Dieter Uh, comments on draft text, provided 2 December 2003, and email to author dated 8 January 2004).

3   At least one expert notes that once the best resources are no longer available, there will be no difference in costs, but countries with pricing systems will have domestic industries while those with quota systems will not (Dieter Uh, comments on draft text, provided 2 December 2003).

4   Note, however, that some US states have increased their RPS targets. In 2003, the year after the RPS was first enacted, the California Public Utilities Commission and California Energy Commission adopted the state's first Energy Action Plan, accelerating the RPS target of 20 per cent from 2017 to 2010 for investor-owned utilities. In September 2005 a draft plan was released that would apply the same standards to non-investor-owned utilities as well (REA, 2005). In July 2005 the Texas legislature doubled that state's previous goal for renewables (2000 additional MW by 2009), calling for Texas to get 5880 MW (about 5 per cent) of its electricity from renewable energy by 2015; at least 500 MW of this total must come from sources other than wind (DE, 2005).

5   Note that China, the Philippines and Guatemala have also enacted tax credits for RETs (Martinot, 2005).

6   Rural households that are off-grid pay US$3–15 per month for energy in the form of kerosene, candles, battery charging and disposable batteries (Reiche et al, 2000).

7   By mid-2005 the number had increased to 230,000 turbines (Martinot, 2005); the author is not clear on whether all resulted from the revolving credit programme.

8   The total project is 2.1 MW, including an existing plant erected in 1997. Construction costs of the new project, of 1.058 MW, are about €5.5 million.

REFERENCES AND FURTHER READING

Asociación de Productores de Energías Renovables (APPA, Renewable Energy Producers' Association) (2002) www.appa.es/ (accessed January 2004)

Babu, Y. D. and Michaelowa, A. (2003) 'Removing Barriers for Renewable Energy CDM Projects in India and Building Capacity at the State Level', The Energy and Resources Institute (TERI) and Hamburgischen Welt-Wirtschaft-Archiv (HWWA), Hamburg, September

Battelle/Pacific Northwest Laboratory (Battelle/PNL) (1991) 'Assessment of Available Windy Land Area and Wind Energy Potential in the Contiguous United States', Richland, Washington (WA), PNL-7789, August 1991

BBC News (2003) '"Minigrids" solve South Asia power crisis', BBC News, 27 October

Beck, F. and Martinot, E. (2004) 'Renewable Energy Policies and Barriers', in Cleveland, C. J. (ed) Encyclopedia of Energy, Academic Press/Elsevier Science, Washington, DC

Berry, T. and Jaccard, M. (2001) 'The Renewable Portfolio Standard: Design Considerations and an Implementation Survey', Energy Policy, vol 29, pp263–277

Bolinger, M. and Wiser, R. (2002) 'Customer-Sited PV: A Survey of Clean Energy Fund Support', Berkeley, May, Lawrence Berkeley National Laboratory

Bravo, I. L. (undated) 'Procedimiento de conexión para inyectar energía a la red eléctrica convencional desde una instalación fotovoltaica', Grupo de Termodinámica y Energías Renovables, Escuela Superior de Ingenieros, University of Seville, Spain

BTM Consult ApS (2003) *World Market Update 2002: Forecast 2003–2007*, Ringkøbing, Denmark, March

BTM Consult ApS (2004) *World Market Update 2003*, Ringkøbing, Denmark, March

BTM Consult ApS (2005) *World Market Update 2004*, Ringkøbing, Denmark, March

Bundesverband Windenergie eV (BWE) (2003a) 'German wind power still flying high', press release, Osnabrück, 22 February

Bundesverband Windenergie eV (BWE) (2003b) 'Down and Out in Denmark', *New Energy*, vol 1, pp20–23, April

California Energy Commission (CEC) (1982) *Wind Energy Program Progress Report*, Sacramento, CEC

CEC (2003), Sacramento, CA, www.energy.ca.gov/renewables/ (accessed January 2004)

Chaurey, A. (2003) The Energy and Resources Institute (TERI), emails to author, 16 October, 3 and 5 November

Choy, M. (2003) 'French law to slow wind power growth – government official', *Reuters news service* (article found on-line), 28 March

Coelho, S. T. (2002) Executive Assistant for the Secretary of State for the Environment, São Paulo, Brazil, interview with author, Washington, DC, 25 July

Coelho, S. T., Bolognini, M. F., Silva, O. C. and Paletta, C. E. M. (1999) 'Biofuels in Brazil: The Current Situation', São Paulo, Centro Nacional de Referência em Biomassa (Brazilian Reference Center on Biomass)

Cox, P. (2003) President, European Union Parliament, speaking at opening session of the European Wind Energy Conference, Madrid, 16–19 June

Centre for Science and the Environment (CSE) (2002) 'Why Renewables Cannot Penetrate the Market', *Down to Earth*, vol 10, no. 23, 30 April, Delhi

Damborg, S. and Krohn, S. (1998) 'Public Attitudes towards Wind Power', Danish Wind Turbine Manufacturers Association, Copenhagen, Denmark

Danish Ministry of the Environment (DME) (Miljøministeriet) (2003) 'GNP, CO_2-emissions and energy', *Nature and Environment: Total Set of Indicators*, Copenhagen, DME

Database of State Incentives for Renewable Energy (DSIRE) (2005) 'Rules, Regulations and Policies', www.dsireusa.org/ (accessed September 2005)

DE (2005), 'Texas Incentives for Renewable Energy', Database of State Incentives for Renewable Energy, North Carolina State University, available at www.dsireusa.org/library/includes/incentive2.cfm?Incentive_Code=TX03R&state=TX&CurrentPageID=1 (accessed 20 March 2006)

de Andrade, A. M., Morato de Andrade, C. A. and Bodinaud, J. A. (1998) 'Biomass Energy Use in Latin America: Focus on Brazil', for workshop on Biomass Energy: Data, Analysis and Trends, International Energy Agency, Paris, 23–24 March, p3, www.iee.usp.br/biblioteca/producao/1998/trabelho/biomass/pdf/ (accessed August 2002)

de Hollanda, J. B. and Poole, A. D. (undated) 'Sugarcane as an Energy Source in Brazil', Instituto Nacional de Eficiência Energética, www.inee.org.br/down_loads%5Cabout/SUGARCANE&ENERGY.pdf/ (accessed May 2005) de Vries, H. J., Roos, C. J., Beurskens, L. W. M., Kooijman-van Dijk, A. L. and Uyterlinde, M. A. (2003) 'Renewable Electricity Policies in Europe: Country Fact Sheets', Energy Research Centre of the Amsterdam, October

Democratic Energy (DE) (2005) 'Renewable Portfolio Standards', www.newrules.org/electricity/rps.html/ (accessed October 2005)

Deutsches Windenergie Institut (German Wind Energy Institute, DEWI) (1998) *Wind Energy Information Brochure*, Wilhelmshaven, DEWI

Eckhart, M., Stone, J. and Rutledge, K. (2003) 'Financing PV Growth: Why Finance May Be the Key to Real Expansion', *Renewable Energy World*, vol 6, no 3, pp112–127

Espey, S. (2000) 'Renewables Portfolio Standard: A Means for Trade with Electricity from Renewable Energy Sources?', *Energy Policy*, vol 29, pp557–566

Environment Daily (2003) 'Europe urged to give renewables more backing', *Environment Daily*, issue 1468, 23 June, quoting Joan Fages, European Renewable Energy Foundation

European Actions for Renewable Energy (PREDAC) (2002/2003) '4 good reasons to favour local investment', www.cler.org/predac/wp1/ (accessed October 2003)

European Commission (2001) 2001/77/EC Directive on the Promotion of Electricity from Renewable Energy Sources in the Internal Electricity Market

European Photovoltaic Industry Association (EPVA) and Greenpeace (2001) *Solar Generation: Solar electricity for Over 1 Billion People and 2 Billion Jobs by 2020*, October, EPVA/Greenpeace

European Renewable Energies Federation (EREF) (2002) '2002 Average Prices (€/MWh) for RES-E, Second Edition', Paris, EREF

European Wind Energy Association (EWEA) (2005) 'Wind Power Installed in Europe by End 2004 (cumulative)', map available at www.ewea.org/ (accessed January 2005)

European Wind Energy Association (EWEA) and Greenpeace (2002) *Wind Force 12*, May, EWEA/Greenpeace

Eurosolar (2003) 'WCRE update November 27, 2003', release to group email list, 28 November

FPL Energy (FPL), www.fplenergy.com/ (accessed October 2005)

Forster, S. (2003) '6,000 MW of Wind Power by the End of 2006', *New Energy*, no 3, pp22–24

French, I. (2003) 'Biogas in Europe: Huge Potential in a Growing Market', *Renewable Energy World*, vol 6, no 4, pp120–131

Frost, T. (2003) 'Novel Funding for Renewable Energy Developments', *Commodities Now*, Asian Energy Supplement, September, p25

Geller, H. (2003) *Energy Revolution: Policies for a Sustainable Future*, Washington, DC, Island Press

German Embassy, United Kingdom (2003) 'Renewable Energy Law to be Retargeted', Environment Information Note 6.03, 3 February, www.britischebotschaft.de/en/embassy/environment/pdf/env-note_03–06.pdf/ (accessed September 2003)

German Ministry for the Environment, Nature Conservation and Nuclear Safety/Deutsches Bundesministerium für Umwelt, Naturschutz und Reaktorsicherheit (BMU) (1994) *Environmental Policy: The Federal Government's Decision of 29 September 1994 on Reducing Emissions of CO_2, and Emissions of Other Greenhouse Gases, in the Federal Republic of Germany*, November, Bonn, BMU

Gipe, Paul (2003) 'Electricity Feed Laws Power European Renewables', *Solar Today*, November/December

Goldemberg, J. (2000) 'Rural Energy in Developing Countries', in UN Development Programme, UN Department of Social and Economic Affairs, and World Energy Council, *World Energy Assessment: Energy and the Challenge of Sustainability*, New York, UNDP

Goldemberg, J. (2002) Secretary for the Environment, State of São Paulo, Brazil, 'The Brazilian Energy Initiative', 25 June

Goldemberg, J., Coelho, S. T., Nastari, P. M. and Lucon, O. (2003) 'Ethanol Learning Curve – The Brazilian Experience', *Biomass and Bioenergy*, vol 26, no 3, pp301–304

Goldstein, L., Mortensen, J. and Trickett, D. (1999) 'Grid-Connected Renewable-Electric Policies in the European Union', National Renewable Energy Laboratory, NREL/TP.620.26247, May

Gonzálvez Hernández, C. (2003) 'Wind Energy Development in Spain', presentation at European Wind Energy Conference, Madrid, 16–19 June

Group of Eight (G8) Renewable Energy Task Force (G8 2001) *G8 Renewable Energy Task Force Final Report 2001*, London

Haas, R. (2002) 'Building PV Markets: The Impact of Financial Incentives', *Renewable Energy World*, vol 5, no 4, p195

Hamrin, J. (2003) 'The Amazing Growth of the US Green Electricity Market', *Renewable Energy World*, vol 6, no 4, pp144–153

Höhn, B. (2003) Secretary for the Environment, Conservation, Agriculture and Consumer Protection, North Rhine-Westphalia, 'Renewable Energy in North Rhine-Westphalia and the Transformation of the Energy Market in Germany', presentation at 'Approaches – Challenges – Potentials: A German–American Roundtable on State Initiatives for Renewable Energy', Heinrich Böll Foundation, 23 October

Hua, Li (2002) 'China's Solar Thermal Industry: Threat or Opportunity for European Companies?', *Renewable Energy World*, vol 5, no 4, p107

Indian Ministry of Non-Conventional Energy Sources (MNES) (2000) *Annual Report 1999–2000*, New Delhi, MNES

Indian MNES (2002) *Annual Report 2001–2002*, New Delhi, MNES

Indian MNES (2003) mnes.nic.in/ (accessed January 2004)

International Energy Agency (IEA) (2003a) *World Energy Investment Outlook*, Paris, IEA

International Energy Agency, Photovoltaic Power Systems Programme (PVPS) (2003b) www.iea-pvps.org/ (accessed January 2004)

International Energy Agency, Photovoltaic Power Systems Programme (PVPS, 2003c) *Japan – National Status Report 2002: Framework for Deployment – Non-technical Factors*, August, Paris, IEA

Inter Press Service (2002) 'Environment: Brazil to Take Renewable Energy Plan to Johannesburg', www.co2e.com/News/story.asp?StoryID=576/ (accessed May 2002)

Iturriagagoitia, N. (2002) 'EHN – A New Energy Model in Action', *Renewable Energy World*, vol 5, no 3, p46

Kammen, D. M. (1999) 'Bringing Power to the People: Promoting Appropriate Energy Technologies in the Developing World', *Environment*, June

Khalip, A. and Blackburn, P. (2002) 'Feature – pioneer Brazil seeks to crank up alcohol motor again', *Reuters*, 3 June

Kittler, R. (2003) 'Renewable Energy in the US: Federal and State Activities to Promote Alternative Electricity Generation', Policy Paper 21, Heinrich Böll Foundation, Washington, DC, April

Krohn, S. (2000a) 'Danish Wind Turbines: An Industrial Success Story', Danish Wind Industry Association, 21 January, www.windpower.org/en/articles/success.htm/ (accessed January 2004)

Krohn, S. (2000b) 'Renewables in the EU Single Market – An Economic and a Policy Analysis', paper presented at the European Small Hydro Conference, Lisbon, May

Ku, J., Lew, D. and Shenghong Ma (2003) 'Sending Electricity to Townships: China's Large-Scale Renewables Programme Brings Power to a Million People', *Renewable Energy World*, vol 6, no 5, pp56–67

Lackmann, J. (2002) 'The German Market Promotion Initiatives for Renewable Energies', Bundesverband Erneuerbare Energie eV (German Renewable Energy Association), presentation at conference of the World Council for Renewable Energy, www.world-council-for-renewable-energy.org/downloads/WCRE_Lackmann.pdf/ (accessed January 2004)

Lamb, C. (2003) 'New Laws Reshape, Fuel Solar Power Growth', *Sacramento Business Journal*, 17 October

Lauber, V. (2004) 'REFIT and RPS: Options for a Harmonized Community Framework', *Energy Policy*, vol 32, pp1405–1414

McPhee, M. (1996) 'Bringing (Solar) Light to the World', *The Energizer*, Solar Energy Association of Oregon, vol 16, no 3, www.solaror.org/ftp/autumn96.pdf, Autumn

Martinot, E. (2002) 'Grid-Based Renewable Energy in Developing Countries: Policies, Strategies, and Lessons from the GEF', Washington, DC, Global Environment Facility, presentation at World Renewable Energy Policy and Strategy Forum, Berlin, 13–15 June

Martinot, E. (2003) 'Renewable Energy in Developing Countries: Lessons for the Market', *Renewable Energy World*, vol 6, no 4, pp50–65

Martinot, E. (2005) *Renewables 2005 Global Status Report*, Washington, DC, Worldwatch Institute

Martinot, E., Chaurey, A., Lew, D., Moreira, J. R. and Wamukonya, N. (2002) 'Renewable Energy Markets in Developing Countries', *Annual Review of Energy and the Environment*, vol 27, pp309–348

Martinot, E. and Reiche, K. (2000) 'Regulatory Approaches to Rural Electrification and Renewable Energy: Case Studies from Six Developing Countries', Working Paper, World Bank, Washington, DC, June

Maycock, P. (2003a) *PV News*, vol 22, no 9, September

Maycock, P. (2003b) 'PV market update', *Renewable Energy World*, vol 6, no 4, pp84–101

Mayer, R. (2001) Bundesministerium für Wirtschaft (BMWi, German Federal Ministry of Economics), email to author, 30 March

Mayfield, K. (2003) '"Fog city" catches a few rays', PowerLight Corporation, news release, San Francisco, 7 January

Menanteau, P., Finon, D. and Lamy, M.-L. (2003) 'Prices versus Quantities: Choosing Policies for Promoting the Development of Renewable Energy', *Energy Policy*, vol 31, pp799–812

Meyer, N. I. (2003) 'European Schemes for Promoting Renewables in Liberalised Markets', *Energy Policy*, vol 31, pp665–676

Meyer, N. I. and Koefoed, A. L. (2003) 'Danish Energy Reform: Policy Implications for Renewables', *Energy Policy*, vol 31, pp597–607

Millais, C. (2003) *European Wind Energy Conference Special Newsletter*, Madrid, 18 June

Mitchell, C. (2003) 'The England and Wales Renewables Obligation', Warwick Business School, University of Warwick, UK, presentation at 'Renewable Energy in America' conference of the American Council on Renewable Energy, Washington, DC, 8–9 July

Møller, T. (2002) 'Government Closes Door in Denmark', *Windpower Monthly*, vol 18, no 7, July

Moore, C. and Ihle, J. (1999) 'Renewable Energy Policy outside the United States', Issue Brief 14, Renewable Energy Policy Project, Washington, DC

Nitsch, J., Fischedick, M. et al (2001/2002) 'Analysis and Elaboration of Long-Term Scenarios for a Sustainable Energy Use in Germany', research study for the German Federal Environment Ministry, Wuppertal Institute, Stuttgart, DLR

Organisation for Economic Co-operation and Development (OECD) and International Energy Agency (IEA) (1999) *Enhancing the Market Deployment of Energy Technology: A Survey of Eight Technologies*, Paris, OECD/IEA

Osafo, Y. and Martinot, E. (2003) 'An Inventory of Renewable Energy Policies in Developing Countries', Global Environment Facility, Washington, DC (working draft)

Panik, M. S. 'Greenhouse Gases Are Global', *Sustainable Development International*, no 4, p112, www.sustdev.org/journals/edition.04/downloaded/ed4.pdfs/sdi4_1111.pdf (accessed 27 July 2002)

Pekic, V. (2003) 'Lula's Resolve Spurs Brazil's Biofuels Market', *Platts Renewable Energy Report*, no 53, July, p33

Perlin, J. (1999) 'Electrifying the Unelectrified', *Solar Today*, November/December

Platts (2003) 'Germany Reaches 100,000 Roof Milestone – but Faces Meltdown', *Platts Renewable Energy Report*, no 53, p33, July

Reiche, K., Covarrubias, A. and Martinot, E. (2000) 'Expanding Electricity Access to Remote Areas: Off-Grid Rural Electrification in Developing Countries', *WorldPower 2000*, London, Isherwood Production, www.martinot.info/Reiche_et_al_WP2000.pdf

Renewable Energy Access (REA) (2005) 'CPUC Plan Extends California Renewable Energy Mandates', www.renewableenergyaccess.com/rea/news/story?id=37180/ (accessed October 2005)

Reuters (2003) 'German 2002 wind power market up 22 pct', *Reuters*, 24 January

Righter, R. (1996) *Wind Energy in America: A History*, Norman, University of Oklahoma Press

Ristau, O. (2003) 'Sunny Prospects on the Costa del Sol', *New Energy*, no 3, June, pp48–50

Sanner, B. and Bussmann, W. (2003) 'Current Status, Prospects and Economic Framework of Geothermal Power Production in Germany', *Geothermics*, vol 32, pp429–438

Sawin, J. L. (2001) 'The Role of Government in the Development and Diffusion of Renewable Energy Technologies: Wind Power in the United States, California,

Denmark, and Germany, 1970–2000', PhD thesis, Tufts University, Fletcher School of Law and Diplomacy, Ann Arbor, UMI

Sawin, J. L. (2003) 'Charting a New Energy Future', in Starke, L. (ed) *State of the World 2003*, New York, W. W. Norton, pp85–109

Shekhar, S., Kumar, A. and Shah, C. (2001) 'Promotion and Financing of Clean Energy Options in India', paper presented at the 9th Greening Industry Network Conference, Bangkok, 24 January

SolarAccess.com (2003) 'Germany Passes Laws to Push Renewable Energy', *SolarAccess.com*, 2 December

SolarAccess.com (2003a) 'Solar Electric Systems Boost California Power Grid', *SolarAccess.com*, 5 November

SolarAccess.com (2003b) 'Trends in European Solar Thermal Market', *SolarAccess.com*, 11 July

SolarAccess.com (2003c) 'UK Boosts Outlook for Solar Energy', *SolarAccess.com*, 6 November

Sprau, P. and Weiss, I. (2002) *National Survey Report of PV Power Applications in Germany 2001*, prepared for the German Federal Ministry of Economics and the Research Centre Jülich, WIP-Renewable Energies Division, Munich, June

The Energy and Resources Institute (TERI) (2003a) 'Public Sector Financing for Wind Power Development', Asia-Pacific Environmental Innovation Strategies, Research on Innovative and Strategic Policy Options, Good Practices Inventory, March, TERI, New Delhi

The Energy and Resources Institute (TERI) (2003b) *Survey of Energy and Environmental Situations in India*, TERI Report 2003IE42

United Nations Development Programme (UNDP) (2000) *World Energy Assessment*, New York, UNDP

United Nations Environment Programme (UNEP) (2000) *Natural Selection: Evolving Choices for Renewable Energy Technology and Policy*, Paris, UNEP

United States Department of Energy (US DOE) (2002) 'An Energy Overview of Brazil', 16 May, Washington, DC, US DOE

Vote Solar Initiative, San Francisco, California, www.votesolar.org/ (accessed January 2004)

Wagner, A. (2000) speaker at presentation at US-European Wind Energy Briefing, sponsored by the Environment and Energy Study Institute, Washington, DC, 22 May (videotape)

Watts, H. (2003) 'Clean energy brings windfall to the village', *Reuters*, 5 November

Weiss, Ingrid and Sprau, P. (2002) '100,000 Roofs and 99 Pfennig – Germany's PV Financing Schemes and the Market', *Renewable Energy World*, vol 5, no 1, January/ February

Wilkins, G. (2002) 'Renewables and Development: Overcoming Barriers to Technology Transfer', *Renewable Energy World*, vol 5, no 4, pp203–215

Wiser, R., Hamrin, J. and Baruch, S. (2000) 'Renewable Energy Policy Options for China: Feed-In Laws and Renewable Portfolio Standards Compared', prepared for Center for Renewable Energy Development, Energy Research Institute, State Development Planning Commission, China, San Francisco, Center for Resource Solutions

Woods, J. and Hall, D. O. (undated) 'Bioenergy for Development: Technical and

Environmental Dimensions', www.eurosolar.org/download/Woods.pdf/ (accessed January 2004)

World Bank and International Energy Association (IEA) (1997) 'Solar Photovoltaic Power: An Overview', Renewable Energy Briefing, Energy Note 10

Wu, L. (1995) 'Inner Mongolia: One of the Pioneers of Chinese Wind Power Development', *Proceedings of the Beijing International Conference on Wind Energy*

5

Renewable Energy and Regulation of Energy Markets

Franco Fugazza and Richard Schlirf

INTRODUCTION

Globally, fossil fuels account for about two-thirds of electricity generating capacity, with the remaining third being almost all large hydro or nuclear, and only about 3 per cent of the total being from renewable energy sources (Martinot, 2002). Given this scheme's generation background, during the past decades two developments have served as catalysts for change in the national energy policies of countries throughout the globe. On the one hand, the steadily increasing demand for electrical energy has created requirements for new capital investments that exceeded the ability of governments to provide, and on the other hand the growing concern about local and global environmental degradation, resulting from the burning of fossil fuels and their effect on global warming and other air pollution problems, has created pressure to spend funds to address these issues.

Renewable energy resources and technologies have as a common characteristic that they are naturally replenishable, or non-depletable. Renewable energy technologies produce power, heat or mechanical energy by converting those resources either into electricity or to motive power.

Large-scale renewable energy (RE) systems, such as windfarms, biomass electricity, hydropower and geothermal, offer considerable economic, environmental and energy security benefits, including long-term competitive price stability, a reduced vulnerability to fuel supply disruptions, flexibility to deliver distributed and household energy to peri-urban and rural populations, minimal emissions of greenhouse gases, minimal local pollutants, the ability to attract investment for domestic infrastructure projects and, being generally modular, the ability to be expanded as demand grows. Renewable resources are environmentally benign, generally having a very modest impact on their surrounding environment (Unit for Sustainable Development and Environment, 2004).

Clean technologies can also produce significant indirect economic benefits; they provide greenhouse gas reduction benefits and enhance the attractiveness of renewable projects in the eyes of investors. Renewable resources promote energy diversification, developing a diverse portfolio of generation assets which reduces both a country's dependence on any one particular form of technology or fuel and its vulnerability to supply disruption and price increases (Armstrong and Hamrin, undated).

Finally, the primary long-term benefit of renewable technologies is that once a renewable project has been constructed, and fully depredated, it becomes a permanent, environmentally clean, and low-cost component of a country's energy system. In effect, the construction of a renewable energy project provides future generations with a low-cost energy facility that produces power with little or no environmental degradation (Armstrong and Hamrin, undated).

That said, and despite the technological advancements of sustainable energy technologies and the growing concerns for energy security and the global environment, widespread use of renewable energy systems for power generation and substantial energy efficiency measures are not expected without key changes in energy markets. In addition to this, other factors may favour conventional fossil fuel power systems, including fuel subsidies offered by many countries, the absence of charges for environmental impacts and the widespread knowledge and general familiarity of conventional technologies. However, with minor changes to the electricity market through policy reform, modern renewable energy and energy-efficient technologies can compete with conventional fossil-fuelled generation, offering long-term price stability (given their independence from fossil fuel price fluctuations) along with other benefits as important contributors to well-diversified and far-reaching energy portfolios (Armstrong and Hamrin, undated).

GOVERNMENT'S REGULATORY ROLE: REGULATION OF RENEWABLES

Although the general goal of regulation is the promulgation and preservation of the public interest (Armstrong and Hamrin, undated), more specifically the objectives of traditional regulation of the electric utility industry have been to ensure reliable power at the lowest price, establishing processes resulting in sufficient revenues to attract additional investment in electricity infrastructure as required to ensure reliable power at a reasonable price, and designing rate structures (tariffs) and incentives (price signals) to encourage the efficient use of electricity.

Traditionally, power utilities have been state-owned monopolies or privately owned monopolies, either regulated by government agencies or self-regulated without much oversight. Their traditional mission has been an engineering one: expanding supply, improving technical efficiency and ensuring or improving reliability and access. As those existing institutions have been unable to create new sources of funding for new electric-sector capacity and new sources of funding to cope with environmental imperatives, governments have been turning to the private sector and have instituted a new electricity structure designed to rely on the open market, namely the unbundled, competitive electricity market, where the electric power industry is divided into three distinct functions: generation, transmission and distribution. This division allows competitive sales of electricity or services among companies in the three sectors. The model typically relies on competitive forces to establish electricity market prices and on private-sector owners to conduct long-term planning (Martinot, 2002).

During the 1990s, waves of restructuring have washed over utilities worldwide, with profound effect on technologies, costs, prices, institutions and regulatory frameworks. Restructuring has changed the traditional mission and mandates of utilities in complex ways, and has had a large impact on environmental, social and political conditions. At the same time, new regulatory approaches are being found for reducing environmental impacts from restructured power sectors. Restructuring is resulting in independent power production and competition in generation; decentralization; privatization; unbundling of generation and transmission; and even competition in distribution (Martinot, 2002).

Along with these changes are a broad variety of new institutional and contractual forms within the power sector which have had both positive and negative implications for renewable energy. The most important are the following.

First, the constitution of competitive wholesale power markets and the removal of price regulation on generation is leading to situations in which utilities no longer have monopolies to produce power, so distribution utilities and large industrial customers gain more choices in obtaining wholesale power. The potential effects of competitive wholesale markets and independent power producers on renewable energy are significant as, in general, traditional utility monopolies have avoided renewable energy sources. On the other hand, competitive power markets may lower wholesale prices, which may stifle renewable energy development (Martinot, 2002).

Second, the privatization and/or commercialization of public utilities is making them behave like commercial corporations even if remaining state owned; they are losing state subsidies and becoming subject to the same tax laws and accounting rules as private firms. The effects of privatization and these trends on renewable energy can be either positive or negative, as private utilities are more likely to focus on costs and less on public benefits, unless specific public mandates exist. On the one hand, privatization might promote renewables by providing new financing mechanisms; on the other hand, transition from public to private may shorten time horizons, increase borrowing costs, and increase requirements for high rates of return, limiting investments in more capital-intensive projects in favour of forms of energy with lower capital costs and higher operating costs (fossil fuels and natural gas in particular) (Martinot, 2002).

Third, there is the unbundling of generation, transmission and distribution, with these functions being parcelled out to different commercial entities and so providing greater consumer incentives to self-generate using a variety of technologies, including renewable energy. In this way, if retail tariffs accurately reflect generation, transmission and distribution costs, customers may face the full costs of centralized generation and delivery, and hence may have more incentive to self-generate and thus avoid transmission and distribution charges – though it can happen too that unbundling creates transmission pricing penalties for intermittent renewable energy sources.

Lastly, the birth of competitive retail power markets and 'green power' sales, where end-users can purchase power from a 'green' supplier, usually at a premium have already been taking place. Although it is not believed that there will be a 'green revolution' in the electric industry, green power markets have begun to flourish in recent years. That trend has been assisted greatly by the exemption of green power from an increasing tax on fossil-fuel generated electricity, which has made green power almost competitive with conventional power (Martinot, 2002).

The primary economic regulatory areas specifically associated with the development of renewable resources include calculation of the total value of renewable resources to the electricity system; analysis and comparison of various types of power within the resource planning process and accounting for the environmental benefits of renewable resources in the planning process; the design of technology-neutral requests for proposals for generation to be developed by independent power producers; the development of standardized contracts for purchasing power from renewable resource facilities; provision of fair and open transmission access for intermittent renewable resources; and assuring quality performance for renewable facilities constructed for rural electrification programmes (Armstrong and Hamrin, undated).

As part of its energy policy, a country will determine its objectives with respect to the use of renewable energy resources. These objectives may be expressed as a commitment to having a certain percentage of the overall energy base come from renewables or may be a commitment to having a certain percentage of the available renewable energy resources developed within a certain timeframe (Armstrong and Hamrin, undated).

The main policy instruments most widely used to achieve targets for electricity produced from renewable energy sources are the quota obligation system, the feed-in tariff system and the tendering system (van der Linden et al, 2005; and see Chapter 4).

In theory, the obligation system is believed to be both effective and cost-effective, though this has not yet been fully demonstrated. It is complex, however, and can only function well if designed carefully. Anyway, the design of support schemes is being improved in that the basic concepts of both the obligation system and the feed-in system have been refined in such a way that the two systems are gradually converging. An important difference between the two types of systems remains, however, namely that an obligation system relies more on market forces whereas the feed-in system is based on greater involvement of the government.

In addition to the main policy instruments for stimulating renewable energy just mentioned, there are other accompanying measures used to achieve the objectives set for specific renewable energy technologies, such as financial subsidies for renewable energy power plants, which often are capital-intensive with relatively low running costs. A major disadvantage of this instrument is the lack of incentive to operate the plant as efficiently as might be the case if production-based financial support were offered. Tax incentives on investment or on production are also widely used, increasing the competitiveness of renewable electricity. Such tax incentives are widespread, probably because they are usually easy to implement and because the cost of such efforts is often opaque and buried in government budgets. Lastly, there is also green marketing aimed at consumers voluntarily paying a premium on their electricity bills to promote renewable electricity, which is used for RE generators to cover the additional generation costs of renewable electricity. This system depends on the consumer's willingness to pay for RE, which is influenced by factors related to consumer environmental awareness and specific market conditions. Of course, the choice between these systems is a political one, and each system can equally be made to function well or can be designed ineffectively (van der Linden et al, 2005).

COUNTRY AND REGIONAL CASE STUDIES

Latin America's Platform Declaration on Renewable Energies

In 2003 the Latin American and Caribbean countries participating in the Regional Conference for Latin America and the Caribbean on Renewable Energies, in Brasilia, agreed to further efforts to achieve the goal of ensuring, on the basis of voluntary actions, that by 2010 the use of renewable energy by the region, taken as a whole, amounts to at least 10 per cent of its total energy consumption.[1] To do so they concurred in fostering the formulation of the long-term public policies needed to further the development of renewable energy sources, in accordance with the regulatory frameworks in place in each country, with a view to their full and effective integration into the energy matrix, and to encourage private-sector investment as well. This includes:

• fostering the adoption of regulatory and institutional frameworks capable of incorporating instruments internalizing social and environmental benefits of renewable energy;
• encouraging the development of renewable energy projects;
• encouraging the creation of markets for green tags and carbon emission credits;
• executing tax incentive programmes and formulating public policies that encourage the development of renewable energy markets.

Wind power in Costa Rica

In Costa Rica a significant private-sector wind power industry has emerged (Martinot, 2002). Apparently, early project preparation activities, including institutional and technical feasibility studies, have engendered favourable perceptions and regulatory frameworks for wind (including 'iron-clad' power purchase agreements). Regulatory frameworks in Costa Rica, technology perceptions, and studies addressing non-technical issues have probably been more important than mitigation of perceptions of technical risk through hardware demonstrations.

Brazil's approval of an alternative electricity sources programme

The Brazilian government approved in 2002 an energy scheme called the Alternative Energy Sources Program (Proinfa), designed to stimulate development of biomass co-generation, wind and small hydro generators by guaranteeing power sale contracts to the first 3300 MW of projects that use these technologies, by the end of 2006 (Unit for Sustainable Development and Environment, 2004; GTZ, 2004).[2] Under this programme, similar to a portfolio standard, the government-owned utility (Electrobrás) will purchase electricity produced from the different renewable resources under contracts lasting up to 15 years. Specifically, the scheme provides for the purchase of electricity from plant operators that use renewable energy sources and supply the electricity generated to the interconnected grid. One of Proinfa's objectives is to achieve greater market participation for independent producers who are not governed by concessionaires in the public supply sector. Prices must satisfy certain minimum rates oriented to average electricity tariffs for final consumers and will be limited by maximum ceiling values resulting from the uniform spread of the additional costs among all electricity consumers. Consumers with very low consumption will be exempted from all additional costs. In a second phase, further projects are to be realized in order to ensure that renewable energies (not including large-scale hydropower) account for a share of 10 per cent of annual electricity demand within a period of 20 years. At least 15 per cent of the annual growth in electricity generation should originate from these sources.

Proinfa will be supported by the Brazilian Development Bank through the creation of the Financial Support Program for Alternative Electric Energy Sources. As for results, although it is too early to draw any conclusions, Brazil's electricity sector appears to be expanding. The Brazilian Association of Electric Power Distributors calculated that the electrical sector would amass a record number of public and private investments in 2005, around US$2.54 billion, nearly 60 per cent more than was invested in 2004. Of course, a great part of the funds, around 40 per cent, are coming from the federal government. It is true that this growth owes more to a programme seeking to ensure universal access to electricity by 2015 than to renewable energy strategies. Nevertheless, it must be noted that wind energy within the framework of the Proinfa programme has stimulated the intensification of investment in wind energy after a number of political approaches towards this objective had failed in the past. Attention is now being focused on the period up to the end of 2006, in which 1100 MW of wind power is due to go on to the grid. By mid-2004, altogether 161 wind projects had received electricity generation licences for a total of some 8560 MW. Applications were made by a number of different operating firms, although some of these licences have already expired or were returned by the operators.

Wind and small hydro power in India

In India, support for wind power occurred in parallel with the explosive market growth that emerged in the mid-1990s, fuelled by favourable investment tax policies and a supportive regulatory framework (Martinot, 2002). Besides investment tax credits, transparent power purchase tariffs, transmission wheeling, third-party sales, guarantees for local utility power purchase contracts and power 'banking' all contributed to the development of the market. By 2000 more than 1000 MW of wind capacity had been installed in India, virtually all of it by the private sector. In addition, dozens of domestic wind turbine manufacturers had emerged, many of them joint ventures with foreign partners. Exports of turbines began and high-technology turbine designs with variable-speed operation were being produced. The capabilities of the Indian Renewable Energy Development Agency (IREDA) were strengthened to promote and finance the private-sector investments that financed many wind and mini-hydro projects. Awareness among investors and banking institutions of the viability of wind power technology increased, which helped those lobbying for lower import tariffs for wind systems. During the 1990s, many financial institutions decided to offer financing for windfarms, a key project goal. In this regard, investment tax credits have been a powerful stimulus to technology transfer and market development while the credits existed. But longer-term market sustainability may rest on the awareness, capabilities, supportive regulatory conditions and commercial financing arrangements put in place by the assistance of international agencies such as the World Bank.

The Renewable Portfolio Standard in Japan

In 2003 the Japanese Renewable Portfolio Standard (RPS) was approved (van der Linden et al, 2005). It placed an obligation on the electricity suppliers to supply 1.4 per cent of total electricity consumption from renewable sources by 2010. Intermediate targets set for 2003 were exceeded that year. The most important renewable sources there are bioenergy, small hydro and wind. The experience with the Japanese RPS indicates that the way the system has been designed has a number of weak points – the most important of them being that the target is too low to attract substantial investments in new renewable energy technologies – so resulting in poor system performance. That is, the lack of necessary incentives; the non-transparency of the certificate market, which creates uncertainties among potential investors; and the strategic behaviour of electricity suppliers, which adds to the instability of the certificate price – all these factors serve to hamper investments in new renewable technologies.

Small hydropower in Sri Lanka

Before an energy services delivery project was implemented in Sri Lanka, all mini-hydro development was done by the national electric utility (Martinot, 2002). The project opened up the market to third-party mini-hydro developers. Many of these, in the form of independent power producers (IPPs), were financed, and regulatory frameworks for IPPs have been developed, including standardized 'non-negotiable' power purchase tariffs and contracts (PPAs). This project provided enough incentive for the national utility to adopt IPP frameworks and agree to PPAs, which together with demonstration effects of prior mini-hydro installations and new incentives for developers (such as import duty waivers and income tax concessions) spurred the market. A problem in Sri Lanka has been that variable power purchase tariffs might have hindered market development. Tariffs had been tied to short-run avoided utility costs based on the international price of oil, but because of the downturn in oil prices later, all development essentially stopped in 1999. This fluctuation has seriously hurt the longer-term interest of private mini-hydro developers in Sri Lanka.

Australia

Australia was one of the first countries to introduce an obligation system (van der Linden et al, 2005). In April 2001 the Mandatory Renewable Energy Target

(MRET) was introduced as an instrument to increase the contribution of renewable sources to Australia's electricity demand. The MRET is aimed at reducing both greenhouse gas emissions and developing the renewable energy industry. In 2003 the MRET was reviewed by an independent panel that concluded that MRET was meeting its objectives. The interim targets for the first two years (2001, 2002) had been exceeded and the industry was well on the way to meeting the targets for the third and fourth years. In a response to the review, the Australian government reconfirmed its commitment to the MRET scheme at its current level, not increasing the target beyond 2010 as recommended by the review panel. As a result, some in the renewable energy industry have questioned whether the existing system can support a growing and thriving renewable energy sector. Other changes were also announced as a result of the review. They were intended to enhance market transparency, increase opportunities for bioenergy and solar technologies, improve business certainty, and encourage innovation through recognizing emerging renewable electricity generation technologies.[3]

European Union: general trends

The directive on the promotion of electricity produced from renewable energy sources (RES) in the internal electricity market is the main legislation affecting renewable electricity at EU level (van der Linden et al, 2005). This directive aims at facilitating a significant increase in renewable electricity from 13 per cent in 1999/2000 to 21 per cent in 2010 for the current 25 member states (the EU25) as a whole. The directive also establishes indicative targets for the penetration of renewable electricity for each member state. The main policy instruments currently used by the member states for promoting electricity production from renewable energy sources are the quota obligation system or renewable portfolio standards (RPS) feed-in tariff system and the tendering system. In addition, a variety of accompanying measures are being applied in the member states for achieving the objectives of particular renewable energy technologies.

Sweden

During the 1990s the Swedish electricity market was reformed in several stages. Sweden now has a liberalized electricity market (van der Linden et al, 2005). The objectives of the reform have been to increase the freedom of choice for electricity consumers and to create conditions for greater pressure on prices and costs in the electricity supply. A new public authority, the Swedish National Energy Administration, was set up in 1998 with responsibility for implementing most of the energy policy programmes and coordinating the work of restructuring the energy system. The electricity market regulation was based on the principle of unbundling and regulated Third Party Access (TPA). The production and sale of electricity take place on competitive terms, while network operations (transmission and distribution), which in practice form a natural monopoly, are regulated and monitored. In order for these two functions not to be intermingled, a company trading in or producing electricity cannot also run network operations. The electricity networks throughout the country must be open to all players on the market who have paid a connection charge somewhere in the country. Grants are available for investments that provide a new contribution to electricity generation.

The Swedish quota obligation system came into force in 2003. The policy is part of the national energy policy targeting a transition towards a more sustainable energy system where nuclear power is phased out and fossil fuel use is limited. The system was intended to replace all previous investment and production subsidies for renewable electricity technology. The overall objective of the scheme is to increase the production of electricity from eligible renewable sources from 6.5 per cent (estimated for 2002) to 16.9 per cent in 2010. The objective was also to introduce competition between different RES eligible for certificates. As expected, this has resulted in more electricity from biomass-based combined heat and power (CHP), whereas the installation of wind power turbines has slowed. Under this system all electricity consumers (except energy-intensive industry) are required to buy a certain proportion (8.1 per cent in 2004) of the electricity they use from RES-E (Electricity from Renewable Energy Sources). A long-term stable playing field, the possibility of internationalizing, avoiding distortion to the function of the electricity market, creating reasonable conditions for existing plants, stimulating technology development and cost efficiency, and promoting new investments have been thought of as objectives of the obligation system which have resulted in positive practical experiences so far.

United Kingdom

The United Kingdom has provided support for renewable energy developments for many years (Agnolucci, 2005; Parliamentary Office of Science and Technology, 2001; van der Linden et al, 2005). The Non-Fossil Fuel Obligation (NFFO) consisted in competitive tendering for the development of renewables energy projects and was the main policy to promote electricity from RES for almost a decade in the United Kingdom. Public electricity suppliers (PES) were required to purchase a certain amount of electricity produced from non-fossil fuels and were compensated for being obliged to do so. The background of the NFFO was the privatization of utilities carried out by the Conservative government in the United Kingdom, when a vertically integrated monopoly was split and auctioned to private companies. The NFFO was generally considered a successful policy instrument. However, while the instrument has driven down costs, it has not resulted in a scale of deployment comparable to that which occurred in other

European countries. The government therefore decided to replace this scheme and introduced a new, more market-driven support mechanism, which places an obligation on electricity suppliers to ensure that a minimum percentage of the power they sell comes from renewable energy sources. This obligation system is known in England and Wales as the Renewables Obligation (RO). The obligation in the United Kingdom is placed on licensed electricity suppliers, which have to surrender tradable Renewable Obligation Certificates (ROCs) or pay a penalty to comply with the legislation. The Office of Gas and Electricity Markets (Ofgem) administers the scheme and issues the ROCs.[4] The obligation level started at 3 per cent of electricity supplied in 2002/2003 and rises to 10.4 per cent in 2010/2011 and to 15.4 per cent by 2015/2016. The obligation levels are ambitious, and the system design guarantees high ROC prices only if there is a large shortfall compared to the target. It is impossible to say with great certainty the extent to which the RO is leading to the development of additional renewable capacity or renewable electricity generation, though capacity and generation have increased markedly as the annual growth rate has significantly increased since the introduction of the obligation.

Denmark's energy policy in the 1990s

Since 1990, diverse policies have been introduced to establish the goal of sustainable development in the energy sector (ESRU, undated). The main overall target of these policies was to reduce CO_2 emissions in 2005 by 20 per cent in comparison to 1988 levels and to make renewable energy constitute almost 14 per cent of the total energy supply by that date. Energy taxes were introduced on fossil fuels and electricity mainly in the household sector consumption of oil to reduce the use of fossil fuels, or at least keep it at the same level, and increase the share of renewable energy while keeping gross energy consumption constant. Renewable energy was exempted from taxes while at the same time subsidies were introduced for environmentally friendly forms of electricity production. To ensure that previously outlined targets would be reached, green taxes became more widely used in all sectors of the economy. As trade and industry green tax schemes included subsidies for energy saving and reductions to labour taxation, not only were energy saving and efficiency encouraged, but also renewable energy became an important role in these sectors' strategies as a means of meeting economic goals.

Concomitantly, the promotion of solar cell systems and the increased use of household wind turbines encouraged the development of these latter technologies by using demonstration projects and small-scale installations and by making them more economically viable. Other important initiatives have been the land-based wind order and the offshore wind order, resulting in the 2005 target for the expansion of on- and offshore wind power being almost reached as early as the end of 1998. The next year an electricity reform took place with the goal of setting up an internal green market, obliging consumers to ensure that 20 per cent of their energy consumption came from renewable energy sources. This purchase obligation would ensure a stable demand for renewable energy and stimulate its growth. In 2000 the government enacted measures aiming at a comprehensive climate strategy. Among other measures, an offshore windfarm order established a concrete basis for fixing the settlement prices, making it possible for offshore to be as economically feasible as onshore wind.

In conclusion, energy policy proved to be effective in developing renewable energy and allowing it to be economically competitive with traditional forms of power generation. Total gross energy consumption did not increase, in part because of the efforts made to increase energy efficiency in energy supply.

CONCLUSIONS AND POLICY IMPLICATIONS

Countries need long-term energy solutions, and in order to obtain these they may elect to design policies which create a market that is receptive to the needs of long-term investments such as renewable technologies (Unit for Sustainable Development and Environment, 2004; Martinot, 2002; van der Linden et al, 2005; Armstrong and Hamrin, undated). Sustainable energy technologies, including renewables and energy efficiency, have the potential to significantly reduce the amount of fossil fuels that are consumed for the production of electricity. It is clear that there are many benefits to be accrued by diversifying power generation resources to include these technologies. While the benefits – including power system price stability, infrastructure security, environmental protection and rural economic development – may be convincing, policy and regulatory reform is required to help level the playing field with regard to conventional fossil-fuelled systems.

Regulatory reforms should be considered to ensure that renewable energy projects could feed into power grids, fostering the reduction or removal of incentives for fossil-fuelled systems and adopting targeted measures that address specific barriers to renewable energy and promote investments in such systems. Policies that have been implemented or ought to be implemented to enhance the possibilities of renewable energies include renewable portfolio standards requiring a minimum percentage of renewables to be part of the overall energy supply portfolio; systems benefit charges in which taxes are collected from all power services to support renewable energy and energy efficiency developments; exemptions from taxes to stimulate investments in renewable energy projects; exemptions from systems charges allowing renewable providers to be exempt from many of the systems charges that conventional power generators must pay; and, lastly, renewable energy resource laws (Martinot, 2002).

Experience suggests that national-level policies for technology market development and industry incentives may partly depend first on technical demonstrations and greater awareness on the part of policy-makers. But market development takes a long time, and a large and growing domestic industry is required to deal with the regulatory, contractual, technical and operational challenges of grid-connected renewable energy. This means that policy must focus explicitly on the medium term as well as the short term and ensure that sustainable regulatory mechanisms, financing, and adequate skills and manpower are developed (Martinot, 2002).

As for obligation and feed-in systems, the former can encourage cost reduction and competition, but its design is more complex than the latter. An obligation system will generally stimulate the lowest-cost and least risky renewable technologies, thereby allowing a set target to be met in an efficient way. Moreover, the total costs of an obligation system can be capped by the size of the quota and the level of the penalty. Short-term tradable renewable energy certificates (TREC) markets can lead to overcompensation of existing capacity or low-cost production, but this can be countered in several ways by technology-specific (investment) subsidies, creating a more level playing field in the TREC market or by developing technology or vintage tiers under the obligation system. Furthermore, the cost-effectiveness of an obligation system can be improved by using penalty revenues to stimulate renewables, instead of letting them flow to the state; and/or by encouraging a high level of compliance by setting a relatively high penalty. Equity aspects should be duly taken into account when designing quota obligation systems to ensure sufficient long-term support from the main stakeholders. Obligation systems seem to be less suited to small producers of renewable energy sources. Other important aspects are the exemptions of particular groups from the system, and the creation of a level playing field for renewable energy source technologies. A sufficient level of certainty for the renewable energy industry is imperative for a well-functioning obligation system (van der Linden et al, 2005).

To conclude, there are a number of specific policies for incorporating renewable energy within power sector restructuring that are observed practice or policy in many countries (Martinot, 2002). Experience and lessons from most of these policies are just emerging, however, and many effects remain poorly documented. Many of these policies have been seen in developed countries, but not yet in developing countries. The majority of them are listed and commented upon below.

- *Enactment of stable frameworks for independent power producers.* Private-sector involvement and investment in the power sector are greatly facilitated by establishing a transparent and stable framework and rules governing competition (both on price and access to customers). Establishing these conditions can assist in promoting renewable energy market development and scale-up. For grid-connected renewables in many countries, utility regulatory frameworks that allow fair competition for electricity generation by independent power producers, including power purchase agreements and a transparent and stable tariff-setting regime, are an essential first step towards creating private markets for renewable energy. In addition, rules and institutions for bidding and transacting power purchases are also essential elements of a power market.

- *Reduction of subsidies to fossil fuels.* If conventional generation remains subsidized, these subsidies should be reduced to create a more level playing field. Explicit or implicit subsidies for traditional forms of generation are prevalent in many countries. Implicit subsidies may exist, for example, if tariffs do not incorporate full capital replacement costs of ageing fossil units or if environmental standards are not being enforced.
- *Open-access provision to transmission.* An open-access transmission system must allow power wheeling between buyer and seller that provides open access to customers. Transmission services should not discriminate against or give unfair advantage to specific ownership or certain types of generation.
- *Comparable environmental standards enforcement on all generators.* Existing facilities, even if old, should face the same environmental standards as new plants, even if this means they must be decommissioned because of prohibitive retrofit costs.
- *Attention to environmental policy at the same time or restructuring.* Emissions standards, monitoring requirements and other aspects of environmental policy can be integrated to strengthen changes in the power sector.
- *Enactment of renewable energy portfolio standards (RPS).* An RPS requires that a minimum percentage of power sold in a given region or service territory is met by renewable energy sources. Usually proposed along with RPS are power trading schemes whereby retail providers may trade their 'renewable energy' generation obligations with one another as long as all meet their respective standards, using 'green certificates'.
- *Distribution utilities regulation to encourage distributed generation.* Regulation can encourage distribution utilities to consider the lowest system cost when making decisions about types of service. Regulation of retail electricity suppliers should create economic incentives that promote full consideration of renewable energy technologies for bulk power, distributed generation and demand-side applications. Power sector reforms should ensure that distributed options can compete to provide electricity services.

And as has been mentioned, other measures include the enactment of mandatory purchases at a fixed price of power based on renewable energy, or of competitively bid renewable energy resource obligations; imposition of system benefits charge levies to provide funds for public renewable energy and energy efficiency programmes; and the encouragement of distributed energy.

NOTES

1   Brasilia Platform on Renewable Energies (the Latin American and Caribbean countries participating in the Regional Conference for Latin America and the Caribbean on Renewable Energies, 2003).
2   Brazil-Arab News Agency (ANBA) – Arab Brazilian Chamber of Commerce (CCAB), www.anba.com.br/ingles/index.php
3   Australia: www.orer.gov.au/; Renewable Energy Certificates (Office of the Renewable Energy Regulator – Australian Government)
4   See www.ofgem.gov.uk/ofgem/shared/template2.jsp?id=9206.

REFERENCES AND FURTHER READING

Agnolucci, P. (2005) 'Opportunism and Competition in the Non-fossil Fuel Obligation', Tyndall Centre Working Paper 78, Norwich, Tyndall Centre for Climate Change Research

Armstrong, A. J. and Hamrin, J. (undated) *The Renewable Energy Policy Manual*, US Export Council for Renewable Energy

Byrne, J. and Shen, B. (1995) *Toward a Sustainable Energy Policy Framework: U.S. Experience and China's Opportunities*, Newark, Center for Energy and Environmental Policy, University of Delaware

Energy Systems Research Unit (ESRU) (undated) *Danish Energy Policy*, Glasgow, Institute for Energy and Environment, University of Strathclyde

Gesellschaft für Technische Zusammenarbeit (GTZ) (2004) *Energy-Policy Framework Conditions for Electricity Markets and Renewable Energies*, Eschborn, GTZ, Division Environment and Infrastructure – TERNA Wind Energy Programme

Golove, W., Brown, R. and Holt, E. (2004) *Guide to Purchasing Green Power*, Washington, DC, Energy Efficiency and Renewable Energy Federal Energy Management Program, US Department of Energy

Kozloff, K. and Bailly, H. (2000) 'Renewable Energy Strategies in Developing and EIT Countries under Restructured Electricity Markets', paper presented at the conference 'Accelerating Grid-Based Renewable Energy Power Generation for a Clean Environment', Washington, DC, 8 March 2000

Martinot, E. (2002) 'Grid-Based Renewable Energy in Developing Countries: Policies, Strategies, and Lessons from the GEF', Washington, DC, Global Environment Facility, presentation at World Renewable Energy Policy and Strategy Forum, Berlin, 13–15 June

Parliamentary Office of Science and Technology (2001) *Renewable Energy*, www.parliament.uk/parliamentary_offices/post/environment.cfm

Sterzinger, G. (undated) *Renewable Energy and Regulation – Why Do Bad Things Happen to Good Energy*, Washington, DC, Renewable Energy Policy Project

Unit for Sustainable Development and Environment – Organization of American States (2004) 'Policy Reform for Sustainable Energy in Latin America and the Caribbean', Policy Series 5 van der Linden, N. H., Uyterlinde, M. A., Vrolijk, C., Nilsson, L. J., Khan, J., Åstrand, K., Ericsson, K. and Wiser, R. (2005) *Review of International Experience with Renewable Energy Obligation Support Mechanisms*, Petten, the Netherlands, Energy Research Centre of the Netherlands (ECN)

Wiser, R., Pickle, S. and Goldman, C. (1996) *California Renewable Energy Policy and Implementation Issues*, Berkeley, Energy and Environment Division, Ernest Orlando Lawrence Berkeley National Laboratory, University of California

6

Removing Subsidies: Levelling the Playing Field for Renewable Energy Technologies

Jonathan Pershing and Jim Mackenzie

INTRODUCTION

Modern energy use is increasingly recognized as the culprit in a number of environmental problems. Locally, production and consumption of energy lead to air pollution, water pollution and land degradation. On a regional scale, in the absence of special precautions emissions of sulphur and nitrogen are distributed for kilometres around fossil fuel combustion plants, while nuclear waste disposal requires sites that can be guaranteed to be safe for millennia. Globally, emissions of greenhouse gases lead to climate change, arguably one of the most serious environmental threats facing the world today.

However, energy also provides fundamental services: lighting, heating and cooling, and mobility all require energy use. Without energy, the standards of modern society for health, education and welfare would be impossible to meet. Thus, a balance must be found that meets the challenge of minimizing the environmental damages caused by energy production and use, while still meeting the demand for energy services to power the economy.

Renewable energy provides one potential response to this challenge. Unfortunately, while renewable energy contributes little to environmental damages, its economic costs[1] are generally higher than those of fossil fuels – leading to limited penetration of renewable sources into the global energy market. However, the cost differential between renewable energy and fossil and nuclear energy has been exacerbated by the long-term support provided by governments to the development of these other sources – support that is ongoing. What is now required is that this government support be turned away from fossil fuels and towards renewable energy, leading to a transition to safe and clean alternatives to power the growing demand for energy services.

History provides ample evidence of the ability of governments to shape energy transitions, dating back to the mid-19th century, when UK government subsidies to coal contributed to its rapidly increasing use, and helped fuel the industrial revolution. Similar levels of government support have been found in subsequent energy transitions, including those driving oil and natural gas production and use, as well as nuclear power. In each transition the focus was on increasing reliability of supply, as well as reducing costs and extending the reach of energy services. Direct financial aid was not the only instrument. For many of these transitions, governments provided technical assistance, guaranteed the purchase of new technologies, placed limits on the liability for accidents, and paid for the research and development needed to overcome technical barriers to implementation. Governments have also set the framework of legal and institutional conditions necessary for the private sector to move new energy technologies into the market.

The result of historic – and ongoing – government subsidies has been to skew the energy playing field against 'clean' sources of energy. Collectively, global energy subsidies provided for oil, gas, coal and nuclear power total in the tens of billions of dollars, and help drive the unsustainable fuel mix we see in today's markets.

There is a growing recognition that governments need to reassess their support for certain sources of energy – or at a minimum, that they begin to incorporate the full price to society of the costs that energy use and production impose. With this goal in mind, subsidies that promote the overuse of damaging energy sources, in addition to promoting the use of fuels that emit local pollutants as well as greenhouse gases, must be re-examined. It is clear that if pricing were to fully reflect environmental damages, our pattern of energy use – including both how much is used and what sources were used – would change.

WHAT ARE ENERGY SUBSIDIES AND WHY DO THEY EXIST?

Defining subsidies

According to the World Trade Organization (WTO), the definition of a subsidy contains three elements: (1) a financial contribution (2) by a government or any public body within the territory of a member state (3) which confers a benefit. All three of these elements must be satisfied in order for a subsidy to exist.[2]

While generally subsidies are thought of as cash payments from a government to a producer or consumer, other kinds of subsidies exist as well (see

Table 6.1 *Types of energy subsidies*

Form of government intervention	Example
1 Direct financial transfer	Grants to producer or consumers, low-interest loans
2 Preferential tax treatment	Rebates, exemptions on royalties, tax credit, accelerated depreciation
3 Trade restrictions	Quotas, trade embargoes, technical restrictions
4 Energy-related services provided by government at less than full cost	Direct investment in energy infrastructure, public R&D
5 Regulation of energy sector	Demand guarantees, price controls, market access restrictions

Source: IEA and UNEP (2002)

Table 6.1). Some have a direct effect on price, like grants and tax exemptions, while others act indirectly, for example through regulations that bias the market in favour of a particular fuel or government-sponsored technology research and development (IEA and UNEP, 2002). For a more extensive review of the economics of subsidies, see OECD (1998).

Overconsumption due to excessively low prices distorts supply and demand. Subsidies for energy consumption, which mask the true energy price, lead to higher use (and emissions) for every unit of output. By lowering the price to consumers, subsidies increase import requirements and decrease the availability of fuels for export.

There are a number of factors related to fuels, energy technology and operating habits that collectively contribute to underpricing of energy, or particular forms of energy supply. The common reference to such distortions is to 'a playing field that is not level'.[3] Particularly when the full price of each fuel (incorporating all direct and indirect costs of use and production) is not reflected in its market price, such a distortion can lead to unwelcome outcomes, including reductions in economic efficiency and environmental damages.

Both direct and implicit or hidden cross-subsidies tilt the playing field away from renewables and create significant barriers to the commercialization of renewable energy. Often, implicit or hidden subsidies manifest as system-wide biases toward large-scale, highly centralized energy systems, and disadvantage the smaller-scale, more modular distributed and decentralized energy systems typical of many renewable energy technologies.

The exact consequence of any subsidy is dependent on the specific national circumstances to which it is applied. As the policy context and the technical and institutional characteristics of the electricity system vary, so do the barriers and potential ways of overcoming these. Whether electricity markets are fully liberalized, unbundled and partially liberalized, or vertically integrated monopolies will influence the approach necessary to level the playing field at the national level.

A number of the key forms of market distortions and their consequences have been evaluated by the International Energy Agency (IEA) (IEA and UNEP, 2002, p13):

• Subsidies that lower consumer prices lead to higher energy use and reduced incentives to conserve or use energy more efficiently. At the same time, they reduce incentives for producers to minimize costs, resulting in less efficient plant operation and investments.
• By reducing the price received by producers, subsidies can undermine energy producers' return on investment and their ability and incentive to invest in new infrastructure.
• Direct subsidies in the form of grants or tax exemptions act as a drain on government finances.
• Price caps or ceilings below market-clearing levels may lead to physical shortages and a need for administratively costly rationing arrangements.
• By increasing energy use, consumption subsidies boost demand for imports or reduce the amount of energy available for export.
• Subsidies to specific energy technologies can undermine the development and commercialization of other promising technologies.

Distributional impacts

The social impacts of energy subsidies vary according to the type of subsidy. Many subsidy programmes intended to boost poor households' purchasing power or rural communities' access to modern energy can, paradoxically, leave the poor worse off (IEA and UNEP, 2002, p14). This result occurs because the benefits often go to energy company equipment suppliers and the better-off households (especially in the towns and cities) rather than to the rural poor. Studies, including those of the IEA and the United Nations Environment Programme, have identified three main reasons why energy subsidies tend not to reach the poor:

• The poorest families may not be able to afford even subsidized energy as middle- and higher-income families tend to get the bulk of subsidized energy through favouritism and local political corruption.
• Even if the poor are able to benefit from a subsidy, the financial value to them may be small because their consumption is so modest.
• Consumption subsidies involving caps on prices below market levels may lead to a need for rationing.

Biases in favour of large scale

Energy subsidies often go to capital-intensive projects such as large hydro dams or big oil refineries at the expense of smaller or distributed alternatives. As a secondary inequitable outcome, it is often low-income families living near dams, refineries and gas processing plants who suffer disproportionately from the air pollution and other local damages of these facilities.

An example of the disproportionate attention to large-scale infrastructure development is obtained from a consideration of recent World Bank energy loans: while the total loan portfolio devoted to renewable energy has risen from 4 per cent in 1990 to 14 per cent in 2003 (Saghir, 2003),[4] the overwhelming preponderance of the total energy investment remains focused on large-scale conventional energy supplies and infrastructure, including loans to oil and gas, coal, and transmission grids. However, Bank policy *is* trending in the right direction; the issue is how to continue to increase the share to smaller and renewable energy projects. The Bank's bias is mirrored in the funding of private lending institutions. In a forthcoming study by the World Resources Institute's Julia Philpott, $95 billion in capital flows for new power plants (from 1994 to 2001 in a selected group of developing countries) was heavily biased to coal and gas – with limited amounts for diesel (which tends to be small scale) or renewable energy). The bias was revealed not only in public financing, but also in private flows.

Rationales for subsidies

Notwithstanding these negative consequences to the implementation of subsidies, there are a number of reasons generally given for adopting energy subsidies, including:

- protecting domestic industry and promoting jobs at home;
- reducing imports and improving national security;
- managing risk;
- making energy more affordable for specific social groups;
- protecting the environment.

These reasons often apply particularly to the poor. According to a study released by the World Bank (Barnes and Halpern, 2000), the poor often are unable to obtain services as a result of high access costs, non-availability, inadequate start-up capital for new supply, as well as import restrictions or tax policies.

Subsidies seek to remedy these effects. In short, those providing the subsidies argue that the benefits and welfare gained justify – and often are much higher than – the long-term costs involved in providing the subsidy. Thus, high up-front costs that may be recouped over time, but cannot be justified by low initial revenues (especially for firms with short-term profit goals), or the assistance to poor households in obtaining higher-quality services, offer rationales to today's ongoing subsidies.

Some of the world's largest subsidies have been offered in support of energy security. Energy legislation debated in the US Congress in late 2003 (although not yet passed) provides an example of this. According to the US Congressional Budget Office (CBO), the draft 2003 energy bill would have increased direct spending by $3.7 billion over the 2004–2008 period and by $5.4 billion over the 2004–2013 period (US Congressional Budget Office, 2003). Furthermore, the CBO estimated that the act would reduce revenues by $17.4 billion over the period 2004–2008 and by $25.7 billion over the period 2004–2013. Revenues would be divided between electricity and gas distribution, coal, oil, natural gas production, as well as nuclear power. Only a very small share would go to renewable energy. The emphasis, according to the bill's drafters, was on reducing the US reliance on imported supplies to power its economy.

A compelling case can be made for promoting renewable energy from an energy security point. The United Kingdom's Renewable Energy and Energy Efficiency Partnership (REEEP) and the UN Environment Programme (UNEP) are currently funding a project focusing on how renewable energy can provide energy security benefits (Mitchell, 2004). The work explores the potential for a portfolio planning approach – which shows there may be less risk associated with renewable energy compared to gas.

Similarly, energy security is affected not only by disruptions in the source of supply, but also by failures in the grid and diminished operating reliability of the electricity system. The power system outages in North America and Canada in August 2003 provide evidence of how highly centralized energy systems can be vulnerable. Taking account of the massive impact of large-scale system failure increases the rationale for distributed and decentralized energy systems within the context of energy security.

Despite the economic, social and environmental problems brought on by subsidy programmes, it is politically difficult to discontinue them, partly because of the short-term economic costs that subsidy removal would entail. Increases in energy prices have immediate impacts, especially among the poor, and there can be major reactions if adequate programmes to at least partially compensate for subsidy losses do not exist. Indeed, there are often more efficient ways – such as social security programmes – that can reduce the burdens faced by the energy-poor. While the political barriers to reducing subsidies do not justify the maintenance of these measures, they do nevertheless suggest that subsidy reforms may best be implemented gradually to improve their chances of success.

As the discussion so far suggests, there may be some cases in which subsidies are appropriate policy tools. Those arguing for subsidies have suggested a framework for effective policy and called for subsidies to be assessed according to their efficacy (the subsidy reaches those for whom it is intended), sector efficiency (the subsidy is structured to provide services at least cost) and cost-effectiveness (the subsidy achieves goals at lowest programme cost while providing incentives to businesses that serve the poor) (Barnes and Halpern, 2000).

QUANTIFICATION OF ENERGY SUBSIDIES AND THEIR EFFECTS

A subsidy by its very nature involves a complex set of changes in economic resource allocation through its effects on costs and/or prices (IEA and UNEP, 2002, p12). These shifts inevitably have economic, social and environmental implications. Indeed, the reason subsidies exist at all is to support some economic, social or environmental goal. However, artificially low energy prices also contribute to poor economic performance on the part of many state-owned energy companies (IEA and UNEP, 2002, p8). This poor performance reduces the ability of companies to invest to meet increasing demand. It also discourages private and foreign investment in the energy sectors of these countries.

Quantifying these different effects, in terms of costs and benefits, is extremely difficult and usually involves a subjective judgment on the part of the analyst. This is especially true when one is measuring the social and environmental benefits.

Economic costs

The change in the price of energy significantly affects energy consumption. As Figure 6.1 makes clear, the higher the prices, the lower the consumption. Subsidies, which act to depress the prices consumers see, thus, perversely, increase national demand – and reduce overall resource availability. Prices reflecting the real value of the resources employed in the generation of electricity ensure that consumers receive the correct signals to use electricity in the most efficient possible way.

Energy subsidies are widespread, though they vary greatly in importance and type according to the fuel and country (IEA and UNEP, 2002). Few studies have been carried out to quantify the energy subsidies for the world as a whole. A 1992 study by the World Bank, however, estimated global subsidies from the underpricing of energy to amount to around $230 billion per year (mostly related to below-market pricing in the former Soviet Union (FSU) and developing countries). This can be compared with GDP in 1992 of about $27 trillion. This was confirmed in a 2001 paper by Andre de Moor of the Netherlands; the results of his study are reproduced in Table 6.2.

Other, more recent studies confirm that energy subsidies are much larger in non-OECD countries. In 1997 the World Bank estimated annual fossil fuel subsidies at $48 billion in 20 of the largest countries outside the OECD and $10 billion within the OECD (IEA and UNEP, 2002, p11).

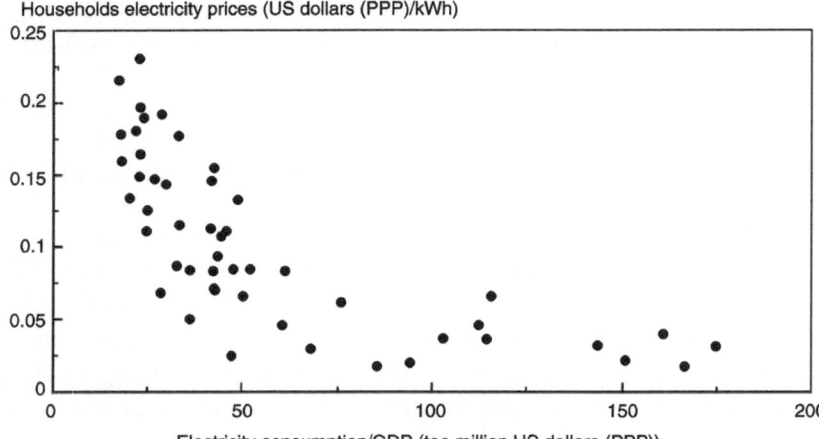

Source: IEA (1999a)

Figure 6.1 *Electricity prices and electricity use per GDP, 1996 (27 OECD and 22 non-OECD countries)*

Table 6.2 *The cost of annual energy subsidies, 1995–1998 ($US billion)*

	OECD countries	Non-OECD countries	Total
Coal	30	23	53
Oil	19	33	52
Gas	8	38	46
All fossil fuels	57	94	151
Electricity	a	48	48
Nuclear	16	nil	16
Renewable and end-use	9	nil	9
Non-payments and bail-out[b]	0	20	20
Total	82	162	244
Global energy subsidies	34%	6%	100%
Per capita subsidies ($/cap)	88	35	44
Per capita GDP ($/cap, 2000)	23,132	3903	7316

Notes:
a   Subsidies for electricity in OECD countries have been attributed to fossil fuels according the shares.
b   Subsidies from non-payments and bail-out operations have not been attributed to energy sources.

Source: de Moor (2001); World Resources Institute (2003).

While globally comprehensive studies are not generally available (nor are those that are available considered particularly accurate), a number of regional analyses have been made at a detailed level. In a World Bank paper considering the FSU, China, Poland, India, South Africa, Czechoslovakia, Mexico, Brazil,

Argentina, Venezuela, Indonesia, Saudi Arabia and Egypt, it was concluded that total economic efficiency losses produced by subsidies in these countries amounted to $33 billion annually (Larsen and Shah, 1992). An IEA study (IEA, 1999a) evaluating China, Russia, Indonesia, Iran, South Africa, Venezuela and Kazakhstan suggested economic efficiency losses of substantially less: about $17 billion (due to reduced subsidies in the intervening period, as well as to a different set of countries being evaluated). Table 6.3, prepared by Helen Mountford of the OECD, drawing from a range of studies, shows how extensive the subsidization is in a variety of OECD countries (Mountford, 2000).

More specificity is available for some fuel subsidies, particularly for coal: the IEA provides a regular update of coal subsidies in those member countries subsidizing this fuel. As can be seen from Table 6.4, while these levels have been decreasing with time, they are still significant.

Table 6.3 *Results of OECD and other studies on energy production subsidy removal*

Study	Subsidy or group of subsidies removed	Monetary equivalent of distortion (US$ million, various years, 1988–1995)	Other economic effects of removing subsidies
DRI (1997)	Coal PSEs in Europe and Japan	5800	Job loss in coal industry, increased coal trade.
Böhringer (1995)	Coal in Germany	6700	Nearly 1% GDP increase. Job loss in coal industry, increased coal trade. Cost of using subsidies to maintain jobs is 94–145,000 DM per job per year. Reduces cost of meeting CO_2 target
Michaelis (1997)	Grants and price supports for coal and nuclear producers in the UK	2500	
Shelby et al (1996)	DFI (1993) analysis of US Federal subsidies	8500[a]	
	DJA (1994) analysis of US Federal subsidies	15,400[a]	GNP increased 0.2% if revenue used to reduce capital taxes
Anderson and McKibbin (1997)	Coal subsidies and import restrictions in Western Europe and Japan (to 2005)	NQ	Increased coal trade. GNP of transition and developing economies increased, while Australia's GNP is lowered

Note: NQ = not quantified; PSEs = producer subsidy equivalents
a   The two studies analyse different sets of energy supports and use slightly different estimates for some of them.

Table 6.4 *Total support to coal production in selected IEA countries (US$ million)*

	1991	1992	1993	1994	1995	1996	1997	1998 (p)[a]
France	397	402	422	361	83[b]	118	c	c
Germany	6919	7885	6873	7930	8502	7165	5880	4982
Japan	815	739	878	918	884	729	469	n.a.
Spain	709	735	595	959	1110	1043	956	929
Turkey	408	613	394	165	267	195	158	171
UK	1991	1898	292	318	204	179	293	0

Source: IEA (1999b)

Notes:
a (p) = preliminary data, subject to revision.
b Note that loans taken out by Charbonnages de France have increased since 1994, as have production costs.
c Aid for the French coal industry has not been authorized by the European Commission for the years 1997 and 1998.

The results of such declining subsidies can be seen directly in the production numbers shown in Figure 6.2. Inasmuch as coal has been replaced by other, cleaner fuels, there is a net environmental benefit to such subsidy reform. However, in many countries the coal that had been produced domestically has been replaced with imported coal – providing little environmental benefit and perhaps a loss in domestic energy security.

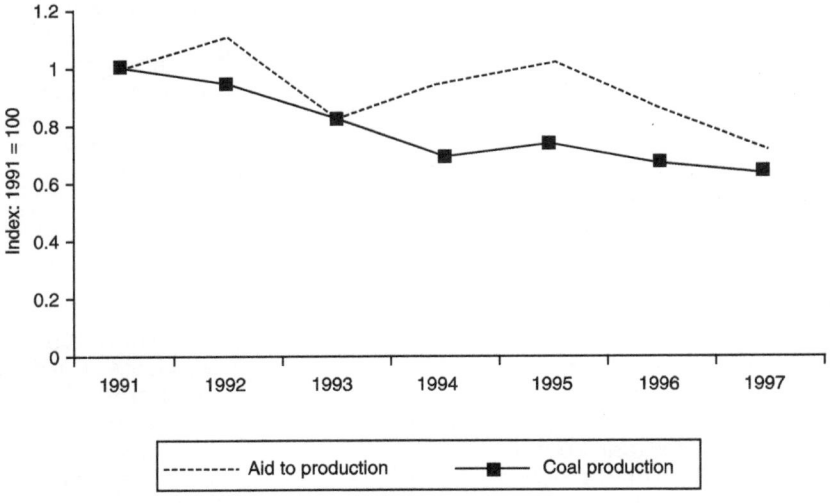

Source: IEA

Figure 6.2 *Coal production and aid to production (subsidies) in selected IEA countries, 1991–1997 (index: 1991 = 100)*

The results of subsidy reform can have significant effects on national – and global – economies as well as the environment. According to the 1999 IEA study, removing subsidies in the energy sector of eight key countries would:

*   reduce primary energy consumption in these countries by 13 per cent;
*   increase the countries' GDP through higher economic efficiency by almost 1 per cent;
*   lower CO_2 emissions by 16 per cent;
*   yield domestic environmental benefits, including reduced local air pollution.

Positive effects on a global scale are to be expected. Subsidy removal in all eight countries would cut energy consumption by 3.5 per cent at world level, thus improving world energy intensity significantly.

Notwithstanding these still-large subsidies, the trend is improving. The overall size of subsidies has fallen sharply since the 1980s, mainly due to economic reforms in the former communist bloc. According to the World Bank, subsidies dropped by more than half in the five years ending in 1996. Additionally, in most OECD countries, gross energy subsidies are more than offset by taxes. In the four largest European states, for example, revenue from special duties and taxes on sales of oil products alone amount to almost $160 billion. This compares to perhaps $20–30 billion per year of energy subsidies for the OECD as a whole (IEA and UNEP, 2002, p12).

Environmental implications

While subsidies that encourage the production and use of fossil fuels often have detrimental environmental impacts, such generalizations are not always accurate. For example, subsidies that lead to switching from firewood and dung to oil products (such as propane) could greatly reduce indoor air pollution, a particularly acute problem in developing countries. Some countries also promote policies to encourage the use of renewables for environmental and security objectives; for example, Brazil, Denmark and the US all subsidize the production of fuels derived from agricultural feedstocks or biomass with these goals in mind. However, more frequently, consumer subsidies that lower the price of fuels or the cost of using them lead to environmental damages at the local level, including water and air pollution through additional use (and misuse) of energy, particularly fossil fuels.

A number of studies have sought to quantify the effects of subsidies on the environment, including on emissions of conventional pollutants such as SO_2, NO_x and particulates as well as carbon dioxide. Table 6.5, from Helen Mountford, shows the results of several studies examining the effects of subsidy removal on CO_2.

A number of studies have shown the harmful effects of various types of fossil fuel subsidies. For example, according to the OECD, global CO_2 emissions

Table 6.5 *Energy production subsidy removal: effects on CO_2*

Study	Subsidy or group of subsidies removed	Decrease in annual CO_2 emissions relative to reference scenario resulting from reforms by 2010	
		Million tonnes	Percentage in sector concerned
DRI (1997)	Coal producer subsidy equivalents (PSEs) in Europe and Japan	10 (DRI estimate); >50 (OECD estimate)	1
Michaelis (1997)	Grants and price supports for coal and nuclear producers in the UK	0–40	0–8
Shelby et al (1996)	DFI (1993) analysis of US Federal subsidies	40	0.7
	DJA (1994) analysis of US Federal subsidies	235	4
Anderson and McKibbin (1997)	Coal subsidies and import restrictions in Western Europe and Japan (to 2005)	13,000 (worldwide)	8 (worldwide)

would be reduced by more than 6 per cent and real income by 0.1 per cent by 2010 if all subsidies on fossil fuels used in industry and the power sector were removed everywhere (IEA and UNEP, 2002, p16). The IEA's 1999 study concluded that the removal of consumption subsidies in 8 of the largest non-OECD countries would reduce primary energy use by 13 per cent, lower CO_2 emissions by 16 per cent and raise GDP by almost 1 per cent in those countries as a whole. These same subsidy reductions would have global impacts including a reduction in energy consumption by 3.5 per cent and global emissions by 4.6 per cent. Other benefits too would accrue from subsidy reduction, such as improved international security from reductions in oil imports from unstable parts of the world. The removal of coal subsidies generally yields the biggest environmental benefits.

Subsidy reform and renewable energy: levelling the playing field

While subsidies often have detrimental impacts, governments can have a clear and pressing policy goal for which subsidies are an appropriate instrument. The existence of market barriers to the introduction of climate-friendly technologies provides an example where subsidies may be justified. Renewable energy sources, which often have high initial (capital) costs and concomitant risks, may be subsidized by governments wishing to encourage investments in new capacity or in research to meet environmental or social goals.

Lowering the unit costs of emerging renewable technologies like solar PV or wind requires experience – which comes from building and operating plants. The time needed to gain this experience may be too long for the market to bear without a degree of government support. The facts bear this out. Few energy technologies have reached maturity without substantial public sector investment.

Identifying a 'good' subsidy is essential to the task of reform. Experience shows that good subsidies should be:

- well targeted so that only a fairly narrow group of consumers or producers are recipients;
- soundly based, so that incentives to provide the service efficiently are not undermined;
- practical, so that the financial resources are adequate and the management of the programme is feasible and affordable;
- transparent, so that the financial costs and the provisions of the programme are clear;
- in effect only for a limited lifetime (i.e. have a sunset provision) and be predictable, so that recipients can plan for future phase-down and phase-out;
- close to market (technologies that are 'deserving' of subsidies should not be oversubsidized lest the subsidy stifle commercial discipline and competition; the subsidy provided should be only equal to the magnitude of the externality);
- competitive, so that subsidies are provided through competitive mechanisms so as to ensure that excess 'rents' are dissipated.

Adopting or removing subsidies must take these properties into account. In addition, the phasing in of reforms can be done over a period of time to reduce the resistance and pain. If new subsidies reduce the economic status of a specific group, compensating measures can be (and may need to be) introduced if such a goal is considered desirable. For example, if subsidies directly supporting energy prices are to be removed, they may be offset by loans for introducing new, energy-efficient appliances. Thus, while energy costs rise (and demand falls), the final price to the consumer may stay level – or even decline. Such multi-goal strategies must be developed and implemented – which, in turn, requires that politicians communicate to the public the overall benefits of subsidy reform.

Subsidies for renewable energy may be usefully focused in three areas: (1) reducing technical barriers; (2) overcoming market impediments (including through the internalizing of externalities); and (3) addressing administrative barriers and social and environmental constraints.

Actions aimed at technical barriers

The barriers to renewables vary as new technologies move into the marketplace. In the initial stages of development, technical impediments usually predominate.[5] They can best be addressed through improved, strengthened and extended international collaboration in R&D in the renewables area.

There is some indication that an increasing emphasis has been placed on R&D in the renewables area. According to the IEA, while the overall levels of energy-related R&D have been steadily declining over the past decades (in 2000 they stood at less than 50 per cent of 1980 levels), the support globally for renewable energy has increased slightly over the past decade – which means it has been absorbing an increasing share of the total R&D.[6] However, even with this support the reduced overall levels of spending have been inadequate to address the serious technical hurdles that remain to competitive pricing in the renewable sector. Figure 6.3 shows the declining levels of support, as well as the dominant share of the total devoted to nuclear power.

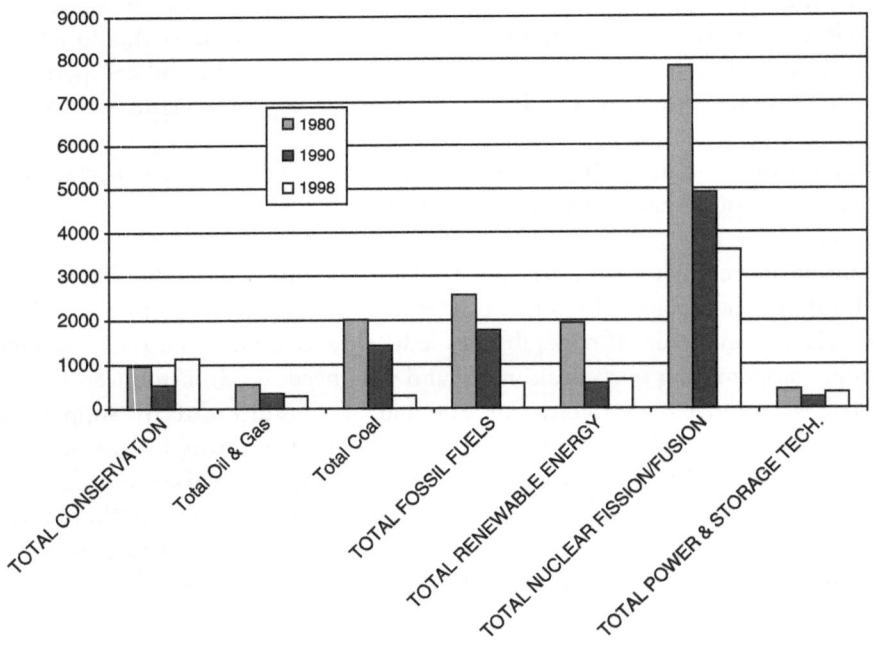

Source: IEA, www.iea.org/dbtw-wpd/textbase/stats/rd.asp

Figure 6.3 *R&D investment by IEA member countries, 1980, 1990 and 1998 (million US$, 2002 prices and exchange rates)*

Actions aimed at market impediments: internalizing externalities

The broadest and most economically sound approach to reform markets lies in the 'internalizing' of externalities. In short, this means that price signals are provided to value any damages – or conversely, to recognize specific benefits. Such price signals can be explicit (in the form of new taxes, user fees or surcharges), or implicit (e.g. in the form of caps on use). If designed in a sufficiently flexible manner, these tools can allow the market to implement the desired outcomes at the lowest cost.

Perhaps the most promising of these are options that use trading systems, allowing the market to efficiently allocate the costs among energy producers and consumers. Two examples offer insight into this policy choice: the market for renewable energy certificates, and the market for greenhouse gas emissions.

According to the IEA,[7] nine IEA member countries (as well as several US states) have established renewable portfolio standards – which allow for trading of renewable energy certificates (RECs). The initial commitments to renewable energy were established to promote their use in order to decrease dependency on foreign imports, to promote long-term R&D, to reduce greenhouse gas emissions (GHG) from fossil fuels, and to provide incentives for the development of new, indigenous industry. While the details vary, each of these systems requires a minimum share of electricity to be generated by renewable energy. This obligation in turn allows the market to establish a price at which renewable energy would trade – and prompts energy producers to invest accordingly. At present, REC permits trade at different prices in different markets: 500 kWh can be purchased in the US market for $7.50,[8] or permits can be traded at a much higher range of £45–48 per megawatt-hour in the United Kingdom.[9]

A similar system of caps (and subsequent trading) allows for the externality cost of climate to be added into energy prices. With fossil fuels emitting considerable carbon dioxide during the combustion stage, any pricing of carbon raises the price of those fuels relative to non-carbon alternatives. Currently a number of countries, responding to their commitments under the UN Climate Convention's Kyoto Protocol, have developed pilot programmes to cap GHG emissions – and the markets have responded with trading and prices for emissions offsets. In 2003, according to information from Natsource, CO2e.com and PointCarbon (three international brokerage firms), the volume of tons trading exceeds 200 million, and the price ranges widely (between $1 and $17). The existence of such prices gives a general sense of the cost to the market of complying with the caps on emissions that have been set to date.

Clearly, while markets are being established in renewable energy, full values of market externalities are not yet assigned. If a technology is to become cost-effective, market impediments such as hostile pricing structures often need to be overcome. This can be achieved through the removal of subsidies to competing non-renewable sources and through measures to ensure full-cost pricing among all forms of energy. This will ensure that energy prices reflect their social and environmental costs as well as their short-term production costs. Full-cost pricing is difficult to achieve because of uncertainties in the magnitude of external costs and the resulting political problems.

In order to take into account the full social, environmental and economic benefits, the full value price for renewable electricity should be estimated and used. Such estimates can be made only with coordinated research on the full valuation of renewable electricity and through the dissemination of experience with approaches such as net metering and exploring voluntary agreements between regulators and utilities. Opportunities in niche markets worldwide should be published to enhance awareness of them and their attractiveness to investors and consumers. Commercial finance should be harnessed to develop technical standards to increase the confidence of investors and developers. At the same time, certified methodologies for assessing resources should be standardized. Provisions should be adopted to support insurance instruments, thereby bringing the renewable industry and the financial community together to develop suitable financing packages.

Actions aimed at administrative barriers and social and environmental impediments

In addition to economic and technical barriers, reforming the energy sector will require removing system-wide institutional barriers. Renewable energy will only become competitive with larger-scale more traditional energy technologies if:

- Cost-reflective pricing and least-cost planning values and rewards for small-scale and modular generation are embedded within the distribution system.
- There is a longer-term move towards performance-based regulation whereby energy suppliers or utilities are (financially or managerially) rewarded for quality of service provided as opposed to quantity of electricity sold.
- Changes are made in the analysis of investment risk, whereby reduced fuel risk and lower capital investment risk are considered and incorporated into financial packages for new energy systems.

Development of guidelines on marketing and public education and certification schemes to build credibility will also aid in reducing administrative barriers. Transactions costs may be minimized by establishing regional energy centres allowing for some market aggregation. They may also be reduced through the establishment of effective financial and market intermediaries (such as funds, banks, dealer networks and concessions) that promote capital investment and liquidity. The development of international consensus on the desirability of removing impediments to new entrants to the energy market would provide a further impetus to new investment as well.

Other measures to help renewables include the creation of internationally recognized academic and vocational qualifications in renewable energy technology design, installation and maintenance. Also desirable would be provision of specialized courses on renewables and a consensus with academic institutions to include a stronger coverage of renewable energy technologies in traditional academic courses.

There is a need to reduce the environmental impacts of renewables. For example, some communities have begun to resist new wind turbine siting, while others have expressed concern at the potential land use requirements for large-scale solar arrays. However, a combination of coordinated research, development and dissemination of results can lead to improvements in public awareness, and planning guidelines can be developed to outline frameworks to take into account the benefits and costs of renewables. In this context there is a need to organize training courses for local authorities to help them identify opportunities for renewables, and to establish technical advisory centres. Incorporating renewables early in the planning processes would also facilitate their use and widespread dissemination: this might be done through the voluntary adoption of planning guidelines calling for assessments of the potential for the use of renewables for every new development. Renewable technologies could be further enhanced if methodologies for cost–benefit analysis were developed and implemented by the power sector, by governments, by funding agencies and by the wider planning community. If such programmes and policies are implemented, the learning curves that apply to new technologies suggest that renewable energy could be competitive with existing fossil alternatives within the next 20–30 years. However, if such policies are not adopted, the business-as-usual energy scenario, in which fossil fuels continue to supply the vast majority of the world's energy needs and emit concomitant quantities of toxic waste and greenhouse gases, remains the scenario for the future.

RECOMMENDATIONS

The following recommendations emerge from this study – as well as from a host of previous analyses and workshops on energy sector reforms (see, for example, IEA and UNEP, 2001 or ECE, 2001).

- While considerable analysis has been undertaken to date, further work is required to accurately identify, particularly at a local and national level, the environmental externalities related to energy production and use, and to the environmental damage caused; identification of the specific forms for each country will be required to evaluate their importance and to determine whether (and, if so, how) they might be removed. It is clear that fossil fuels have historically benefited from such subsidies, leading to a distortion in the market.

- Any negative social impact of a given subsidy or tax reform should be identified and measures to alleviate such an impact considered. The most distortional (from the market point of view) and environmentally damaging subsidies and tax provisions should be identified so that they can be abolished first. From an environmental perspective, subsidies to fossil fuels can and should be removed expeditiously.
- To provide for appropriate full life-cycle costs, environment-related taxes or charges should be introduced and environmentally harmful energy subsidies removed. Renewable energy subsidies, which meet public goods goals for environment and security, should be adopted.
- Timing matters for subsidy reform: sudden shocks can destabilize economies. Thus, energy price reform should be announced in advance to allow producers and consumers to adapt their behaviour accordingly and to create a reliable investment climate, and the relevant legislative and regulatory instruments and institutions should be phased in.
- Transparency of a process improves the final product, and creates political buy-in. Thus, public and private stakeholders should be involved and a broad consensus on energy price reform sought.
- Internationally coordinated action can facilitate the process of removing environmentally damaging subsidies. However, unilateral actions may still be appropriate in the absence of international agreements.

While there is generally broad agreement on these principles, it is equally clear that although such policy recommendations have been consistently made for decades, they have yet to be undertaken by the majority of countries.

There are, however, lessons from some successful subsidy reform efforts. In these examples, reforms seem to be linked to other cost-reducing activities. Successes tend to be introduced gradually, allowing changes to take place over a period that does not cause economic dislocation. Often, the decision process has been taken up front, and day-to-day operational decisions are removed from the purview of political winds. Policies that lead to negative impacts are offset with 'safety nets' for those affected and there is a conscientious effort made both to assess (in advance) the nature of these impacts, and to inform affected populations of programmes to ameliorate their costs. In short, the success stories are available – and indicate that with clear, careful and rational planning, reforms are possible.

Thus, perhaps the key issues for consideration at the conference are twofold: (1) *why* such strongly endorsed and broadly agreed changes have not been undertaken; and (2) *how* to improve the likelihood that such measures will be taken in the future. It is clear that success in the area of subsidy reform will generate the twin benefits of economic growth and environmental protection described in this short chapter. If the session of the International Conference for Renewable Energies on which this book is based is able to make progress in understanding barriers and improving commitments to subsidy reform, it will indeed be a success.

NOTES

1 'Economic costs' here refers only to those used in standard accounting. Thus, in suggesting that renewable energy is more expensive, no value is assigned for environmental damages or for fuel security values – both of which might make renewable energy more competitive. In fact, as discussed later in this chapter, such economic accounting itself provides a form of 'subsidy' to the existing energy regime.
2 See www.wto.org/english/thewto_e/whatis_e/eol/e/wto04/wto4_24.htm#note3.
3 The underlying image is a football or soccer field that is tilted towards one end giving one of the teams an unfair advantage over its opponents. In a similar manner, energy subsidies promote one fuel at the expense of others; in this case, fossil fuels and nuclear power at the expense of renewable energy.
4 Jamal Saghir was Director, Energy and Water, at the World Bank at the time of the study (October 2003).
5 The following discussion is taken from 'Penetration of renewable energy in the electricity sector' (OECD, 1998b).
6 It might be noted that the share of R&D devoted to nuclear power (both fission and fusion) has absorbed a much larger share of the total R&D budget than any other element of the R&D budget. According to the IEA (www.iea.org/dbtw-wpd/textbase/stats/rd.asp), nuclear energy (including both fission and fusion) was given approximately 51 per cent of the more than $90.5 billion devoted to R&D between 1988 and 1998, while fossil fuels were given 13 per cent and renewables only 7 per cent.
7 See www.iea.org/dbtw-wpd/textbase/pamsdb/re_webquery.htm.
8 See www.eadenvironmental.com/buy_credits_now/.
9 See www.natsource.com/markets/index.asp?s=103.

REFERENCES AND FURTHER READING

Anderson, K. (1995) 'The Political Economy of Coal Subsidies in Europe', *Energy Policy*, vol 23, no 6, pp485–496
Anderson, K. and McKibbin, W. J. (1997) 'Reducing Coal Subsidies and Trade Barriers: Their Contribution to Greenhouse Gas Abatement', Seminar Paper 97–07, Centre for International Economic Studies, University of Adelaide, Australia
Barnes, D. and Halpern, J. (2000) *The Role of Energy Subsidies*, World Bank, Washington, DC

Böhringer, C. (1995) 'Carbon Taxes and National Policy Constraints: The Case of German Coal Subsidies', draft report, Institut für Energiewirtschaft und Rationelle Energieanwendung, Universität Stuttgart

Commission of the European Communities (2002) *Inventory of Public Aid Granted to Different Energy Sources*, Brussels

De Moor, A. (2001) 'Towards a Grand Deal on Subsidies and Climate Change', Natural Resource Forum, JNRF: 25.2, May 2001

DFI (Decision Focus Incorporated) (1993), 'Analyzing Energy Subsidies and Greenhouse Gas Emissions', Task 6 Memo for EPA Contract 68-W1–0009, Work Assignment # 131, DFI, Washington, DC

DJA (Dale Jorgenson Associates) (1994), 'The Elimination of Federal Energy Subsidies: Environmental Gains, Tax Considerations and Economic Consequences', report of results and findings prepared for the OECD under contract to US EPA, Cambridge, MA

DRI (DRI/McGraw-Hill) (1997) 'Effects of Phasing Out Coal Subsidies in OECD Countries', in Organisation for Economic Co-operation and Development (OECD), *Reforming Energy and Transport Subsidies: Environmental and Economic Implications*, Paris, OECD

Economic Commission for Europe, Economic and Social Council (2001) *Reforming Energy Prices for Sustainable Energy Development*, Committee on Sustainable Energy and Committee on Environmental Policy, ENERGY/2001/6, CEP/2001/11, 3 September 2001

International Energy Agency (IEA) (1999a) *World Energy Outlook, Looking at Energy Subsidies: Getting the Prices Right*, Paris, Organisation for Economic Co-operation and Development (OECD) and IEA

IEA (1999b) *The Role of IEA Governments in Energy: 1999 Review*, Paris, OECD and IEA

IEA and UNEP (2002) *Reforming Energy Subsidies*, Paris, OECD and IEA

Larsen, B. and Shah, A. (1992) 'World Energy Subsidies and Global Carbon Emissions', World Development Report 1992, Background Paper 25, Washington, DC, World Bank

Michaelis, L. (1997) 'Electricity-Related Supports in the United Kingdom', in Organisation for Economic Co-operation and Development (OECD), *Reforming Energy and Transport Subsidies: Environmental and Economic Implications*, Paris, OECD

Mitchell, C. (2004) 'Energy Security and Sustainability', available at www.reeep.org/index.cfm?articleid=35&iid=884

Mountford, H. (2000) 'Experiences with Reforming Energy Subsidies', paper presented at UN-ECE/OECD Workshop 'Enhancing the Environment by Reforming Energy Prices', 14–16 June 2000, Prague Czech Republic

Organisation for Economic Co-operation and Development (OECD) (1997) *Reforming Energy and Transport Subsidies: Environmental and Economic Implications*, Paris, OECD

OECD (1998a) *Improving the Environment through Reducing Subsidies: Part II, Analysis and Overview of Studies*, Paris, OECD

OECD (1998b) *Penetration of Renewable Energy in the Electricity Sector*, Paris, OECD

Philpott, J. (2004) 'Keeping It Private, Going Public: A Money, Carbon and Clean Air

Analysis and Reporting Framework for Power Sector Investment and Finance', World Resources Institute, forthcoming

Saghir, Jamal (2003) 'The Role of the World Bank Group in Renewable Energy and Energy Efficiency', available at www.worldbank.org/energy/pdfs/mso337.pdf

Shelby, M., Shakleton, R., Shealy, M. and Cristafaro, A. (1997), 'Climate Change Implications of Eliminating US Energy Subsidies', *OECD, Reforming Energy and Transport Studies: Environmental and Economic Implications*, OECD, Paris, pp77–89

US Congressional Budget Office (2003) Letter from the Director of the CBO to the Chairman of the US House of Representatives Committee on Energy and Commerce, available at http://energy.senate.gov/legislation/energybill2003/cbo_report.pdf

World Resources Institute (2003) 'Climate Analysis and Indicators Tool', available at: http://climate.wri.org/project_description2.cfm?ProjectID=93

7

International Institutional Arrangements in Support of Renewable Energy

Achim Steiner, Thomas Wälde, Adrian Bradbrook
and Frederik Schutyser

INTRODUCTION

This chapter updates the analysis and proposals contained in the Thematic Background Paper (TBP) for the 2004 International Conference for Renewable Energies (hereafter Renewables 2004) 'International Institutional Arrangements. Bundling the Forces – but How?'. The TBP's point of departure was that renewable energy (RE) is an essential component of sustainable development, and should provide a bigger proportion of the world's energy supply.

The TBP provided an overview of the international institutional landscape and identified key constraints that need to be addressed, most efficiently at the international level. The paper concluded that that the existing RE landscape did not seem optimal to enable RE to grow. Two pathways were suggested that could increase the development of RE as a source of complementary/alternative energy in the future. The first consists of developing key principles and guidelines for 'best practices', compiled in a *Code of Conduct/Code of Best Practices on Promoting Renewable Energy* in the overall context of global sustainable energy development. Second, the TBP suggested a *new type of international organization* to ensure that RE becomes an essential element of global efforts at sustainable energy development.

These pathways are distinct yet complementary and mutually reinforcing processes, and both involve multi-stakeholder participation, considered essential to bundling the forces in a highly dispersed institutional landscape. It is essential not to isolate and compartmentalize renewable energy, but to design policies in such a way that the interaction, competition and substitution between the various energy sources and energy industries are properly taken into account, and guided towards an overall more sustainable form of global energy. Single-minded promotion of renewable energy is less relevant than a policy and institutional context that favours a sustainable energy mix for the global community.

In this field the creation of the Renewable Energy Policy Network for the 21st Century (Ren21, www.ren21.org) is a key outcome of Renewables 2004. We shall briefly recall the main issues related to institutional arrangements for RE, and revisit our earlier recommendations in the light of Renewables 2004 and Ren21.

EXISTING INTERNATIONAL INSTITUTIONS AND ACTIVITIES: A BRIEF OVERVIEW

This section begins with a brief typology of the existing international landscape, and then proceeds to examine issues that should be addressed by these international actors for RE to grow. A third section will indicate shortcomings of the current situation.

Stocktaking: an overview of the international RE landscape

The brief stock-taking exercise produced an extensive list of activities undertaken by international actors on RE. The following typology is one of the many according to which these numerous and varied institutions and activities can be distinguished. The RE landscape is very dynamic and constantly changing. The overview in Table 7.1 is therefore not exhaustive and merely provides some examples.

Action required for the promotion of RE

The overview of actors on the one hand shows that many actors deal directly or indirectly with RE, from the project to the international policy level. On the other hand, no one existing institution brings together all stakeholders at a global level, which is arguably where many constraints hindering RE growth should be addressed.

Some of the key areas where coordination is needed for RE to be able to grow are the following:

- *Information*. Research is needed into the environmental aspects and costs of every energy option. This is a precondition for the implementation of measures that are often proposed to increase the share of RE (e.g. 'internalize environmental costs into the price of energy').

Table 7.1 *Overview of the institutional renewable energy landscape*

Intergovernmental organizations, whose primary activity is energy related	Examples include the International Energy Agency (IEA), the Organización Latinoamericana de Energía (OLADE) and the Energy Charter Conference and Treaty. On the one hand, these organizations have expertise, a governmental support base, and in some cases the authority to make binding rules. On the other hand, membership of most of these organizations is limited geographically or otherwise, and none have RE as a main focus
The World Bank Group, and the Regional Development Banks	These are significant players, with an important RE impact in developing countries. They finance a significant number of RE projects throughout the world, ranging from technological assistance to energy sector reform, sometimes with private sector co-financing
Regional organizations	Examples include the European Union (EU), the Association of Southeast Asian Nations (ASEAN), the Southern African Development Community (SADC), and Asia-Pacific Economic Cooperation (APEC)
The Global Environmental Facility (GEF)	The GEF (through implementing agencies) operates more than 100 programmes for the promotion of energy production and consumption from RE (backed by private sector development and sometimes by energy sector reform), mainly with a domestic scope. Projects do not address issues such as taxation, subsidies or trade law on a global scale
The UN system	The UN Regional Economic Commissions play an important capacity building role (e.g. United Nations Economic Commission for Europe (UNECE) or the United Nations Economic and Social Commission for Asia and the Pacific (UNESCAP)). Globally, the United Nations Environment Programme (UNEP) and the United Nations Development Programme (UNDP) are important actors (cf. the Global Network on Energy for Sustainable Development, the UNDP Initiative for Sustainable Energy (UNISE), and the World Energy Assessment). Many other specialized UN agencies have addressed RE within their niche (for example the United Nations Department of Economic and Social Affairs (UNDESA), the World Health Organization (WHO), and the Food and Agriculture organization of the United Nations (FAO)). The Commission on Sustainable Development (CSD) includes energy as a major component of its work plan for the coming years. An interagency group 'UN-Energy' has been created to help ensure coherence in the UN response to the WSSD in general and its energy-related outcomes in particular
The World Summit on Sustainable Development (WSSD) and its Plan of Implementation (and the resulting 'type II' partnerships)	The WSSD Plan of Implementation, while not binding, is the broadest international instrument with the most extensive references to renewable energy and energy efficiency yet produced by the world community. It focuses on development, implementation, technology transfer and rapid commercialization of RE. It sees energy as key to the eradication of world poverty, and to change of unsustainable consumption and production patterns. More than 20 type II (public–private) partnerships are active in RE, for example the Renewable Energy and Energy Efficiency Partnership (REEEP). The Political Declaration resulting from Renewables 2004 reaffirms the WSSD commitment

Non-Governmental Organizations (NGOs)	The NGO community ranges from green advocates (most environmental NGOs have a work programme on energy and climate change), to NGOs focusing specifically on energy, to consumer interest groups. Examples are the World Energy Council, the World Council for Renewable Energies, the World Wind Energy Association, the International Network for Sustainable Energy. Some charitable foundations also support RE activities
The research community	This group includes a wide variety of actors, ranging from fundamental research at universities to applied research to technology development specifically for commercial purposes
The private sector	Individual companies involved in energy supply (utilities, increasingly working in more than one country), technology supply and research and development (R&D), but also groups such as industry associations (e.g. Eurelectric) and the World Business Council on Sustainable Development

- *Demand.* There needs to be a substantial and sustained demand for renewable energy at economic conditions that make investment in supply a long-term viable business proposition; at present, such demand is hindered by factors that make RE not competitive with other energy sources. Institutional arrangements to support demand can be conceived: for example, regulating the cyclic demand for energy to match the typically varying available RE supply, or prescribing RE inclusion in building standards.
- *Markets.* RE can grow if there are functioning – and competitive – national and regional markets where the advantages of renewable energy can compete effectively against the advantages of incumbency (in terms of established suppliers and network operators, established and proven technologies, established regulatory and fiscal conditions and established political and institutional alliances and support). Such RE markets need an enabling regulatory framework that encourages RE, which in turn depends on political will. Expanded markets in industrialized countries reduce the cost of technologies and therefore accessibility for developing countries. Incumbents (such as existing network operators, in particular in countries where renewable energy has not as yet been fully embraced in actual practice and mentality by existing energy monopolies) may obstruct interconnection, or access to domestic and international markets; industry associations naturally reflect vested positions rather than a pro-RE perspective. The same applies to transmission charges, which as a rule are not geared towards easy and economically practical feed-in of renewable energy into grids. Equally important is the competitiveness of RE markets themselves.
- *Investment, taxation and subsidies.* Investment in renewable energy facilities must rely on a favourable, stable framework in terms of international law, national regulation and taxation, a framework that minimizes the much

greater risk of long-term RE investments in particular against instability and revocation of committed subsidies and support systems. Taxation and state support/public subsidies regimes must create a level playing field between renewable and conventional energy, with some element of an initial 'affirmative action' to counter the handicaps of renewable energy. Targeted and efficient subsidies that are limited in time but stable may be a tool to encourage RE investment (such subsidies must be designed so as to encourage further research and development; compatibility with trade rules is a delicate issue, and demands a coordinated international approach). Regulatory and tax regimes that encourage RE investment should be subject to external, enforceable disciplines that affect their credibility and stability. 'Greening' of investment protection is one of the very few methods available that is likely to enhance the investment climate for RE investment and RE entrepreneurs.

- *Economic incentives.* Developing and applying innovative technologies requires economic incentives. New technologies as a rule cannot be imposed by state action, but have to emerge out of entrepreneurial action in markets – and that requires a favourable legal, regulatory and fiscal setting. There may have to be privileges for developing countries, but they must not undermine the economic incentive for developing and widely applying technological innovation.
- *Infrastructure.* Available infrastructure should be accessible – technically and legally – to renewable energy. The creation of new infrastructure (interconnectors) necessary for RE trade should be encouraged and some public assumption of long-term commercial risk needs to take place.
- *Trade.* Cross-border trade in increasingly integrated and liberalized energy markets needs to be structured so that renewable energy is not penalized by domestic regimes which encourage – and protect – only domestic production of renewable energy. Cross-border trade rules also need to be responsive to the competition between environmentally more favourable and environmentally less favourable energy. The issue is a contentious one within the World Trade Organization (WTO), but given that different countries are endowed with RE opportunities to a different degree, international trade in, for example, electricity generated by RE must be specifically addressed.
- *Governance system.* The many elements constituting the governance system of the global economy (mainly international trade, investment, environmental treaties) should not create obstacles for RE investment and trade, but rather provide a protective, encouraging and facilitating effect. An important issue for international cooperation is protection of intellectual property, protecting investments in R&D until RE can break out of high-cost niche markets. The protection of intellectual property may be a particular concern in certain developing countries where the opportunities for RE are deemed to be high.

Current situation: constraints

This brief overview shows a highly compartmentalized RE landscape with little systematic pooling of information, analysis and coordination at an international level. There is a lot of engagement for RE but it is mostly dispersed and diffuse. All activities are relevant, but not sufficient to create more efficiency and synergies (e.g. sharing lessons, establishing common standards). One important consequence of this is that there is no comprehensive assessment yet of the effectiveness of the different policies and activities, or of the impact they have on each other (though assessments of effectiveness of policies are under way). Nor is there a clear view on what the overall financial commitments for RE projects are, how they evolve and how they compare with measures for conventional energy (e.g. coal subsidies, or support for decommissioning of nuclear plants).

Box 7.1 Example of effectiveness assessment: the Renewable Energy and International Law Project

The Renewable Energy and International Law Project (REILP) is supported by the UK Foreign Office's Renewable Energy and Energy Efficiency Partnership, the law firm Baker & McKenzie and several universities (Yale, the Centre for Energy, Petroleum and Mineral Law and Policy at the University of Dundee, and others). Other project partners include the World Conservation Union (IUCN – through its Environmental Law Programme), the United Nations Environment Programme (UNEP), the Australian and US governments, and the secretariat of the North American Free Trade Agreement (NAFTA). The project is looking at ways in which international law can be used as a tool to support the development of renewable energy, and, conversely, ways in which it may currently be impeding that development.

It has not been possible to identify, in a clear and convincing way, which initiatives work, which do not, and why. While success and failure are always relative, and conditioned upon the particular context, a successful case has a potential to serve as a model, and a failure has the potential to provide some pointers towards the need for a different approach. An assessment by an independent, respected and expertise-based institution or institutionalized processes appears imperative to find out how to make promotion of renewable energy work in practice, and not just in organizational public relations.

No international agency identifies wholeheartedly with the issue of global sustainable energy (in particular, energy conservation, energy efficiency, renewable and climate-friendly energy) and focuses on it – in terms of agenda setting, initiatives for international negotiation of principles, rules and standards, setting up a global stakeholder consultation process and relationship-building with all relevant actors, including the private sector. Sustainable energy does not have an 'international home'. This fact may contribute to uncertainty about many initiatives' lasting effect, and to the fact that many primarily look like experimental showcase activities.

The survey also provides little evidence that RE issues have permeated non-RE energy (or related) activities, for example the negotiation, amendment and application of international treaties related to trade, investment, environment and energy, or technical and financial assistance in other areas of energy. It may well be that many environmental treaties and initiatives are being promoted which in effect do not favour, and may hinder, the emergence of a viable RE activity. It is only now that the specific impact of WTO rules (favourable and unfavourable) on RE industries is being analysed in more depth. For example, renewable energy would benefit from an open cross-border energy market, but also needs protection from competition of energy produced under lax safety conditions or without internalization of external costs – the 'energy dumping' issue.

The compartmentalized nature of the response of international actors to the RE challenge may be due to their internal organization: most of these agencies are classic intergovernmental organizations with only a very limited participatory role for non-state actors (in RE, primarily energy companies, but also NGOs, banks, professional firms and associations, and academic institutions). This suggests that it would be inappropriate to call for a 'World Energy Agency' in the classic model, but rather for an international institutional platform that is 'owned' by all significant stakeholders. It should be the institutionalized form, secretariat and focus of a network that combines all relevant interest, influence, expertise, financing and regulatory power.

There is a need for instruments (both institutional and of the nature of rule- and standard-setting) that can:

- assess national/regional/global impacts of RE policies;
- exercise expertise-based 'agenda power' by launching new initiatives encompassing stakeholder consultation;
- act as forum, vehicle and secretariat for the negotiation and continuous adaptation of principles, rules and standards for designing RE policies, assessing RE initiatives and integrating RE objectives into other relevant organizational activities.

In particular, trade is an issue that cannot be dealt with successfully by a national or sectoral approach. In consultation between the custodians of trade rules (mainly the WTO, the NAFTA secretariat, Asian and Latin American trade organizations, and the European Union), it is necessary to develop, based on already available precedent, a set of guidelines and principles that helps to apply available trade rules (non-discrimination – justification for non-protectionist, guideline-covered renewable energy; subsidies; dumping and state aids – against energy produced below international standards) in a way that facilitates cross-border RE trade, combats RE-based protectionism (by purely national schemes) and creates a truly level playing field in the competition between RE and conventional energy and energy products.

The functional deficits of existing actors can be summarized as follows:

- *Area of activity.* There do not seem to be real gaps in substance (e.g. market reform, capacity-building, specific RE sources) covered by existing actors, but current coordinated research is insufficient to fill the information gaps that prevent coordinated policy-making. It is important to keep all options open, including through support of research into future sources that may at present be in a very early stage of development.
- *Assessing RE potential.* Here a clear gap can be identified. The numerous existing initiatives and projects indicate what is being done to promote RE, but this activity hides the fact that many countries do not have any RE activities, or may not even be aware of the specific RE potential present within their territory. Whereas many institutions try to influence decision-makers, none is specifically mandated to consult governments on realizing the RE potential of their country.
- *Multi-stakeholder participation.* No single institution has been found where all stakeholders are represented and are able to contribute effectively in terms of their expertise and agenda-setting capacity.
- *Structures and finance.* Many activities are designed around specific studies, projects, reforms or experiments. There does not seem to be an institution that is able to act on RE on a global scale, independent of projects, and with stable, possibly governmental, funding.

ADDRESSING THE CONSTRAINTS: TWO POTENTIAL PATHWAYS FOR PROGRESS

The overview of existing activities at the international level shows that there is no need for a new body to *invest* in RE, but there is a need for sharing experience, expertise and information. RE needs an institutional platform at the international level and a relevant international code of best practices in standards and policies – for governments, independent energy regulators, international organizations, the private sector, academic institutions and NGOs – on designing, pursuing and integrating RE objectives in overall energy, environmental, investment, fiscal and trade policies and strategies.

Two parallel processes can be envisaged to address these issues, one leading to an institutional platform, the other leading to a compilation of guidelines, standards and best practices. The two are interrelated: the proposed negotiation of a Code of Conduct, or Code of Best Practices, on Renewable and Sustainable Energy would bundle the relevant forces in a consultative process; it should be serviced by an interim secretariat. The proposed outcome – the Code of Conduct – would then lend itself as one of the key mandates of what is proposed: a global institutional platform for sustainable energy.

A coordinated international approach must, however, take into account that energy policy remains a largely national prerogative. New international instruments would need to reflect national sovereignty considerations and respect the principle of subsidiarity. They would only come into play where action at the international level is more efficient.

An institutional platform

We argue that for RE and sustainable energy to realize its full potential, a permanent institutional platform would be required. Two possibilities can be distinguished. They are not mutually exclusive and may happen sequentially.

The possibility should not be excluded that an existing institution could take the lead on global RE issues. At the same time, any existing institution would need to be restructured if it were to serve the purpose of convening all stakeholders as suggested. Any institution should not be a conventional intergovernmental agency, but rather an institutionalized and stable focal point of a network, grouping together the main stakeholders. The governing board should not be composed exclusively of state representatives, but rather of existing international agency representatives with a stake and expertise in RE (for example, the International Energy Agency (IEA), the World Bank, the United Nations, Asia-Pacific Economic Cooperation (APEC), the Organización Latinoamericana de Energía (OLADE), the Association of Southeast Asian Nations (ASEAN), the International Atomic Energy Agency (IAEA)), of governments, of private financial institutions, energy companies, energy professionals, industry associations, NGOs and civil society.

One option, possibly a necessary first step, would lead to the creation of a *specialized RE agency*. Such an agency would be the 'international home' for renewable energy. Such an organization had been proposed prior to Renewables 2004 and was provisionally named IRENA (the International Renewable Energy Agency). It could be created rather quickly, and it would be able to perform functions such as coordination of policies, agenda-setting and the monitoring of a Code of Conduct or Code of Best Practices. At the same time, it would remain a specialized RE agency, and the challenge would be to integrate its work within the broader context and, ultimately, policy decisions on sustainable energy choices. One important task for the institution would be to inform and advise governments about RE options. It should play a subsidiary role, and focus on countries that have an underdeveloped RE potential.

A second outcome could be an *energy for sustainable development agency*. This would be a global organization concerned with setting agendas, developing expertise, organizing networks and providing technical assistance in the field of energy. It is suggested that energy, as the world's most important industry, deserves and requires a separate international agency. Its mandate would have to be focused on promoting a sustainable model of energy supply with a primary focus on energy conservation, energy efficiency and renewable energy promotion. This may be a distant target, but arguably it is a logical endpoint of true integration of new energy sources in global energy policy.

The value added by any new institution to the range of existing institutions would be twofold. First, RE would be a key element of the institution's work from the start, instead of a small component, as is often the case in existing models. Second, the institution would have a global perspective, bring together all stakeholders and have an explicit mandate for sustainable energy.

Code of Conduct/Code of Best Practices related to Renewable Energy

A non-binding code could be an important catalyst for coordinated action by providing a source of *de facto* rules and standards for energy policy and development – an enabling framework for actors in RE. Such a code should be developed through a mechanism that involves all stakeholders: existing institutions active in the field of energy (e.g. the IEA, the European Union, but perhaps also important actors that traditionally do not deal with RE, such as the Organization of the Petroleum Exporting Countries (OPEC) or the IAEA), the research community, national agencies, the private sector, energy utilities, the World Bank and the regional Development Banks, consumers, banks and civil society.

The aim at this stage should thus not be a multilateral treaty. The experience with such treaty-making is patchy. It is much more difficult to achieve a treaty, and once it is completed, ratifications are as a rule very hard to get – unless the treaty itself is of little practical significance.

Codes and other non-binding instruments have been criticized for being too weak, and without real impact. However, a comprehensive text that has been developed and agreed by stakeholders can give a feeling of ownership, which makes it more likely that policy-makers or governments would draw on the content of the code. It may therefore be more feasible to start with an instrument that is not legally binding, but may acquire a persuasive authority – and some legal effect – over time. The condition for success is that such an instrument represents a progressive consensus of the main stakeholders while being suitable for realistic and practical application. If such a pre-codification works, the next step may include the conversion of a code into a legally binding instrument.

Issues that could be covered in the code include:

- phasing out of subsidies to non-sustainable energy systems;
- protection of investments: guaranteed subsidies for a limited time; initial

'affirmative action'; full internalization of external effects by carbon and related taxes;

- mandatory prior assessment of the environmental effect of RE investments;
- mandatory prior assessment of the economic effect of RE investments, and the effect on utilization and development of other energy sources;
- the development of an RE-friendly trade regime (e.g. cross-border electricity trade involving 'green electricity');
- monitoring of the impact of RE projects and lessons learned (failure and success; which policies work, and under what conditions);
- facilitation of technology transfer, while protecting investors' rights for a certain period;
- investment in research capacity in developing countries, creating an enabling environment that makes RE projects self-sustaining;
- differentiated standards for developing countries, with a further distinction between energy supply to the growing urbanized centres and to remote rural communities not connected to the grid.

Importantly, the development of a Code of Conduct or Code of Best Practices is a mechanism by which to channel the work, experience and expertise of existing initiatives and institutional capacities. Recent examples of multi-stakeholder processes of this kind include the work of the World Commission on Dams, the World Health Organization Commission on Macroeconomics and Health, the 'Berlin Guidelines on Mining and Sustainable Development' and the Mining, Minerals and Sustainable Development Project (MMSD).

A code drafted and endorsed by the major stakeholders could serve significant legal functions:

- as justification, under WTO/GATT (General Agreement on Tariffs and Trade) rules, of import restrictions if necessary to protect RE investment;
- as an authoritative argument to protect RE investment if threatened by abrogation of governmental support and protection upon which the investor has relied;
- as a balancing weight when other environmental objectives pursued by single-focus treaties create an impediment to RE investment.

THE OUTCOMES OF RENEWABLES 2004 AS A RESPONSE

In this section we shall briefly discuss how the outcomes of Renewables 2004 address the constraints identified earlier in the institutional landscape.

The political declaration resulting from Renewables 2004 provided the mandate for the creation of a Renewable Energy Policy Network for the 21st Century, Ren21. Its goal is 'to allow the rapid expansion of renewable energies in developing and industrial countries by bolstering policy development and decision-making on sub-national, national and international levels'. Ren21 will also play a role in monitoring the 197 actions and commitments contained in the International Action Programme that resulted from Renewables 2004. The political declaration also supports policy analysis, and technological assessments and the sharing of good practice.

The fact that just over a year since Renewables 2004, Ren21 was created, and commitments towards the establishing of a secretariat have been confirmed indicates the international political will to make significant progress for RE. Many of the policy barriers we identified can be addressed by the policy network. The proposed clearinghouse being developed with REEEP will advance information-sharing significantly.

We would like to suggest some ideas for the further evolution of Ren21, recognizing that it may become the international vehicle for sustainable energy. The authors did not participate in the debates among stakeholders in general and within the steering committee in particular. The points below are therefore no criticism of the work accomplished so far, but merely suggestions for possible future developments.

1   While an *informal policy network* is an important step to bring together RE stakeholders, we think it should not be the final goal. Ren21 has a secretariat, but there is a strong feeling that it should not create a new institution. While some fatigue in the field of international organizations is perhaps understandable, we believe that a permanent, institutionalized multi-stakeholder organization is required to have an influence on the international political agenda, or on RE commitments that can be monitored and are more than voluntary.

    Many partnerships have been created, especially since WSSD, and they often bring together stakeholders in a way that did not happen before. This model could also be expanded to the governing board of a more stable focal point, more than a 'network'. A stable institutional platform as a focal point may better facilitate coordination and cooperation among existing actors.

2   Related to the issues of organization is that of *membership*. Ren21 currently proposes a loose network with the option of perhaps formalizing membership at a later stage. The range of stakeholders already in the network is impressive, and Ren21 explicitly aims at bringing together the environment and development communities. The main point we would emphasis in this context is that we feel any network or institution should have as broad as possible a scope in the field of energy. Our original TBP argued for an 'energy for sustainable development' agency. We still believe that mainstreaming of RE may be better achieved if an organization contains stakeholders from outside the RE community, such as traditional energy companies, as well as those from inside it.

3 Ren21 plans an impressive range of *activities* to share information and knowledge (cf. also the thematic background papers planned, including a global status report on RE). This will no doubt strengthen the RE community. One important function of any institutional mechanism to strengthen RE could be to carry out what have been termed 'RE audits', as discussed in the preparations for Renewables 2004. This function could be developed further within the context of Ren21. Perhaps the focus should be less on 'making the case for RE', since this case seems to be widely accepted. Rather, countries should be able to find assistance in determining the RE potential of their country, and the different options for sustainable energy more broadly. The Code of Conduct/Code of Best Practices we suggested could fit within Ren21's knowledge exchange activities.

4 A *challenge* for Ren21 or any organization in whatever form that promotes sustainable energy is to influence non-energy agendas, and not to operate in isolation. Bringing the development and environment perspectives together is a major step forward (only a few years ago, when the world agreed on indicators to monitor progress towards achieving the Millennium Development Goals, no indicator on RE, such as biomass, was included, despite its major importance for the rural poor). A parallel may be drawn with environmental protection and biodiversity conservation. International commitments and achievements since the Rio Earth Summit in 1992 have been impressive. But today, mainstreaming the environment into sectors that have a major influence, such as trade, still remains a challenge.

CONCLUSION

The promotion of RE requires cohesion, focus, coordination and, ultimately, critical mass. Traditional models of developing international policy and organizational frameworks through exclusive intergovernmental processes would not provide an effective response to this need. A multi-stakeholder approach is far more likely to harness the significant potential of non-state actors to promote RE in partnership with governments by drawing on their expertise, resources and capacity to influence and respond to consumer choice. The effectiveness of governments in establishing policy and regulatory frameworks for the promotion of RE will in large part depend on their capacity to mobilize the potential of researchers, investors and consumers.

While the future role of RE depends on the research and development of technological options, the economic, institutional and development policy frameworks will determine whether RE can establish itself as a viable source of complementary/alternative energy. The establishment of a new multi-stakeholder platform and the codification of an emerging consensus on RE are considered two pathways for strengthening the institutional framework for RE at an international level. The logic of this approach points ultimately towards the need for a global institutional platform for sustainable energy development.

The creation of a multi-stakeholder network after Renewables 2004 is an important step towards the promotion and mainstreaming of RE. We have made suggestions for the future development of Ren21, in terms of both organization and substantive activities. The current international momentum for RE promises long-overdue developments that will depend on appropriate institutional support. We believe that an informal multi-stakeholder network is an essential first step, and could evolve into a more permanent institution.

NOTE

The authors wish to thank Gavin Longmuir for comments.

PART THREE

FINANCING

8

Mobilizing Finance for Renewable Energies

Virginia Sonntag-O'Brien and Eric Usher

INTRODUCTION

Rising oil prices, growing energy security concerns and the human and environmental devastation caused by adverse weather events attributed to climate change are increasing the attractiveness of alternatives to conventional fossil fuel-based energy sources and adding reasons to rethink the world's energy future. Investment is at the core of the transition to a sustainable energy future, as massive amounts of new capital will be required to cover the world's growing energy demand and to meet commitments to large carbon emissions reductions.

Renewable energy (RE) has a significant potential to mitigate global climate change, address regional and local environmental concerns, reduce poverty and increase energy security. The challenge is to provide the right policy frameworks and financial tools that will enable RE to achieve its market potential and move from the margins of energy supply into the mainstream.

Renewable energy is, in fact, a multi-billion-dollar industry and the most dynamic sector of the global energy market. Still, the renewable energy sector remains by far the smallest segment of the world's energy industry. Various finance-related risks and barriers are hindering faster growth. RE is site specific and most sites are still not cost-competitive with conventional fossil fuel energy sources in the short to medium term. Most renewable energy projects have high up-front capital costs relative to competing technologies and low rates of return. Many investors therefore are not ready or willing to invest, or only under very unfavourable terms for the project developer. This is particularly the case in developing countries, where access to affordable finance is extremely difficult and often reliant on targeted subsidies.

Policy-makers thus have a mandate to take action that focuses on creating enabling frameworks and finance mechanisms for technology R&D, commercialization and investment. Policies are needed that create markets, reduce risk, provide acceptable rates of return for investments and create conditions for a sustainable and profitable RE industry. This entails long-term regulatory frameworks supported by financial mechanisms that address specific barriers to and gaps in technology or project development.[1]

This chapter focuses on the potential for market development and for creating investment opportunities by pointing out the dynamic growth of market trends in renewable energies. It explains how RE projects are financed and describes the barriers that make financing RE difficult. Finally, it analyses the most important financial instruments that are still weak or missing for the promotion of renewable energies. These address the issue of financial risk, and risk management instruments are described that can mitigate the perceived risks associated with RE and affect conditions for investment in RE projects.[2]

THE CLIMATE FOR INVESTMENT IN RENEWABLE ENERGY

Within the energy sector renewable energies have to compete with the other conventional segments of the industry. Generally, the market for RE is improving and is the fastest growing in the energy sector. Though market and investment conditions vary according to technology (size, capacity, on- or off-grid), the market drivers for RE are the same: improved economics (in some cases); energy security; global, regional and local environmental benefits; economic development; and consumer support.

RE investment grew from $7 billion in 1995 to $30 billion in 2004, and today renewable energy receives about one-third of the global investment in the power sector (REN21, 2005). Market growth responds to a number of factors, the most significant of which is cost reductions. Renewable technologies are improving all the time and are becoming less costly to manufacture and operate. Wind and solar photovoltaics (PV) are one-tenth of the cost they were in the early 1980s, and additional cost reductions of about 5 per cent per year are expected in the near term. The volatility of the fossil-fuel market – in particular, the rising price of oil and gas and the ratification of the Kyoto Protocol[3] – is making RE energy sources more and more competitive, and attracting new forms of investment. Watching decreasing returns on traditional investment vehicles is luring non-utilities to the RE market, such as Goldman Sachs. GE expects its wind-energy revenues to grow to more than $2 billion in 2005. Institutional investors are turning their attention to RE, especially wind, because of the long-term timeframe for investment.[4] A number of renewable energy companies have raised capital through the stock markets in 2004 and 2005, particularly in Frankfurt and through the London Stock Exchange's Alternative Investments Market (AIM).[5] The World Bank has calculated that the commitments made by developing countries to renewable energy will require investment of around US$10 billion per year (Saghir, 2005).

Specifics of financing renewable energies

Renewable energies represent a major step-change innovation as compared with existing energy supply options. In terms of scale, capacity, energy resource characteristics, points of sale for output, status of technology and a number of other factors, RE technologies are usually markedly different from conventional energy systems. The differences are not lost on financiers, as financing an RE plant is different from financing conventional fossil-fuelled power plants and requires new thinking, new risk management approaches and new forms of capital.

Financing renewable energies is new to financiers. Considering investing in the RE sector for the first time is an investment in itself. To become more effective at placing capital in RE markets, financiers must travel up a learning or experience curve. Market failures impede this learning process and create barriers to entry into the market. To operate effectively, markets rely on timely, appropriate and truthful information. In perfect markets this information is assumed to be available, but the reality is that energy markets are far from perfect, particularly those that, like the RE market, are in technological and structural transition. The information that enables a correct assessment of a project's viability is generally lacking, and there is limited economic justification for any single market participant to produce such information. As a result of insufficient information, underlying project risk tends to be overrated and transaction costs can increase.

Compounding this lack of information are the issues of financial structure and scale. Renewable energy projects typically have higher capital costs and lower operational costs than conventional fossil-fuel technologies. The external financing requirement is therefore high and must be amortized over the life of the project. This makes exposure to risk a long-term challenge (which also has political risk implications in terms of changes in government policy). Since RE projects are typically small, as for example solar PV and mini-hydropower, the transaction costs are disproportionately high compared with those of conventional infrastructure projects. Any investment requires initial feasibility and due diligence work, and the costs for this work do not vary significantly with project size. As a result, pre-investment costs, including legal and engineering fees, consultants and permitting costs, have a proportionately higher impact on the transaction costs of RE projects. Furthermore, the generally smaller nature of

RE projects results in lower gross returns, even though the rate of return may be well within market standards of what is considered an attractive investment.

Developers of RE projects are often underfinanced and have limited track records. Financiers therefore perceive them as being high risk and are reluctant to provide non-recourse project finance. Lenders wish to see experienced construction contractors, suppliers with proven equipment and experienced operators. Additional development costs imposed by financiers on undercapitalized developers during due diligence can significantly jeopardize a project.

Financiers perceive many RE technologies as being commercially unproven. With the exception of onshore wind, financiers regard the full cost and long-term performance risks of RE technologies as being higher than with conventional technologies. As with most new technologies, a vicious circle exists, with financiers and manufacturers reluctant to invest the capital needed to reduce costs as long as demand is low and uncertain. But unless there is investment, demand stays low, because potential economies of scale cannot be realized at low levels of production.

Developing financial solutions for renewable energy

Eventually, market forces will be the best way to determine how and where RE are used. This ideal solution, however, assumes mature technologies, efficient markets and full internalization of environmental and social costs – conditions that do not exist in any country at present. Public interventions are therefore needed to help accelerate RE development, commercialization and financing.

Financing RE, whether on- or off-grid, in developed or developing countries, is part of a larger value chain that exists in different degrees of completeness, depending on the market, the technology and the infrastructure available to bring the technology to market. The completeness of this value chain in any location will determine whether the financial sector will become engaged.

The RE sector is marked by incomplete value chains. Financing is but one part of the value chain, and it is the part that engages only after all the other parts are in place. Therefore, to attract finance it is necessary to assess the state of a specific value chain, identify the missing steps and design public interventions to effectively and efficiently bridge the gaps.

To be successful, public interventions must:

- address specific market barriers or failures;
- be removable (not create dependence);
- reward innovation (improve the technology/service offering and bring costs down);
- be cost-effective (leverage private-sector capital).[6]

Box 8.1 How Financiers Make Investment Decisions

Financiers make lending and investment decisions based on their estimation of both the risks and the returns of a project. In considering a project a financier will usually prepare a risk/return profile, as is shown in Figure 8.1. The analysis involves assessing each individual risk and the means of mitigating its potential impact on the project. Assessing the returns involves verifying the cost and revenue projections and then comparing the financials of the project with the cost of financing to be used.

A lender will specifically focus on the ability of the borrower (or, in the case of project finance, the project) to make loan repayments. An equity investor, who shares in the upside of the project, will base his or her decision on an estimation of the risk-adjusted return of the project, which graphically (see the smaller diagram in Figure 8.1) means deciding whether the project falls above or below the investors' risk/return yield curve. For on-grid RE the returns are usually easy to assess; it is the risks that can be difficult to assess or manage. When it comes to off-grid business models, a financier usually has difficulty understanding both the risks and the returns.

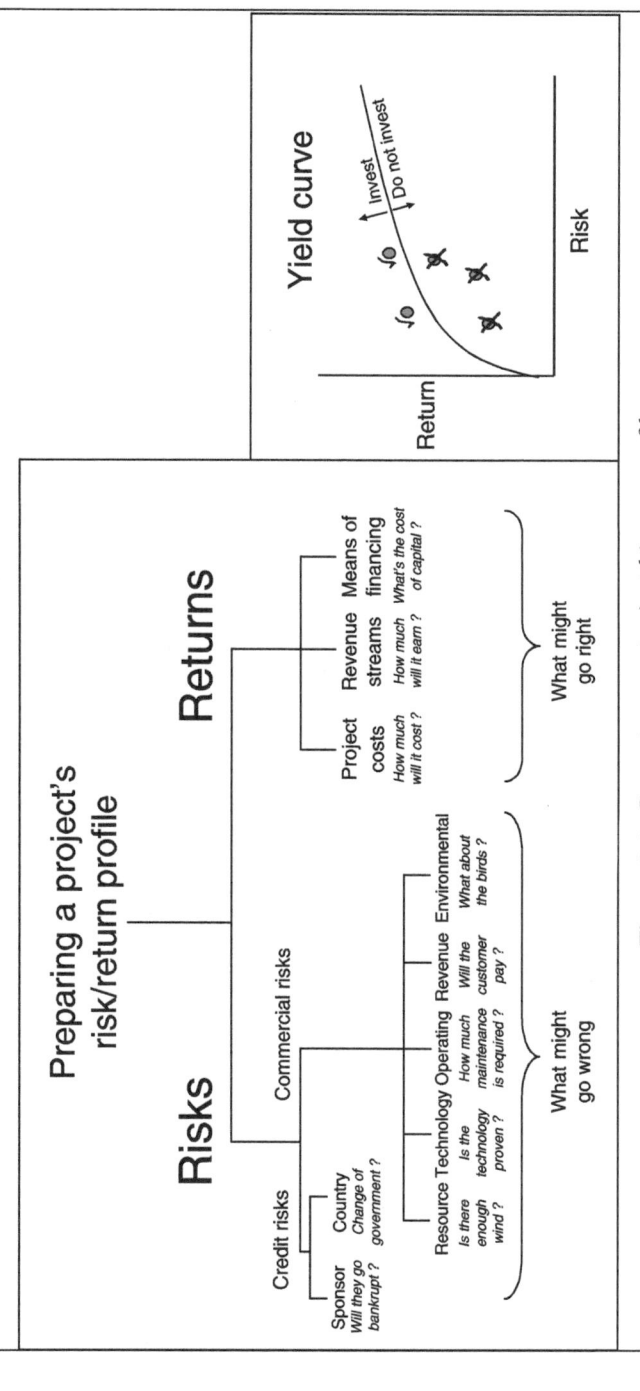

Figure 8.1 *Preparing a project's risk/return profile*

Financing sustainable energy technology innovation

Technology innovation is one of the catalysts behind sustainable-energy market growth. The technology innovation process progresses from R&D to demonstration, to pre-commercialization through to commercialization. Each stage is influenced by technology push and market pull forces that need to be managed and balanced by technology innovators, governments, financiers and eventually end-users.

Supportive regulatory and tax environments are key drivers of the development and financing of new technologies. To lead technology innovation towards the market and to engage commercial investment in the RE sector, governments are starting to implement a range of new finance mechanisms capitalized by public sources. These include technology and business incubators, contingent grants, convertible loans, public-backed venture capital, and revenue support.

Developing financial solutions for on-grid RE projects

Generally, large and medium-scale renewable energy projects need to operate within the same financing rules as are applied to conventional fossil-fuel energy projects.[7] A key issue for financing on-grid renewables investments is how to create a price support mechanism that provides stability and predictability over the medium and long term. Such mechanisms reduce the risk premium in the cost of capital, which will increase the amount of investment in renewables and lower the price that consumers have to pay for renewable energy. Policy interventions are taking a range of forms, including market-based mechanisms such as carbon emissions trading and renewable obligation arrangements, and fixed-price schemes such as the feed-in laws in Germany and Spain.

No one approach will be equally appropriate in all markets and regulatory environments. To be effective, however, they all must create financial incentives for investors to change the pattern of investment away from carbon-emitting conventional technologies in favour of large-scale investment in renewable/non-carbon-emitting technologies.

Bridging the gaps in the finance continuum

Numerous forms of capital are involved in the finance continuum of a grid-connected RE plant, as is shown in Figure 8.2.[8] The conventional power sector finance continuum includes the following sources of capital:

- equity provided by the companies involved in the project, as well as in some cases by institutional and strategic investors;

- corporate or project-financed loans provided by commercial banks or special purpose funds;
- guarantees perhaps provided by an export credit agency (ECA) to cover specific cross-border risks;
- insurance provided by an insurer or insurance broker;
- key parties to the transaction, such as fuel suppliers or power purchasers, who have entered into long-term contracts with the project.

For RE on-grid projects the finance continuum is actually quite incomplete, and the gaps can often only be filled with niche financial products, some of which already exist and some of which need to be created. Figure 8.2 shows which types of finance are often secured today by mid- to large-scale on-grid projects, which types are occasionally secured, and the current gaps and barriers in the continuum. It also proposes some interventions that might be supported by public sources to close the gaps.

Project development capital

Project preparation for on-grid RE projects is generally carried out either by large energy companies or specialized project development companies (as is the usual case in Germany). Energy companies finance project preparation from operational budgets. Specialized companies finance project development work through private finance, capital markets or with risk capital from venture

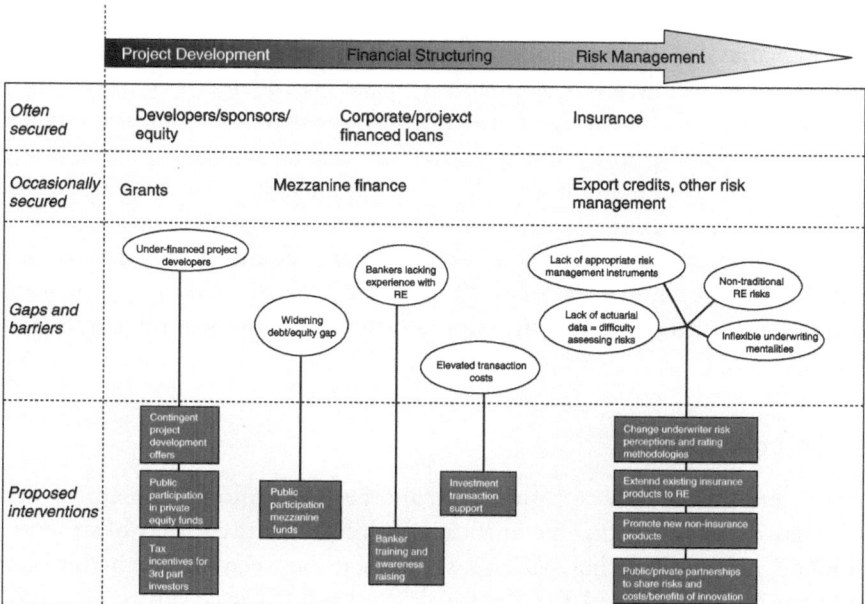

Figure 8.2 *On-grid financing continuum*

capitalists, private equity funds or strategic investors (e.g. equipment manufacturers).

Once a project developer has prospected out a site and has assessed the resource potential, the access to the grid and the expected power purchase price, he or she will prepare a pre-feasibility analysis. If the analysis is promising, the developer will then begin the engineering design, the permitting and the environmental assessment processes.

These steps can take several years and significant resources to complete. Facilities that can share some of the costs of development on a (contingent) grant basis can effectively help move RE projects forward.[9] These facilities need to be carefully structured to target the right projects and align interests on project development. For example, grant programmes are increasingly being combined with loan instruments to shift the focus from early stage 'prospecting' to later stage project engineering and development.

Equity finance

If the concept successfully passes through the development stages, the project developer is then in a strong position to attract external financing. To reduce the expected time for plant commissioning and to minimize project complexity, smaller developers may wish to sell the greater part of their ownership in the project to an entity that has sufficient resources to serve as sponsor and carry the project through financial structuring and contract negotiating. To secure loans, developers and their equity sponsors will generally need to provide between 25 per cent and 50 per cent of the capital required for a project in the form of shareholder equity. As the risk (real or perceived) associated with a project increases, lenders will require that equity play a larger role in the financing structure. This not only strains a developer's capital resources but also raises the cost of the entire project, since the cost of equity capital is always higher than the cost of debt capital. Therefore, innovative structures are needed that can fill the widening gap between the equity and debt available to a project. One institution willing to invest straight equity is the Connecticut Clean Energy Fund, which takes an early-stage participation in renewable energy projects with the specific aim of catalysing later-round commercial financing.

Mezzanine finance

Further along in the finance continuum, another option to fill this equity/debt gap is quasi-equity or mezzanine finance, which constitutes a variety of structures positioned in the financing package somewhere between the high-risk/high-upside equity position and the lower-risk/fixed returns debt position. It most commonly takes the form of junior debt (paid only after senior debt claims have

---

BOX 8.2 THE FIDEME MEZZANINE FUND

FIDEME is a €45 million public–private investment partnership that provides mezzanine finance to renewable energy companies in France. The French Environment and Energy Management Agency (ADEME) provides both capital and a first loss guarantee, which increases the risk-adjusted returns for investors and retail banks.

Innovation: ADEME is taking a mezzanine position within a mezzanine fund (see the next subsection), therefore its capital is double leveraged.

Source: www.ademe.fr/

BOX 8.3 COMMUNITY-OWNED WIND FUNDS

Community-based collective investment in RE is growing in countries such as Germany, Denmark, Finland and Canada, especially in wind projects. In Germany most commercial-scale wind turbines are financed by community-owned funds, a participation scheme that raises awareness and local acceptance and allows everyday citizens to make sound ethical investments. The funds receive tax benefits.

Innovation: Tax-incentivized community-owned funds.

---

been satisfied), coupled with the option to purchase shares in the company at a predetermined level. Public participation in mezzanine funds, if structured appropriately, can buy down the risks or buy up the returns for commercial investors. A number of RE mezzanine funds now exist in both developed and developing countries.[10]

Debt finance

The bulk of the financing provided to a project is usually in the form of senior debt, which can be structured as on-balance sheet corporate finance or off-balance sheet project finance.

Corporate financing requires a decision by the corporate sponsor to accept the risk and potential reward of a project in its entirety and can only be used by sponsors with a significant base of assets, debt capacity and internal cash flow. In this financing approach, project funding is arranged through the corporate treasury. For small projects internal corporate cash flow will often be a sufficient source of capital. For larger projects the sponsor may need to augment investment from free cash flow by some combination of new corporate debt, equity or bond issuance, or asset disposal. Any debt issuance is underwritten by the overall creditworthiness of the corporation, not the specific project revenues.

Corporate finance is the preferred financing approach for small projects (e.g. less than $15 million), since the financing can be executed more quickly and project revenues may be inadequate to support the transaction costs associated with other financing methods. Costs of legal and arrangement fees can be kept low. With a single sponsor accepting most of the project risk, technical and financial due diligence will typically be accomplished more quickly. Tax incentives, such as accelerated depreciation, and leasing structures, can help improve the financials of RE projects for corporate sponsors.

Off-balance sheet project financing involves the use of a special-purpose financial vehicle to fund a specific power-generation project with only limited recourse to the assets of external investors if the project underperforms or fails. Typical project finance requirements include the following:

- Firm long-term fuel supply and power purchase agreements are obtained with creditworthy parties for all project activities.
- Fixed-price turnkey design and build contracts are placed with experienced contractors.
- Guarantees, warranties or bonds for completion and performance are provided from sponsors and contractors.
- All contracts and insurance polices are assigned to the bank, which allows the lender to take over the project in the event of non-performance by the project company. The ratio of debt to equity is higher when a plant is project financed, and the loan tenor (duration) can be extended for longer periods, so that the loan repayments can match the power purchase agreement revenue streams.

Some public-sector debt instruments exist that are aimed at facilitating corporate and/or project financing to the SE sector. KfW (the German bank for reconstruction and development) has designed a series of soft loan credit lines for renewable energy projects within its Programme for the Promotion of Renewable Energies. Partner banks on-lend the financing and assume the credit risk in return for risk-adjusted margins.

Alternatives to bank financing

Third-party financing is another form of off-balance sheet financing that can be used in place of debt. An example of government-led third-party financing is the Spanish IDAE model, which has been financing renewable projects in Spain since the late 1980s. IDAE identifies a project, provides the capital to a developer to construct it, and recovers its investment plus the cost of its services out of the energy production or savings. At the end of the contract, all capital assets are owned by the project developer and user of the installation. IDAE plays the role of ESCO and has invested €95 million in renewable energy projects and leveraged another €104 million for 144 projects under the third-party finance mechanism.

In OECD markets the largest RE projects are starting to access the bond markets as a means of raising capital. The first known examples of this approach occurred in summer 2003, when FPL Energy in the US and Britain's Scottish Power raised $380 million and $700 million, respectively, through new bond issues for wind-park financing. The scale of these offerings is indicative of the level of maturity of the wind sector, and also the limited role that public support programmes can play in these commercial markets. The US Energy Policy Act of 2005 allows municipal utilities, rural cooperatives and governments to issue *clean renewable energy bonds*, with a national cap of up to $800 million (of which a maximum of $500 million is available to governments). The Federal government will give the bondholder a tax credit for what they would usually receive from the issuer in the way of interest.

Transaction costs

In terms of transaction costs, particularly for project finance, but also more generally for any debt financing, the extra costs associated with satisfying the higher 'burden of proof' that a bank's loan committee would normally apply to the first few RE investments usually fall on the project developer. RE developers are typically undercapitalized and often unable to absorb these transaction costs. Public facilities that share the costs of the investment decision-making and the transaction process can help bring bankable projects through to financial closure. At the same time, building RE awareness and capacity within financial institutions is also important.

Risk management

An integral element of deal structuring, particularly for off-balance sheet projects, is financial risk management. This process entails using financial instruments to transfer specific risks away from the project sponsors and lenders to insurers and other parties better able to underwrite or manage the risk exposure. Among other important factors, financial risk management is one of the keys to deployment of renewable energy technologies. Applied correctly, certain financial risk management instruments can help mitigate the perceived risks associated with RE and affect the degree and terms of investment into such projects. However, there are currently constraints on the availability of such risk management instruments, which relate to factors such as the willingness and capacity of insurance and capital markets to respond.

A number of insurers had bad experiences with wind power in the 1980s and early 1990s, and although the industry has undergone enormous growth since then and the technology has matured considerably, many insurers are still

---

BOX 8.4 BUILDING CAPACITY IN THE BANK

When one is building renewable energy investment capacity within a financing institution (FI), the approach needs to be flexible, as different institutions follow different 'product development' paths. To enter a new sector, some FIs first focus on creating the right policies or strategies, while others focus on training personnel. Other approaches include learning 'hands on' by taking first investments or developing specialized funds or loan portfolios.

Pursuing change in a financial institution takes time and commitment at all levels. To be successful across the institution, changes in the incentive structure are often needed. Although the CEO may be interested in renewable energy investment activity for its policy implications, loan officers often focus on narrower targets, such as simply meeting the traditional benchmarks of rapid loan disbursement with minimal risk. Without stronger incentives, loan officers may pay only limited attention to renewable energy investments.

Changing the way a financial organization considers new investments therefore requires both better information and new mandates to combine social and environmental factors – both risks and returns – as integral measures of economic performance.

The United Nations Environment Programme's Sustainable Energy Finance Initiative (SEFI), for example, has been developing a Transaction Support Facility (initially run through a number of banks in North Africa) for helping financiers evaluate and transact first-time loans/investments in the renewable energy and energy efficiency sectors. SEFI support comes in the form of direct or contingent grants to investors or lenders evaluating RE or EE transactions, usually to cost-share project assessment and advisory work undertaken by third-party experts.

---

reluctant to insure wind projects. There are some, however, who will do so, and a fully financed wind project will usually find cover today. Cover for biomass is available for larger projects; however, what is still needed is a product to cover the security of fuel supply. Financiers want fuel supply insured, but as yet there is no product to do it. Large-scale hydro is well understood and can be insured. Run-of-the-river hydro facilities are also catered for; however, small-scale and micro-hydro developers sometimes have difficulty finding sufficient cover, particularly for contractors' all risks (risk of non-delivery of contracts) (Olivier, undated).

There are still many insurance gaps in the finance continuum. Projects of less than $15 million have difficulty finding insurance cover and, as a result, financing. Only niche insurance operations with low overheads are able to service small-scale developers, and even then there is a steep learning curve and an indeterminate risk reward ratio for many projects. For emerging markets, targeted enhanced political risk insurance is needed that covers the risk in the case of default in performance of obligation by government or other entity. Such insurance could come from governmental or from public–private entities (Worenklein, undated).

There are a number of key barriers to the development of risk management instruments. Many risks associated with RE projects are non-traditional and hence uninsurable. It can be difficult to diversify risks, and actuarial data are not available to properly assess the risks (e.g. offshore wind construction risk). Underwriters have limited understanding of RE projects and associated risks, and have difficulty in aligning strategies for dealing with them. Underwriting mentalities are therefore generally rigid and inflexible.

Efforts in the risk management area should aim to help:

- extend existing energy insurance product lines with similar operations or facing similar risks to include standard RE projects and, as and where possible, prototypical RE projects;
- change underwriting risk perceptions to increase the availability of risk transfer products;
- develop new markets through convergence of insurance and capital markets;
- develop appropriate new underwriting rating methodologies;
- develop new risk management instruments to bundle heterogeneous risks;
- aggregate projects to create portfolios of scale and risk diversification.

Table 8.1 lists emerging risk management instruments that are at various stages of development. The table describes what risks the instrument covers and highlights the transfer of risk from one party to another where relevant.

Public–private partnerships can be developed to help move many of these instruments forward. Partnerships need to be developed on a risk-sharing basis in a manner that is equitable.

On-grid financing in developing countries

Most forms of financing used for on-grid projects in developed countries are also applicable to developing countries; however, the less mature financial markets make the gaps in the continuum significantly larger. The problem is exacerbated by market volatility and economic instabilities, which have caused foreign direct investment to decrease in recent years. It is a challenge to get capital markets more engaged, especially private-sector institutional investors such as pension funds and insurance companies that manage large pools of capital and must invest conservatively.

A lot of innovation in public support instruments for debt financing is actually happening in the emerging markets. A variety of instruments are being used to improve access to long-term financing, including currency swaps to reduce foreign exchange risk,[11] two-step bridging mechanisms to allow project refinancing,[12] lease-financing arrangements to reduce off-take risk, and various other approaches.[13]

To cover commercial and political risks in developing countries, export credit agencies (ECAs) support exports through direct credits/financing, refinancing,

Table 8.1 *Emerging risk management instruments for the renewable energy sector*

Risk mitigation product	Nature	Basic mechanism	Risks covered
Weather insurance/ weather derivatives	Hybrid of reinsurance and traded commodities	Contracts and derivatives including weather-linked financing traded over the counter (OTC). Risks transferred from project owners/sponsors to market through trading companies, banks, reinsurers	RE 'volumetric' resource risks (e.g. temperature, wind, precipitation)
Catastrophe bonds	Synthetic re-insurance	A securitized risk finance instrument based on catastrophe insurance. Some risk transferred from client to re-insurer/institutional investors	Risks related to natural catastrophes (e.g. managing resource supply risk)
Contingent capital	Risk finance	Insurance policy, swap option, hybrid security, debt or preference share provided by (re)insurer to support and/or replace capital. A revenue guarantee	Inability to meet debt servicing caused by defined events; loss of equity capital due to defined event
Captives or other pooling/ mutualization structures	Alternative risk transfer (ART)	Self-insurance programme whereby a firm sets up its own insurance company to manage its retained risks at a more efficient cost than transfer to a third party. Pooling through 'mutualization' or 'protected cell' structures have been used in the wind and geothermal areas	Property/casualty insurance. Can be adapted to include financial risks
REC or emissions reduction delivery guarantees	Insurance	Products provided by insurers and reinsurers to guarantee future delivery of 'credits' or money to purchase credits in spot markets to fulfil contractual requirements. Can be used to monetize future carbon or renewable credit-related cash flows and front-end associated revenue stream	Risks associated with delivery of RECs or emissions reductions, whether performance related or through interference by third parties resulting in contract frustration/ repudiation
GEF contingent finance mechanisms	Grant, loan, guarantee	Contingent grant, performance grant, contingent/concessional loans, partial credit guarantees, investment funds and reserve funds. Transfers a certain level of project risk to providers of such finance	Incremental costs associated with private-sector investment or commercial lending to RE projects
Synthetic securitization guarantees	Credit-backed securities	These structures bundle (credit or other) risk together. Securities issued based on different levels of credit/risk exposure, thus creating a risk transfer and financing conduit based on credit differentials	Could be used to manage a 'pool' of project credit risks that individually would not attract efficient pricing

| Guarantees from MFIs | Guarantee | Partial risk guarantee (covers creditor/equity investors) and partial credit guarantee (covers creditors) by World Bank group (IBRD, IDA, IFC and MIGA), regional development banks (e.g. AfDB, IDB, ADB), etc. Risks transferred from commercial investor/lender to MFIs | Specific political risks (e.g. sovereign risks arising from a government default on contractual obligations) and credit default |
| Export credit guarantees | Guarantee, export credit, insurance | Guarantees, export credits, insurance provided by bilateral export credit agencies | Commercial and political risks involved in private sector trade/investment abroad |

Source: Scoping Study on Financial Risk Management Instruments for Renewable Energy Projects, UNEP SEFI, 2004

> ## Box 8.5 The Bulgarian Energy Efficiency and Renewable Energy Credit Line
>
> The Bulgarian Energy Efficiency and Renewable Energy Credit Line (BEERECL) offers loans, technical assistance and grant support to Bulgarian clean energy projects. The facility is financed with €50 million from the European Bank for Reconstruction and Development (EBRD), into which is blended a €10 million grant facility from the Kozloduy International Decommissioning Support Fund.
>
> *Innovation:* Loan recipients receive grants towards the principal of the loan, up to 20 per cent for renewable energy projects.
>
> *Source:* www.beerecl.com/

interest rate support, aid financing (credits and grants) or export credit insurance and guarantees. ECAs have had limited experience to date with RE support; however, a new renewables agreement has recently been negotiated under the OECD Arrangement on Officially Supported Export Credit, which will allow them to provide extended financing terms (up to 15 years, as compared with 10 or 12 years for conventional projects) for renewable energy projects.

The Energy Future Coalition, a bipartisan energy policy initiative in the US, is looking into the creation of a new class of international debt securities called Global Development Bonds (GDBs) to finance sustainable development. Qualified issuers such as commercial banks might issue such securities to finance portfolios of qualifying projects in developing countries. GDBs would be entitled to certain automatic insurance coverage on political risk and foreign exchange risk from a government agency such as the US Overseas Private Investment Corporation (OPIC) and in return be regulated.[14]

Developing financial solutions for off-grid RE businesses in developing countries

To initiate a new business activity, an off-grid RE entrepreneur needs various sources of capital and business-development support.[15] The capital needs can also be shown along a finance continuum (see Figure 8.3):

- starting upstream with the start-up investment the entrepreneur injects to plan and initiate the business;
- shifting to bank loans as the business gets up and running so that it has operating capital to finance day-to-day operations and business growth;
- finally, downstream, possibly some form of customer or transaction finance (supplier credit, consumer credit, leasing, performance contracting, etc).

Each step in the finance continuum has a distinct role to play and can involve either public, private or public–private (or NGO) delivery channels. Each step also has its own risk/return profile and often requires specialized financiers to effectively deliver.

Owing to the various barriers and market failures cited earlier that inhibit financier participation in the (particularly off-grid) RE markets, there are many gaps today in this continuum, both financial and non-financial. Ultimately these gaps make it difficult to launch a new off-grid RE business or even to expand an existing proven business.

Start-up capital

To start a new off-grid RE business, or expand an existing business into RE, the entrepreneur needs a significant amount of time and capital resources. Innovators are needed that have the capital, the capacity and the entrepreneurship to take risks in developing and testing new business approaches and service offerings. For the entrepreneur it is not so much about understanding a new technology, but rather understanding a market need and being able to package an appropriate technology with a service offering to address this need. In developing countries the offering of RE products and services generally develops in two distinct phases:

1   Developing-country RE industries usually first begin to grow in response to public procurement markets. For solar PV, for instance, this has included applications such as telecoms, vaccine refrigeration and street lighting. The RE companies responding to government tenders act mostly as traders and turnkey contractors, activities that require limited up-front investment and only moderate risk-taking (e.g. timing risk – how long will it take to get paid?). Since these RE trading/contracting companies are now established in

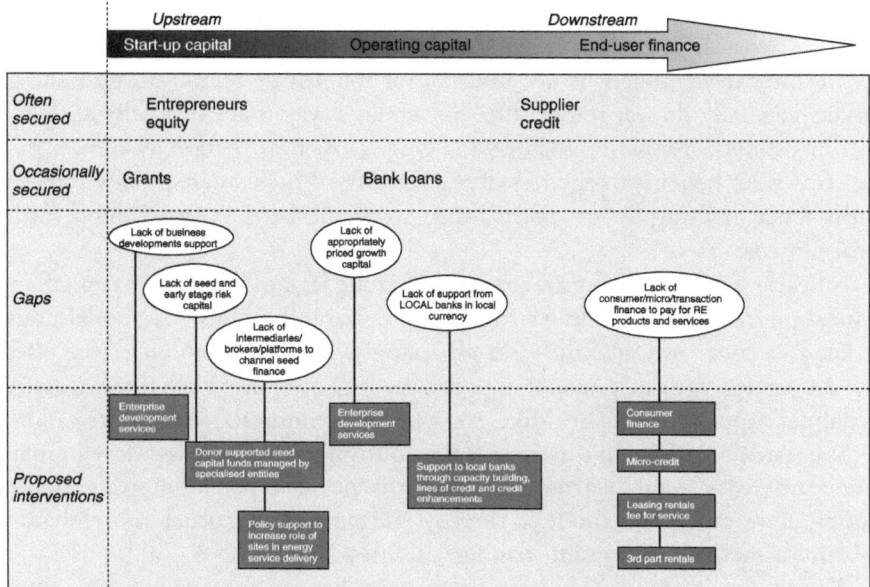

Figure 8.3 *Off-grid financing continuum*

the major urban centres of most developing countries, the technology is available. However, technology availability alone is not sufficient to create substantial RE markets.

2   The second stage of off-grid RE sector development is when companies begin building service infrastructure, thereby directly accessing end-users. Here they take on the role of energy service providers, somewhat akin to decentralized utilities. The business models involved are sophisticated. They have varying financing needs and investment horizons, and they require entrepreneurs and financiers with larger risk-taking appetites. Selco and Shell Solar are examples of such RE innovators, both of which have extensive service infrastructure in place in parts of India and Sri Lanka and are seeing healthy market growth.[16]

The up-front costs of this business model are significant and include:

- increased financing costs (raising capital for a new type of business is time-intensive and costly);[17]
- increased market awareness costs (RE products do not usually substitute directly for conventional products and therefore the consumer must be made aware of the benefits of the clean-energy option);
- increased transaction costs.

To date, the few 'first mover' innovators that have set up extensive service infra-structures have done so mostly with their own resources. Though investing in new business development is not unusual for the private sector, off-grid energy markets generally do not reward first movers in a way that can justify such up-front investment. Resolving sector-wide barriers such as market awareness pro-vides free-rider benefits to the rest of the industry. Much smart money thus sits on the fence, waiting for others to bring down the barriers that inhibit initial sector growth.

An early gap in the off-grid continuum is the lack of early-stage capital and donor support needed to help RE innovators develop their business models, raise market awareness and take the risks associated with new product/service offer-ings. Associated with this financing gap is the lack of appropriate intermediaries to channel support in 'business-like' ways to these young RE innovators.

Suggested solutions in this area of the continuum are business development grants and risk-capital instruments – approaches that align interests around creating new enterprise models, thereby preventing the usual moral hazard problems associated with grant-making activities.

A fairly new form of early-stage finance mechanism for small and medium-size enterprises (SMEs) is seed capital, which is applied as a small initial invest-ment to help an entrepreneur plan and launch a specific business. The seed capital needed can range from several thousand dollars for a new enterprise to simply pilot a new business activity, to a few hundred thousand dollars to take a new enterprise from a tested approach to a proven commercial business. The RE finance leader for the past decade in the seed finance area is the energy investment company E+Co.[18] The willingness to take more risk than with conventional sources, combined with the provision of enterprise development services, constitute the main concessional aspects of the seed-financing approach.

Box 8.6 Rural Energy Enterprise Development

Initiated in 2000 by the United Nations Environment Programme (UNEP), E+Co and a number of country partners, and backed by the UN Foundation, the Rural Energy Enterprise Development (REED) initiative supports sustainable energy enterprises that use clean, efficient, renewable and affordable energy technologies to provide energy services to rural and peri-urban customers in seven developing countries. REED offers rural energy entrepreneurs a combination of start-up financing, enterprise development services, such as business planning, management structuring and financial planning, and assistance in securing later-stage financing. Thirty-six enterprises financed to date in Ghana, Mali, Senegal, Tanzania, Zambia, China and Brazil include crop drying, charcoal production, biofuels, wind pumps, solar water heating and efficient cook stoves.

Innovation: Using seed finance to capitalize early-stage clean energy SMEs.

Source: www.areed.org; www.energyhouse.com

Donor aid to date has mostly taken the lead in the seed-finance area.[19] Donor agencies today are more willing to provide grant funding for demonstration projects than for the up-front costs of developing a new clean energy business. A key difference between the two approaches is the nature and extent of risk the donor takes. In a technology demonstration programme the risk is that the *technology* will fail, while with the seed-finance approach the risk is that the *entrepreneur* will fail. It is a basic fact of business that not all ventures succeed. It is also a fact that risk is an inherent and accepted element of doing business. The risk of failure – and the potential for profit – is what drives entrepreneurs to use their resources efficiently and to constantly seek new ways to deliver an improved product or service.

In industrialized countries the risk capital needed to start new companies is now often provided by venture capitalists, who are willing to invest both business development support and capital in risky ventures in return for the prospect of high returns. Although many elements of the 'hands-on' investment model of venture capitalists are very appropriate for off-grid developing-country SMEs, the prospect of elevated returns in this sector are limited, for the reasons stated above, and therefore the seed capital business will need to be principally donor supported.

Operating capital

As a new company scales up its operations, it will usually need a second injection of capital to manage its growth. The more successful a young company is, the higher its capital requirements will be. These 'second-stage' growth capital requirements include both short-term needs (e.g. working capital) and long-term needs (investing in service infrastructure), both of which should normally be financed with loans from local banks.

BOX 8.7 THE SOUTH AFRICAN EMPOWERMENT THROUGH ENERGY FUND

The Empowerment Through Energy Fund (ETEF) was launched by the Shell Foundation in partnership with GroFin, ABSA and IDC in 2003 with ZAR51 million (US$6.3 million) to help back empowered SMEs in the energy sector, including those offering affordable modern energy services to poorer communities. ETEF provides pre-investment business development assistance, start-up and/or early growth stage capital, as well as post-investment business development assistance. As of July 2005, 14 deals had been approved, resulting in the creation of almost 200 new jobs.

Innovation: The Shell Foundation provides a grant to cover initial fund management costs as well as risk capital contribution to the fund.

Source: www.shellfoundation.org/energise/etef; www.grofin.com

For SMEs in the RE sector, however, this is rarely possible. Whether loans are actually accessible depends on a combination of three factors: cost (interest rate), collateral requirements and, for larger loans, exchange-rate risk. The first two of these factors are based on bank risk assessment of the company and the project or product. Very few commercial lenders today are financing off-grid RE SMEs in developing countries. If they do, they usually provide funds in the form of private or corporate finance, and therefore access to these loans has less to do with the business model than with the size of the owner's asset base and the owner's willingness to provide these assets as security.[20]

Public and donor funding mechanisms and capacity development are needed to help engage commercial lenders and investors in financing the operating capital needs of off-grid RE companies. Typically, these support mechanisms have taken three forms: lines of credit, credit enhancements for loan provision and SME growth capital funds.

Lines of credit are a common approach for development finance institutions (DFIs) to support the creation of credit windows in national or local banks for specific areas of lending, including RE enterprises. When local financial institutions start out in a new sector, this form of refinance is the quickest way to engage their participation and provides the most latitude for the donor/DFI to set the financing terms. On the other hand, treating local financial institutions only as intermediaries eliminates their risk exposure, which can make them less motivated to manage the credit portfolio effectively. The key to success for this approach is thus finding ways to align DFI and FI interests.

Credit enhancements are a variety of subsidies provided by DFIs or other donor programmes aimed at softening loan financing, either for the lender or for the borrower. The concessionality comes in the form of risk-sharing or interest rate reductions. Partial risk guarantees ensure debt-servicing payments to selected lenders or other investors in a project, usually for specific time periods or exposure levels.[21]

Partial credit guarantees act to extend loan repayment periods, thus improving the project's cash flows. Both forms of guarantee can motivate banks to lend for projects they perceive as risky. Buying down the risk can mean lower costs of financing for the borrower or decreased security requirements. Guarantees are most effective at addressing elevated banker perceptions of risk; once a bank has gained experience in managing a portfolio of RE loans, it is in a better position to evaluate true project risks. Guarantees, however, do nothing in the long term to reduce an RE technology's true project risk.

Interest rate subsidies, which are another form of credit enhancement, lower the cost of financing for the borrower and can be an effective means of helping banks build their loan portfolios in specific RE sectors. Because the bank assumes the entire credit risk, its interests are fully aligned with those of the donor, in terms of both minimizing defaults and continuing lending activity after the donor support has been phased out.[22] This approach is subtle, however, and

---

Box 8.8 The Global Renewable Energy Fund of Funds

Sponsored by the European Commission, the Global Renewable Energy Fund of Funds (GREFF) is designed to provide subordinated non-commercial capital to a number of investor-financed funds. Each fund will invest some form of patient equity or quasi-equity into RE businesses or projects in developing countries and economies in transition. The capitalization target is €75 million public/donor funds, which is expected to leverage €300 million investor capital in the individual funds.

Innovation: Public-backed equity funding with less demanding return expectations than pure commercial private equity.

---

therefore will only work in larger RE markets where banks can be confident of building sizeable loan portfolios quickly. In less developed markets transaction costs might outweigh the benefits for the bank.

More public-supported credit enhancement programmes are needed that build on successful approaches applied to date.

A number of RE growth capital funds (also termed patient capital funds) have been created in the past few years to help SMEs grow or expand businesses that are still not financeable with commercial loans. In contrast to seed capital funds, which are generally purely donor-financed, growth capital funds are financed with a mix of donor and commercial capital. As with mezzanine finance funds (to which growth capital funds are quite similar), this blending is used either to buy down the risks or to buy up the returns for commercial investors. Financing is provided either as equity or as debt, although the trend is towards debt, owing to the difficulty of exiting equity positions. Experience with these funds has been mixed, as some have not managed to meet their investors' expectations and consequently have been dissolved. As compared with technology support programmes, investment funds are usually more technology-neutral and therefore can be more effective at responding to a specific market opportunity or need. Public support for growth capital funds must be redoubled, building on the approaches that have worked to date. As with seed capital, the intermediary fund manager is key, and support to develop this area of finance expertise is necessary.

End-user financing

There are a number of models for financing RE transactions through the end-user, all of which have been implemented in various developing countries with varying degrees of success.

The *supplier credit model* is the most basic means for an RE enterprise to finance transactions with end-users and is usually short term (3–12 months). It requires that the enterprise either finance its purchases or receive manufacturer credit, which on occasion extends to six months, but seldom longer. This model also requires that the RE enterprise manage a credit portfolio, something that requires a different sort of expertise from that of running an RE business.

The *consumer credit (or micro-credit) model* is perhaps the most developed means of facilitating individual household purchases of renewable energy systems. Loans are made by local banks or entities that specialize in originating small-scale loans for solar or other renewable energy systems, at either the household or the village scale. The transaction between the end-user and the RE enterprise is commercial and does not require that the enterprise supply credit to the end-user. Notable examples of consumer micro-credit for solar home systems have emerged in South Asia, including Grameen Shakti in Bangladesh, Sarvodaya in Sri Lanka, and Syndicate and Canara banks in India. The two forms of credit enhancements – guarantees and interest rate softening – can also be applied with consumer credit.

The *fee-for-service model*, whereby customers pay for an energy service, is an approach that makes RE products and services more affordable, while minimizing the long-term risks for the customer inherent in the credit models (technology risk, rural electrification risk, etc.). The basic grid extension model for rural electrification is fee-for-service. Soluz has applied this model to PV in the Dominican Republic and Honduras. Though it is easier to build a customer portfolio with this model, experience has shown that managing customer turnover and system removal costs can be a challenge. Government concessions can be used to apply the fee-for-service model, usually with off-grid equivalents of grid extension subsidies. The national utilities in Morocco (ONE) and South Africa (ESKOM) have followed this approach.

The *lease model* is similar to the fee-for-service model in that the lessor retains ownership of the equipment and hence responsibility for maintenance and equipment replacements. Leasing is normally provided by a specialized financial institution and can include tax benefits that lower system costs.

CONCLUSIONS AND RECOMMENDATIONS

Social, environmental and energy security concerns, coupled with improved renewable energy technologies, are increasing the momentum for support for renewable energy. The challenge is to introduce the right policy frameworks and financial tools to enable RE to achieve its market potential. This is particularly crucial in developing countries, where investment is endangered by geopolitical, economic and regulatory risks and where the lack of developed financial markets and products leaves the risks resting solely on the shoulders of the lender or investor.

A free market does not function according to rules of social responsibility and therefore needs to work within certain boundaries that serve social and environmental as well as economic goals. Policy-makers have a social responsibility to set those boundaries. Legislators and regulators have it in their control to shift the future energy portfolio if they choose to decrease demand by improving energy efficiency and to push energy generation towards renewable energy.

Financiers have a social responsibility to develop and deliver market solutions to the challenges of building a sustainable energy future. That energy future must take the form of a free energy market with less carbon, less fuel and fuel-price risk, and more and better access for the poor.

The following recommendations to address barriers to investment in RE are for policy-makers as well as the financial sector.[23]

Table 8.2 *Recommendations for policy-makers*

Goals	Recommendations for policy-makers
Have coherent and consistent policies in place to create genuine incentives for investors	• Introduce policy interventions that raise the output price and/or reduce the investment cost to ensure an adequate and predictable return on investment • Price in environmental benefits of clean technologies and environmental costs of fossil fuels • Encourage fixed-price schemes and market-based instruments • Provide incentives that reflect the various stages of development from R&D (e.g. public funds) to commercial use (e.g. production tax credits). Structure incentives to address specific market barriers, be removable, reward innovation and be cost-effective
Lower front-end barriers to project development	• Create public facilities to share some of the costs of project development on a grant or contingent grant basis • Improve the creation and sharing of information on RE plants to help financiers better understand and manage project risks and decrease risk perceptions
Strengthen the equity base of project developers	• Implement public interventions that give third-party investors incentives to take early participation in RE projects (training, tax benefits, soft capital for fund capitalization)
Close the debt/equity gap	• Increase public support for mezzanine-type funds to buy down the risks and buy up the returns of commercial investors
Mobilize corporate finance	• Introduce tax incentives and leasing structures that improve the financial of RE projects for corporate sponsors
Lower loan/investment evaluation and transaction costs	• Develop public facilities to share the costs of investment decision-making and the transaction process
Mobilize the finance sector to improve/expand risk management tools	• Build public–private partnerships aimed at moving new risk management instruments forward

Table 8.3 *Recommendations for financial institutions*

Goals	Recommendations for financial institutions
New financial products	• Develop and test innovative financial products tailored to the RE sector
Develop/improve/expand risk management tools	• Extend existing energy insurance product lines with similar operations or facing similar risks to include standard RE projects and, where possible, prototype RE projects • Change underwriting risk perceptions to increase the availability of risk transfer products • Develop new underwriting rating methodologies • Develop new risk management instruments to bundle heterogeneous risks • Aggregate projects to create portfolios of scale and risk diversification • Develop new risk transfer markets through convergence of insurance and capital markets • Develop the actuarial data sets needed to assess project risks
Increase awareness, information and skills	• Increase awareness of the threats of climate change and other environmental impacts of energy production and use • Initiate and carry out more work on financing renewable energy • Develop the skills to evaluate renewable energy project risks and revenue streams • Provide information to customers and clients on the opportunities of investing in RE • Increase analyst coverage of listed RE companies

Table 8.4 *Recommendations for policy-makers and multi- or bilateral institutions in developing countries*

Goals	Recommendations for policy-makers and multi-/bilateral institutions
Support business development	• Support companies that ensure reliable access to energy, provide local income generation and improve living standards • Help project developers fill the gaps along the finance continuum. This can be in the form of support for feasibility and due diligence work, as well as for business planning
Support new risk capital approaches for enterprise development	• Support the creation of early-stage seed capital funds, providing capital and enterprise development services to innovative clean energy entrepreneurs • Support local intermediary capacity as a more efficient way to deliver seed capital and to support SMEs with business development services

Support growth capital approaches that help proven SMEs scale up their businesses	• Finance growth capital funds using blended arrangements that buy down the risks and buy up the returns for commercial investors
Improve access to, and affordability of, credit markets	• Provide credit enhancements to share the risks (guarantees) or buy down the financing cost (interest rate softening) of commercial loans
Partner with the private sector	• Build effective financing partnerships with the private sector on a risk-sharing basis. This is necessary as long as the renewable energy industry in developing countries continues to rely on government and public funding

Table 8.5 *Recommendations for financial institutions in developing countries*

Goals	Recommendations for financial institutions
Greater role in financing RE in developing countries	• Develop risk management tools to support RE infrastructure and institutional capacity in developing countries • Develop innovative financing strategies that ensure quantifiable benefits – in both social and financial terms – for both the public and the private financier
Partnership with the public sector	• Co-invest with IFIs in the RE sector to diversify risk and increase effectiveness and efficiency of financing • Play an active role in managing the execution and financing of publicly funded projects
Greater role of ECAs in RET export support	• Design and market tailored products, jointly with RE sector • Create risk management products inspired by and responding to Kyoto Mechanisms • Develop processes to bundle smaller RE projects

Note: ECA = export credit agency

NOTES

The authors gratefully acknowledge Akanksa Chaurey (TERI), Martina Otto (UNEP), Michaela Pulkert (Hypovereinsbank), Gareth Hughes (Climate Change Capital), Charles Donovan (BP Alternative Energy) and Sandra Makinson (Canopus Foundation) for their valuable contributions.

1   *Public Finance Mechanisms to Catalyze Sustainable Energy Sector Growth* – a guide that traces the investment stages of technology innovation and sustainable energy project development and highlights the public-sector mechanisms that address the funding gaps and leverage private capital; UNEP/BASE SEFI in collaboration with the Clean Energy Group, October 2005.

2   For the sake of brevity, this chapter does not cover all RE technologies and financing models in detail, but rather uses some generalizations to describe the basic forms of renewable energy finance. Two of the most grievous generalizations are (1) focusing

mostly on electricity sector RE examples; and (2) grouping together larger projects as 'on-grid' and smaller projects as 'off-grid'.

3 The so-called Kyoto effect is discernible in the Global Energy Innovation Index (GEIX). During the first quarter of 2005 the 20 constituent companies of the GEIX which are quoted in countries that signed the Kyoto Protocol were up by an average of 21.9 per cent, contrasting with the 30 GEIX constituents quoted in the US and Australia, which were down by an average of 13.5 per cent (New Energy Finance).

4 US and British institutional investors who collectively manage over $3 trillion in assets pledged in May 2005 to invest $1 billion in clean energy companies.

5 The second quarter of 2005 saw the second largest ever Initial Public Offering of a pure-play new energy company, Conergy AG, which completed a €243m float in Frankfurt in March. Conergy is now trading at a market capitalization of around €695 million (New Energy Finance).

6 A more comprehensive list of sustainable energy programme design criteria known as the Pocantico Principles is available at http://newsletter.winrock.org/December03.

7 This section applies to both developed and developing countries.

8 Financial continuum analysis based on work of Philip LaRocco.

9 Two examples are the Massachusetts Pre-Development Financing Initiative and the Canadian Green Municipal Energy Fund, both of which help project developers and public entities develop RE grid-connected generating facilities. Pre-development finance can be provided to support feasibility studies and to cover costs related to siting, environmental analysis and assessment, permitting, facility design, financial and legal analysis, and marketing activities.

10 Some of the US State clean energy funds provide mezzanine-type investment, for example the Sustainable Development Fund of Pennsylvania, which offers junior debt to projects not easily financed on a purely commercial basis. In emerging markets an example is the Central American Renewable Energy and Cleaner Production Facility (CAREC), a mezzanine fund managed by E+Co.

11 For example, DEG, the private-sector arm of KfW, has put in place a cross-currency swap to finance a windfarm in China.

12 The World Bank has offered a liquidity stand-by guarantee to banks providing term loans to mini-hydro plants in Uganda. Although banking regulations limit loan maturities to eight, this form of put option (the banks can sell the loan to the World Bank after eight years) allows banks to stitch together two loans into a 15-year financing package, a term more appropriate for RE infrastructure financing.

13 An in-depth review of these mechanisms is given by Lindlein and Mostert (2005).

14 John E. Mullen (jmullenje@globalnetpartners.com) or Michael Eckhart (meckhart@acore.org).

15 Although this section applies mostly to developing-country enterprises, generic elements of the financing needs/responses are the same for developed-country SMEs.

16 For example, Selco India, a small privately owned PV company, has invested for years in developing the market for PV in southern India and Sri Lanka. Shell Solar,

part of the Shell group of companies, has invested significant capital in developing its solar rural electrification businesses in a number of countries.

17 The typical cost of raising financing for a conventional business is 1–5 per cent of the capital raised. In the RE sector the costs of raising capital can range from 5 per cent to 20 per cent.

18 E+Co has invested in some 90 companies in 28 developing countries, through applying a combination of seed finance and enterprise development services.

19 The two current exceptions are the Seed Capital Access Facility, a mechanism under development by UNEP to shift donor support towards engaging mainstream energy investors in the seed capital area, rather than displacing them, and the Shell Foundation, which has been taking this approach with the GroFin funds.

20 Security requirements can range up to 300 per cent of the loan value.

21 IFC has developed a number of GEF-financed guarantee products for the RE sector, most of which will be offered in the future through its new Environmental Business Finance Programme.

22 This approach is being successfully applied to both solar thermal (IREDA) and PV (UNEP) sectors in India.

23 We also refer the reader to the recommendations made by the G8 Renewable Energy Task Force 2001, as well as the UNEP FI Climate Change Working Group 2002.

REFERENCES

Lindlein, P. and Mostert, W. (2005) 'Financing Instruments for Renewable Energy Instruments, Strategies, Practice Approaches', KFW

Olivier, E. (2003) 'Survey of Contingent Financing & Risk Mitigation Instruments for Clean Infrastructure Projects', PCF Workshop on Tools for Risk Mitigation in Clean Infrastructure Projects

Renewable Energy Policy Network for the 21st Century (REN21) (2005) *Renewables 2005 Global Status Report*, Washington, DC, Worldwatch Institute

Saghir, J. (2005) 'The Global Investment Challenge – Financing the Growth of Renewable Energies in Developing Countries', *Renewable Energy World*, July/August

Worenklein, J. J. (2003) Background Paper, PCF Workshop on Tools for Risk Mitigation in Clean Infrastructure Projects

9

Clean Development Mechanism and Joint Implementation: New Instruments for Financing Renewable Energy Technologies

Axel Michaelowa, Matthias Krey and Sonja Butzengeiger, with contributions from Wendy Soh and Martin Burian

The need for climate policy: the Kyoto Protocol and its 'Kyoto Mechanisms'

Anthropogenic climate change due to emission of greenhouse gases (GHGs) will be one of the major environmental problems of the 21st century. Greenhouse gases are not local environmental pollutants and thus there is no domestic incentive to reduce their emissions. They arise in all sectors of an economy, which means that an efficient climate policy has to be cross-sectoral. Owing to global mixing of greenhouse gases, an efficient climate policy must be applied on a global scale. Renewable energy does not lead to greenhouse gas emissions and thus is a crucial part of a strategy to reduce emissions.

After difficult negotiations, it became possible for a United Nations Framework Convention on Climate Change (UNFCCC) to be signed at the UN Conference for Environment and Development in Rio de Janeiro in 1992. It remained rather general, though, and did not include specific emission targets or binding instruments of climate policy. The UNFCCC entered into force in 1994, and the first Conference of the Parties (COP 1) in Berlin 1995 decided to embark on negotiations of a Protocol with binding targets. The negotiations were crowned with success when in 1997 COP 3 in Kyoto achieved the negotiations of the 'Kyoto Protocol', which has been ratified by 155 countries by

August 2005. However, it took until February 2005 before the Protocol could enter into force, as the US declared its unwillingness to ratify in 2001 and Russia, which was needed to pass the threshold set for entry into force, hesitated to ratify for a long time.

The Protocol sets differentiated, legally binding emission targets for the industrialized countries and countries in transition (Annex B countries). The targets apply to a basket of six greenhouse gases. During the 'First Commitment Period', from 2008 to 2012, Annex B countries are required to reduce average annual emissions to a specified percentage of 1990 levels. Actual national limits range from 92 per cent for the European Union to an allowable increase of 10 per cent for Iceland. The emission targets can be reached via domestic emissions mitigation, by investment in emission reduction projects abroad or the acquisition of emission rights from another country. The latter two options are possible because four so-called Kyoto Mechanisms have been set up that allow trans-boundary cooperation in emission reduction. A large part of the Kyoto Mechanisms rules was agreed in the 2001 Marrakech Accords, as the Kyoto Protocol had only given a general framework. To participate in the mechanisms, countries have to ratify the Kyoto Protocol and fulfil certain reporting requirements. Whether all Annex B countries, particularly countries in transition, will fulfil these rules remains to be seen.

There are four Mechanisms: the Clean Development Mechanism (CDM), Joint Implementation (JI), International Emissions Trading (IET) and bubbles. CDM and JI are project-based whereas the latter two relate to transfers of parts of the national emission budgets. IET is only possible between Annex B countries and consists just of a transfer from one country to another, after 2008. Countries forming a bubble can redistribute their targets internally *ex ante* as long as the sum of the targets is not exceeded. The European Union is the only country group forming a bubble; it has redistributed its target of –8 per cent so that Portugal can increase its emissions by 25 per cent while Luxembourg has to reduce them by 28 per cent, to name the extremes.

The CDM allows countries with emission targets to buy emission credits from projects in countries without targets. It also has the goal to further sustainable development in the latter. Owing to the fact that CDM emission credits are added to the overall emissions budget of Annex B countries, their quality has to be guaranteed. Therefore, emission credits only accrue after independent verification through so-called Designated Operational Entities (DOEs or OEs), which are mainly commercial certification companies and thus are called Certified Emission Reductions (CERs). The Marrakech Accords defined an elaborate 'project cycle' that is overseen by the CDM Executive Board (EB), whose ten members are elected by the UNFCCC Conference of the Parties. It has to check whether projects conform to the rules and formally register them. The 'project cycle' is sketched in Figure 9.1 at the end of this section.

Any institutional arrangement is possible to set up CDM projects: bilateral agreements, multilateral funds or even unilateral activity by the host country. This is helped by the full interchangeability of CERs with other types of emission rights under the Kyoto Protocol. Both host and investor country have to set up an official approval agency, the 'Designated National Authority' (DNA), for CDM projects; the host country defines criteria to check whether the project leads to sustainable development. Up to September 2005, 67 developing and 17 industrialized countries had set up DNAs.

To calculate the amount of CERs of a project, a baseline has to be fixed that describes the situation that would have existed in the absence of the project (OECD, 2000). For a long time, it was unclear how baselines would have to be set up, as the Marrakech Accords only define some principles. Especially contentious was how to check whether a project is 'additional' – that is, whether it would have happened anyway. If business-as-usual projects are accepted, the CERs will create fictitious emission reductions (Greiner and Michaelowa, 2003). Finally it was decided that a 'case law' would be developed. Project developers have to submit a baseline methodology proposal to the EB. The EB sets up a Methodology Panel that evaluates the proposed methodology with the help of independent experts. Up to September 2005, 24 methodologies had been accepted, two of which relate to renewable electricity generation. As the procedure showed that many methodologies were submitted for similar project types, the EB has developed consolidated methodologies on the basis of the submissions. Of four consolidated methodologies, one relates to renewable electricity for the grid. The implications of the methodology decisions will be discussed later in the chapter.

Box 9.1 gives an idea of the challenges in methodology development that are additional to those found in conventional renewable energy project development.

Projects can have a lifetime of ten or three times seven years. They are subject to an in-kind adaptation tax of 2 per cent that is waived for projects in the least developed countries. Another tax will cover CDM administration costs but its rate remains to be specified. Until then, project participants will have to pay a fee for administration that is fairly stiff, ranging from $5000 for the smallest projects to $30,000 for large ones.

Owing to the fear that transaction costs will be prohibitive for small projects (Michaelowa et al, 2003; see also later in this chapter), more lenient rules have been decided for renewable energy projects below 15 MW capacity, energy efficiency projects that save less than 15 GWh per annum and other projects that annually directly emit less than 15,000 tonnes of CO_2. They can use standardized baselines. First evidence suggests that small projects will play a relevant role under the CDM, as 43 per cent of all submitted projects are of small scale.

The term 'Joint Implementation' has acquired a narrow meaning through the Kyoto Protocol; formerly it was the umbrella term for all project-based

Box 9.1 The CDM project underlying the first approved renewable energy methodology: rice husk-fired plant in Thailand

A 20 MW rice husk power plant in central Thailand managed to get the first approval by the CDM Executive Board for a baseline methodology of a renewable energy technology in October 2003. The project participants had previously submitted a methodology that was rejected, but resubmitted a new methodology immediately afterwards.

The plant plans to generate 133 GWh per year, which will be sold through a 25-year power purchase agreement (PPA) with the Electricity Generating Authority of Thailand (EGAT). Within the project's procurement area, it is estimated that there are over 1 million tonnes of rice husk, of which approximately 700,000 tonnes are unused. This large surplus rice husk represents almost five times the plant's requirement. Eight-year fuel supply agreements have been concluded with about 30 rice millers, principally within 80 km of the proposed plant. CER volume is estimated at 83,000 per year. At a price of $7 per tonne of CO_2, the CER revenue will enhance the project's return on investment (ROE) by 7.2 per cent. The project developers argue that it is additional, due to investment and technology barriers. The technology used is state of the art (suspension-fired boiler), much superior to the stoker boilers used by other rice husk power plants in Thailand. This technology has increased costs. Moreover, the absence of a core fuel supplier has made it impossible for the developer to find investors under business as usual. Another barrier is hostility of the neighbouring communities to any thermal power plant. Opinion surveys were conducted among 20 community leaders and 150 villagers. It remains to be seen whether the validator accepts these arguments for project additionality.

The baseline emission factor is EGAT's grid average emissions projected until 2012, which are set to fall from 624 g CO_2/kWh in 2006 to 578 g in 2012 and have to be revised downwards if the actual emission factor is below that. As the plant might prevent other biomass plants from coming on line, owing to competition for biomass supply, a test will be made to check whether the surplus supply of rice husk is at least twice as large as the amount needed to fuel grid-connected rice husk power plants.

Source: Project Design Document NM0019: A.T. Biopower rice husk power project, available at http://cdm.unfccc.int/methodologies/process

---

reductions abroad. It now only applies to projects that take place in Annex B countries, which are, according to the Kyoto Protocol, countries with binding targets. Emission credits ('emission reduction units', ERUs) can only accrue from 2008. JI has two distinct 'tracks'. The first track is very liberal and leaves choice of baselines and project lifetimes to the participating countries. This is due to the fact that ERUs are deducted from the emissions budget of the host country and thus there is no incentive for baseline manipulation. The second track is similar to the CDM and applies if the host country does not fulfil the reporting requirements for Annex B countries; of course, it can also be chosen voluntarily. It is overseen by a 'Supervisory Committee' and the ERUs have to be certified by 'Independent Entities'. It is likely that the rules developed by the CDM Executive Board will be used under the second track. To garner the

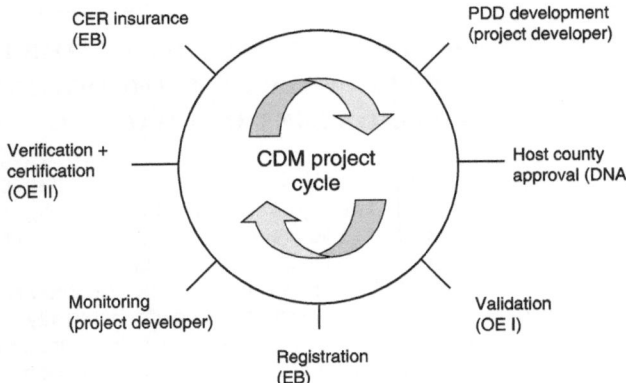

Figure 9.1 *The CDM project cycle*

potential for emission reductions before 2008, some countries are already invit-ing investments into 'early JI' emission reduction projects and granting post-2008 emission rights from their budgets for the pre-2008 reductions.

RENEWABLE ENERGY IN THE CDM AND JI: INITIAL EXPERIENCES AND MARKET PROJECTIONS

The empirical data on CDM and JI projects are increasing. We start with an overview of a 'test phase' from 1995 to 2001 and then look at the current market for CDM and JI projects, and at estimates on how it is going to develop in the future. The section closes with an overview on renewable energy CDM and JI projects that are currently being developed. In the following analysis, energy generated from the collection and burning of landfill and sewage gas is not considered renewable.

Test run: how renewables fared during 'Activities Implemented Jointly'

As the idea of project-based emissions credits surfaced as early as 1992, COP 1 in 1995 decided that the concept should first be tested without accrual of emission credits and called this test phase 'Activities Implemented Jointly' (AIJ). AIJ started relatively slowly and did not lead to the desired convergence of methodologies (Chatterjee, 1997; Dixon, 1999). Reporting was uneven and it was totally unclear which projects were just exercises on paper and which ones were actually implemented (see Table 9.1; see also Michaelowa, 2002; Beuermann et al, 2000). So, there was less experience to draw on than had been expected.

Table 9.1 *The 'Activities Implemented Jointly' pilot phase over time*

	1995	1996	1997	1998	1999	2000	2001
Projects reported	10	16	61	95	122	143	152
Projects implemented[a]	0	3	13	60	86	n.a.	n.a.
Investing countries	3	3	5	8	11	12	12
CDM host countries	2	2	5	14	23	27	30
JI host countries	5	5	7	10	11	11	11
Share of JI countries in all projects (%)	60	50	74	72	65	58	56
Planned emission reduction (Mt CO_2)[b]	23	111	140	162	217	366	442
Share of JI countries (%)	56.5	39.5	32.6	31.3	24.3	15.6	13.6

Notes:
a   These are estimates as no reliable information exists. The projects implemented tend to be small projects in countries in transition.
b   The emission reduction actually implemented is much lower (see previous note).

Source: Michaelowa (2002)

In 1999 only 70 per cent of the planned AIJ projects had been implemented. However, the number of projects implemented steadily increased over time despite the fact that no credits accrued. Hence, it is reasonable to assume that Annex I country governments as well as the private sector that invested in these projects were eager to gain experience with the concept of credit-based trading. It should not go unnoticed that in 41 potential host countries projects have been planned and that those countries in which they have been actually implemented accumulated know-how concerning credit-based trading.

Generally, the development of renewable energy is hindered by relatively high costs and other significant barriers (Wohlgemuth and Missfeldt, 2000; Sathaye et al, 2001; Moomaw et al, 2001). Despite this fact, renewable energy projects contributed a considerable proportion of the total number of AIJ projects planned. This proportion peaked in 1997 at 42.6 per cent and decreased to 34.0 per cent in 2001 (see Table 9.2).

On the basis of these data it is legitimate to assume that renewable energy projects will also play a lead role in CDM and JI in terms of number of projects. However, one clearly sees that when considering the total emission reductions the average size of renewable energy projects was much smaller than the size of other project categories.

CDM and JI market size and prices: initial euphoria, hangover and growing optimism

Before the US pullout from the Kyoto Protocol, estimates of Annex B countries' annual demand for emission reductions through the Kyoto Mechanisms during the period 2008–2012 ranged between 1.7 and 5 billion tonnes (Austin et al,

Table 9.2 *Project types as share of number of planned projects and of total emissions reduction (in parentheses)*

	1995	1996	1997	1998	1999	2000	2001
Forest protection and reforestation	30 (84.5)	25.0 (68.4)	13.4 (58.5)	11.6 (52.2)	9.8 (64.8)	9.3 (38)	8.5 (33.3)
Afforestation	10 (1.3)	6.3 (0.2)	1.6 (0.2)	1.1 (0.2)	1.6 (0.1)	1.4 (0)	2.6 (1.1)
Agriculture	0	0	1.6 (<0.1)	2.1 (1.8)	1.6 (1.4)	1.4 (1)	1.3 (0.7)
Fuel switch	20 (10.7)	12.5 (2.2)	3.2 (1.7)	3.2 (1.8)	5.7 (1.7)	6.4 (2)	6.6 (2.0)
Methane capture	0	6.3 (27.0)	3.2 (21.3)	2.1 (18.6)	3.3 (14.4)	5.0 (30)	5.9 (42.9)
Energy efficiency	30 (2.5)	25.0 (1.7)	34.4 (2.6)	37.9 (4.8)	40.1 (3.5)	41.4 (12)	40.5 (11.3)
Renewable energy	10 (1.0)	25.0 (0.5)	42.6 (15.6)	42.1 (20.5)	37.7 (13.9)	35.0 (17)	34.0 (8.7)

Source: Michaelowa (2002)

1998; Figueres, 1998). These estimates saw a CDM market share between 19 per cent and 57 per cent. Prices were estimated at \$3.5–11/tonne CO_2 and annual financial flows from \$2.8–17.4 billion, i.e. 6–36 per cent of current Official Development Assistance and 1–7 per cent of current Foreign Direct Investment. After 2001, estimates became much more sanguine. Michaelowa and Jotzo (2005) estimate an annual demand of just 1.1 billion tonnes, of which the CDM could capture 33 per cent and JI only 5 per cent. The price would amount to €3.8/tonne. However, over a quarter of this would go towards transaction costs and taxes, leaving just around €2.5/tonne CO_2 for implementation of the project. Annual average revenue from CDM projects would be around €0.7 billion until 2012. Energy sector projects account for 57 per cent of total CERs generated; the remainder come from sinks, gas flaring and landfill gas projects. The share of CERs from renewable energy projects has not been modelled so far.

While originally it had been envisaged that private companies from industrialized countries would be the primary investors in CDM and JI projects, the demand side of the market is currently dominated by programmes that purchase CERs and ERUs run by government or financial institutions:

- The World Bank Carbon Finance unit started demand for CERs and ERUs with its Prototype Carbon Fund in 1999. It now hosts an array of multilateral and bilateral vehicles. The Community Development Carbon Fund and Bio Carbon Fund belong to the former, the IFC-Netherlands Carbon Facility, the Netherlands Clean Development Facility and the Italian and Spanish Carbon Fund to the latter.
- The Dutch government was the first industrialized country government to develop a portfolio of approaches to acquire CERs and ERUs to hedge risks.

From 2001 onwards it organized the CERUPT and ERUPT tenders. It also buys CERs through pipelines generated by banks (Rabobank) and bilateral contracts with governments (Indonesia).

- The Austrian, Belgian, Danish, Finnish and Swedish governments are buying CERs through tender programmes.
- Since late 2003, financial institutions have come up with funds of surprisingly large volumes. The front runner was the Japan Carbon Fund, and a dozen other funds have been added since. It is not always clear whether the reported volume is really available.

The current market size amounts to about €2.1 billion for both JI and CDM (see Figure 9.2). However, several EU governments need to commit much more to cover their Kyoto gap.

Prices for CERs and ERUs have differentiated over time and strongly depend on the project risks. The Prototype Carbon Fund (PCF) initially offered €2.5–3, while the Community Development Carbon Fund (smaller projects with high development benefits) quoted €4–7. Recent World Bank transactions have been made at prices of around €4. In the Dutch ERUPT/CERUPT, renewables go up to €5.5, bioenergy and energy efficiency up to €4.4 and other project types only up to €3.3. In 2003, Japanese buyers offered up to €12 for renewable energy CERs from South-East Asia; in 2005 a number of EU buyers spent up to €15 for CERs when the price for EU allowances reached up to €30.

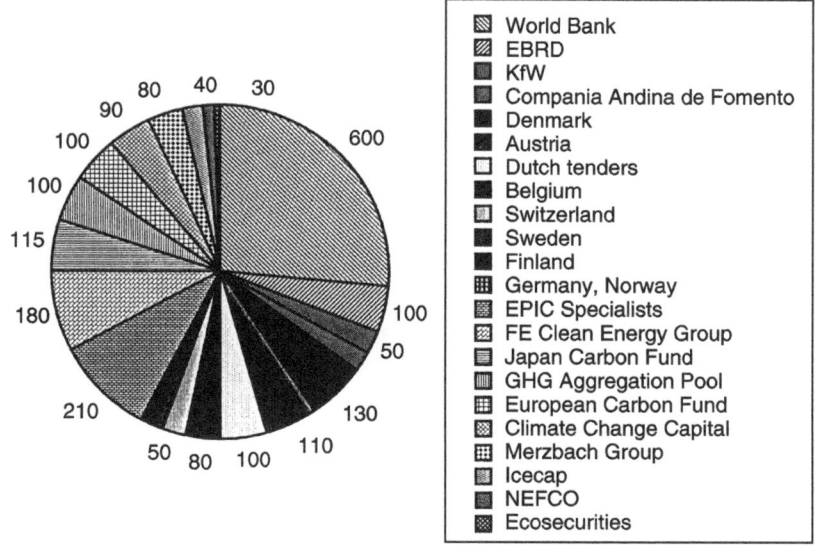

Figure 9.2 *Current market size until 2012 (€ million)*

Supply of CERs and ERUs: renewable energy projects currently in the pipeline

As at 2 September 2005, 19 CDM projects have been registered and 209 have invited public comments, of which 130 are renewable energy projects, i.e. 57 per cent of the total. All project documentations submitted to the EB are available on the UNFCCC website. While for JI there is no central documentation, certifiers publish project documentation for public comments. Moreover, some of the above-mentioned programmes are very transparent and provide detailed project documentation (the World Bank, the Dutch Government).

By 2 September 2005, 180 project documents for renewable energy CDM and JI projects were publicly available. An analysis gives the picture seen in Table 9.3. Generally, it can be seen that the CDM is the preferred mechanism for project developers, as less than a quarter of the projects are JI.

It can be observed that hydro and wind projects are attracting most attention under both JI and CDM, followed by biomass projects. Geothermal projects are rare but large. These projects achieve on average significantly higher annual emission reductions (0.34 megatonnes (Mt)) than the other renewable energy project types, among which the range is 0.07–0.12 Mt.

The share of the wind JI projects in total wind capacity in the host countries is remarkable; essentially all wind projects there are done under JI. Likewise,

Table 9.3 *Shares of different renewable energy technologies in proposed CDM and JI projects*

Project type	No. of projects	Capacity (MW)	Expected CERs/ERUs per year (Mt)	Average CERs/ ERUs per project (Mt)	Share of capacity in respective region[a] (%)
Biomass CDM	60	1135	2.98	0.05	8
Biomass JI	12	60	1.51	0.13	n.a.
Geothermal CDM	2	120	0.28	0.14	3
Geothermal JI	1	5	0.01	0.01	n.a.
Hydro CDM	50	900	1.90	0.04	0.2
Hydro JI	22	170	0.68	0.03	n.a.
Solar CDM	1	0	<0.1	<0.1	n.a.
Wind CDM	17	1050	1.93	0.11	45
Wind JI	15	505	1.44	0.05	935
Total CDM	130	3205	7.09	0.05	0.9
Total JI	50	740	3.64	0.07	n.a.

Note:
a   For JI, only countries in transition are counted. Data for biomass, geothermal and hydro are for 2000 (Martinot et al, 2002), for wind for 2002 (AWEA and EWEA, 2003).

Sources: Websites of UNFCCC, World Bank Carbon Finance, SENTER, Det Norske Veritas, KPMG, SGS, TÜV Süd; own calculations.

CDM wind is making up a significant share of the original wind capacity of developing countries; the picture would, however, be less impressive if capacity growth since 2002 were to be taken into account. Biomass CDM is starting to make up an appreciable share of total biomass energy in developing countries, but is still not dominant. As regards hydro, the impact is still minimal.

Figure 9.3 provides an overview of the share of the renewable energy technologies according to their expected amount of CERs generated each year.

Box 9.2 provides an analysis of the first registered CDM hydro project.

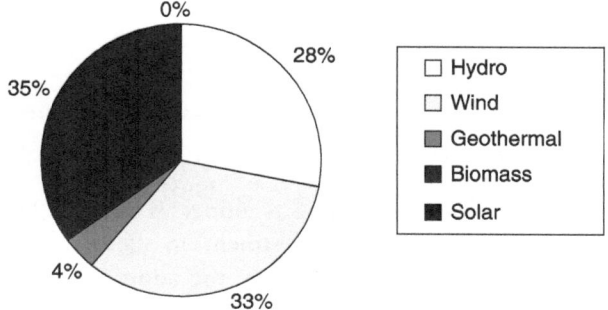

Figure 9.3 *Share of renewable energy technologies in the overall Certified Emission Reduction volume projected from CDM and JI projects*

Box 9.2 The Rio Blanco small hydropower project in Honduras

Rio Blanco was the first renewable energy project and the second overall to be registered under the Clean Development Mechanism (CDM). Its host country, Honduras, aims to reduce its dependence on oil imports. The run-of-the-river hydro project is a decentralized, privately owned, small-scale project activity according to the CDM framework and also according to international hydro power standards. It is projected that 32 GWh/year and thereby 17,800 CERs will be generated. The extra revenues helped to overcome significant financial barriers. Construction costs of US$8 million could only be financed by a loan by the Central American Bank of Economic Integration (CABEI) under relatively bad conditions with an interest rate of 12 per cent. Through the CERs, IRR is increased by about 1 percentage point, which made it possible to take up the loan. Moreover, Rio Blanco uses revenues from CERs to finance reforestation activities in the project area. Five million trees – a combination of timber and fruit trees – are planted in cooperation with local communities and thereby create another source of sustainable income for the future.

REVENUE FROM CDM AND JI FOR RENEWABLE
ENERGY PROJECTS

The internal rate of return is most commonly used to determine the viability of any investment. It is usually calculated on the basis of a cash flow analysis which rests on a considerable number of parameters that are project specific. This chapter cannot cover all of those parameters comprehensively and therefore focuses on the following questions. First, what factors does the additional revenue from CDM and JI projects depend on? Second, what does this mean for renewable energy projects under CDM and JI in particular? Finally, estimates for the magnitude of additional revenue from the project-based mechanisms are presented.

Factors on which the revenue from CDM and JI depends: implications for renewable energy projects

Investment in renewable energy projects under JI and CDM is physically inseparable from the conventional investment in renewable energy projects (Laurikka and Springer, 2003). By contrast, the additional revenue generated from the 'GHG mitigation component' of CDM or JI projects can be distinguished from the conventional income stream of the project. The total revenue from the 'GHG mitigation component' depends on the revenue from the sale of CERs and the transaction costs incurred by the project developer (see Box 9.3).

It can be seen that the revenue from the 'GHG mitigation component' of any project type will depend on a variety of factors. $q_{credits}$ will be very project specific. The baseline emissions, for example, depend on the baseline option chosen, as well as on the project type, as illustrated in Box 9.4. They will also depend on the host country. $p_{credits}$ will depend on the price the buyer is prepared to pay for the CERs.

Specific TAC are mainly determined by the amount of CERs generated, as most of the transaction costs are more or less fixed (Michaelowa et al, 2003). Table 9.4 shows the effect on projects of different sizes. At current market prices, all projects below 5000–10,000 CERs per year are not viable, unless transaction costs are subsidized. There is indeed a tendency to subsidize development of Project Design Documentand validation though public money.

Generally, the above findings apply to any type of GHG mitigation project, renewable or not. Apart from large hydro power installations, no renewable technology has the potential to generate such a high amount of CERs that transaction costs become negligible.

However, renewable energy projects can potentially achieve higher CER prices than other technologies. They come with a number of sustainable development

Box 9.3 Revenue from sale of credits and influencing factors

Specific revenue 'GHG mitigation component' (e.g. €/kWh or €/tonne CO_2) = specific $R_{credits}$ – specific TAC

where $R_{credits}$ is the revenue from the sale of credits/value of credits and TAC are the transaction costs that accrue from the 'project cycle' and from potential market transactions (e.g. finding a buyer for the credits).

$R_{credits} = q_{credits} \times p_{credits}$

where $q_{credits}$ is the total amount of credits generated until the end of the (last) crediting period and $p_{credits}$ is the price of each credit/value of each credit.

$q_{credits}$ depends on:

- baseline emissions;
- project emissions;
- crediting period.

TAC depend on:

- project complexity;
- host country;
- maturity of the GHG market.

Table 9.4 *CDM project size, types and indicative specific transaction costs*

Size	Type	CERs (t CO_2/year)	Transaction costs (€/t CO_2)
Very large	Large hydro, industrial gases, gas power plants, large combined heat and power (CHP) plants, geothermal, landfill/pipeline methane capture, cement plant efficiency, large-scale afforestation	>200,000	0.1
Large	Wind power, solar thermal, energy efficiency in large industry	20,000–200,000	1
Small	Boiler conversion, demand side management, small hydro	2000–20,000	10
Mini	Energy efficiency in housing and small and medium enterprises, mini hydro	200–2000	100
Micro	Photovoltaics	<200	1,000

Source: Michaelowa et al (2003)

Box 9.4 CALCULATING THE BASELINE FOR
GRID-CONNECTED RENEWABLE ELECTRICITY PROJECTS

Baseline rules have been defined by the CDM Executive Board for renewable energy projects below 15 MW as well as general renewable electricity for the grid (consolidated methodology). The baseline here is set as follows:

The average of the 'operating margin' and the 'build margin' is taken, where: (i) the operating margin is the weighted average emissions (in kg CO_2/kWh) of all generating sources serving the system, excluding hydro, geothermal, wind, low-cost biomass, nuclear and solar generation; (ii) the build margin is the weighted average emissions (in kg CO_2/kWh) of recent capacity additions to the system, defined by the most recent 20 per cent of plants built or the five most recent plants (whichever has had a higher generation in the year with the most current data). For small projects, the weighted average emissions (in kg CO_2/kWh) of the current generation mix can also be used.

A numerical example:

Your hydro plant of 10 MW generates 70 GWh per year. The grid it serves has the following characteristics:

- 5000 MW hydro, generating 35 TWh per year;
- 10,000 MW coal, generating 70 TWh per year with an emissions factor of 1.1 kg CO_2/kWh;
- 3000 MW gas, generating 15 TWh per year with an emissions factor of 0.5 kg CO_2/kWh;
- 2000 MW oil, generating 6 TWh per year with an emissions factor of 0.8 kg CO_2/kWh.

The last 4000 MW built has the following characteristics:

- 1000 MW hydro, generating 7 TWh per year;
- 2000 MW coal, generating 14 TWh per year with an emissions factor of 0.9 kg CO_2/kWh;
- 1000 MW gas, generating 6 TWh per year with an emissions factor of 0.4 kg CO_2/kWh.

The baseline is calculated as follows:

The operating margin is $\dfrac{70 \cdot 1.1 + 15 \cdot 0.5 + 6 \cdot 0.8}{91} = 0.981$ kg CO_2/kWh

The build margin is $\dfrac{7 \cdot 0 + 14 \cdot 0.9 + 6 \cdot 0.4}{27} = 0.556$ kg CO_2/kWh

The average of the two is 0.769 kg CO_2/kWh.

The small-scale option gives:

$\dfrac{35 \cdot 0 + 70 \cdot 1.1 + 15 \cdot 0.5 + 6 \cdot 0.8}{126} = 0.709$ kg CO_2/kWh

Baseline emissions are 70 GWh × 769 tonnes CO_2/GWh = 53,830 tonnes CO_2

In many countries, collection of these data will involve a substantial effort.

benefits compared to conventional energy technologies. Evidence such as from prices paid in CERUPT and Community Development Carbon Fund and individual transactions suggests that such projects achieve a premium as a result of their social and environmental benefits (see earlier in the chapter, and Springer and Varilek, 2004). Another chance to increase the CER price of a renewable energy CDM project is to validate it under the 'Gold Standard' of the NGO community (WWF, 2003). The Gold Standard was unveiled in July 2003 and is intended to define best-practice CDM projects. It excludes fossil fuel generation, efficiency improvement and fuel switch projects and thus promotes renewable energy with the exception of large hydro.

Estimates for additional revenue for renewable energy projects from CDM and JI

As mentioned, the revenue from CDM and JI is project specific. Table 9.5 provides an overview on either the additional revenue generated by the sale of credits or the impact of the additional revenue on the IRR. The projects listed in Table 9.5 are either case or desk studies. CER prices assumed usually range from €1 to €10/tonne CO_2.

It can be seen that the sale of credits can lead to an increase in IRR by 1–2.4 per cent and additional revenue in the order of magnitude from 0.2 to 0.5 ct/kWh. As the studies were mostly based on prices of €3/CER, the recent price increase would mean a tripling of the revenue per kilowatt-hour to about 1 ct/kWh; in cases of extremely favourable baselines (1500 g CO_2/kWh) and premium CER prices (€15/tonne CO_2), up to 2.2 ct/kWh could be reached. However, it should be highlighted that the figures do not factor in transaction costs.

Table 9.5 *Additional revenue from CDM and JI*

Project	IRR (%)	IRR with credits (%)	Additional revenue (ct/kWh)
Wind farm (Brazil)	6.7	7.5–8.5	0.20
Wind farm (Morocco)	11.3	13.6–17.9	0.25
Wind (desk study)	–	+1	0.25
Small hydro (Uzbekistan)	11	11.2–3.8	–
Small hydro (Uzbekistan)	>12	>12.4–14.8	–
15 MW hydro (desk study)	–	–	0.45
PV (Brazil)	8.4	8.7–10.2	–
PV (desk study)	–	–	0.50
Biomass (Zimbabwe)	18.3	18.4–21.7	0.40

Source: Langrock et al (2003); Bode and Michaelowa (2003); Michaelowa et al (2003)

NECESSARY CONDITIONS TO GENERATE NEW AND ADDITIONAL FUNDS FOR RENEWABLE ENERGY PROJECTS THROUGH CDM AND JI

Domestic instruments in Annex B countries

Without incentives, there will not be any investment in the Kyoto Mechanisms and thus also no promotion of renewable energy projects through them. A necessary condition for private involvement in CDM and JI is the link with climate policy instruments in their home country (Michaelowa, 2004). These instruments can take the form of emissions trading, emission taxes, subsidies or regulation. In a system of domestic emissions trading, CERs or ERUs could be imported and used. Tax concessions could be granted and should be proportional to the emission credits achieved by a CDM/JI project. It would in principle also be possible to subsidize emission reduction projects at home and abroad in general. While the European Union in principle allows use of CERs/ERUs in its emissions trading system, other countries still have to follow suit.

Project additionality determination

In the case of the CDM no cap exists for the host country, and everybody would profit from an overestimate of reductions: the investor who gets more CERs and the host who can sell more. The problem is that there are many emission reduction opportunities that are profitable for a company and thus will be implemented anyway. Under the CDM, they would not be 'additional'.

Since the Marrakech Accords did not specify a specific additionality test, most project developers and analysts believed that any project could pass, provided its greenhouse gas emissions were lower than average emissions in the country for production of the same product. They were shocked when in April 2003 the CDM Executive Board rejected eight proposed baseline methodologies outright and returned another six for revision (for a detailed analysis see Jotzo, 2003). The EB cited lack of additionality tests as one of the main reasons for this result. In October 2004 the CDM Executive Board published a consolidated additionality test with the following steps:

- Projects that started before registration: proof of CDM influencing the decision;
- Identification of alternatives to the project;
- Investment analysis to determine that the project is not the most economically or financially attractive; or

- Proof of prohibitive barriers. Unfortunately, the EB did not specify a list of 'accepted' barriers, which it did in the case of small scale projects (UNFCCC, 2003);
- Common practice analysis – that such projects are not normally carried out in the proposed area of implementation;
- Impact of CDM registration on overcoming barriers.

For JI projects of the first track, additionality determination is not mandatory, but host country governments will make sure that the projects are additional. Every non-additional ERU sold has to be made up with reductions in other parts of the host country economy, and that will cost money.

Capacity-building plays a key role

The issue of who pays for capacity- and institution-building in the CDM and JI context deserves attention. It has rightly been feared (Srivastava and Soni, 1998) that ODA could be diverted to such uses, as currently many industrialized countries fund capacity-building up to the development of PDDs. All large CDM procurement programmes have a capacity-building component, and the UN organizations are competing against each other to see which can offer more capacity-building activities. The importance of information, training, appropriate capacity and focal institutions for the development and approval of CDM projects cannot be overestimated (Michaelowa, 2004).

CONCLUSIONS

The project-based Kyoto Mechanisms can improve financing of renewable energy projects but will not provide a panacea for large-scale renewables promotion. However, an increase in the market price for greenhouse gas reduction credits means that the incentive per kilowatt-hour can reach around 1 ct/kWh under normal baselines. Nevertheless, CDM and JI currently only promote renewable energy technologies whose costs are not much above those of fossil fuel technologies. Thus, wind and hydro make up a large share of current CDM and JI projects. The Kyoto Mechanisms will definitely not be a vehicle to promote photovoltaics. In the best locations for wind, hydro and biomass, problems with additionality determination may arise as the renewables projects would have gone ahead even without the CDM revenues. For project developers, the lengthy CDM project cycle will generate transaction costs that make CDM projects viable only if they generate more than 5000–10,000 CERs.

REFERENCES AND FURTHER READING

Austin, D., Faeth, P., Seroa de Motta, R., Ferraz, C., Young, C., Zou Ji, Li Junfeng, Pathak, M., Srivastava, L. and Sharma, S. (1999) 'How much sustainable development can we expect from the Clean Development Mechanism?', Climate Note, Washington, DC, World Resources Institute

American Wind Energy Association and European Wind Energy Association (AWEA and EWEA) (2003) *Record Growth for Global Wind Power in 2002*, Washington, DC

Baumert, K. (1999) 'Understanding Additionality', in Goldemberg, J. and Reid, W. (eds) *Promoting Development while Limiting Greenhouse Gas Emissions: Trends and Baselines*, World Resources Institute, New York

Beuermann, C., Langrock, T. and Ott, H. (2000) 'Evaluation of (Non-sink) AIJ Projects in Developing Countries', Wuppertal Papers 100, Wuppertal, Wuppertal Institute

Bode, S. and Michaelowa, A. (2003) 'Avoiding Perverse Effects of Baseline and Investment Additionality Determination in the Case of Renewable Energy Projects', *Energy Policy*, vol 31, pp505–517

Chatterjee, K. (ed) (1997) *Activities Implemented Jointly to Mitigate Climate Change – Developing Country Perspectives*, New Delhi, Development Alternatives

Dixon, R. (ed) (1999) *The U.N. Framework Convention on Climate Change Activities Implemented Jointly (AIJ) Pilot: Experiences and Lessons Learned*, Dordrecht, Kluwer

Figueres, C. (1998) 'How Many Tons? Potential Flows through the Clean Development Mechanism', in World Resources Institute, Foundation for International Environmental Law and Development and Center for Sustainable Development in the Americas (eds) *The Clean Development Mechanism*, World Resources Institute, Washington, DC

Greiner, S. and Michaelowa, A. (2003) 'Defining Investment Additionality for CDM Projects – Practical Approaches', *Energy Policy*, vol 31, pp1007–1015

Grubb, M., Vrolijk, C. and Brack, D. (1999) *The Kyoto Protocol*, Eigenverlag, London

Heller, T. (1998) 'Additionality, Transactional Barriers and the Political Economy of Climate Change', Venice, Fondazione Eni Enrico Mattei Nota di lavoro 10.98

Intergovernmental Panel on Climate Change (IPCC) (2001a) *Climate Change 2001 – The Scientific Basis*, Cambridge, Cambridge University Press

IPCC (2001b) *Climate Change 2001 – Impacts, Adaptation, and Vulnerability*, Cambridge, Cambridge University Press

Jotzo, F. (2003) 'CDM and the Additionality Question', *AETF Forum*, August/September, pp4–5

Jotzo, F. and Michaelowa, A. (2002) 'Estimating the CDM Market under the Marrakech Accords', *Climate Policy*, vol 2, pp179–196

Langrock, T., Michaelowa, A. and Greiner, S. (2003) 'Defining Investment Additionality for CDM Projects – Practical Approaches', HWWA Discussion Paper 106, Hamburg

Laurikka, H. and Springer, U. (2003) 'Risk and Return of Project-based Climate Change Mitigation: A Portfolio Approach', *Global Environmental Change*, vol 13, pp207–217

Martinot, E., Chaurey, A., Lew, D., Moreira, J. and Wamukonya, N. (2002) 'Renewable Energy Markets in Developing Countries', *Annual Review of Energy and the Environment*, vol 27, pp309–348

Michaelowa, A. (2002) 'The AIJ Pilot Phase as Laboratory for CDM and JI', *International Journal of Global Environmental Issues*, vol 2, pp267–280

Michaelowa, A. (2004) 'CDM Incentives in Industrialized Countries – The Long and Winding Road', *International Review for Environmental Strategies*, vol 5, pp217–231

Michaelowa, A. and Jotzo, F. (2005) 'Transaction Costs, Institutional Rigidities and the Size of the Clean Development Mechanism', *Energy Policy*, vol 33, pp511–523

Michaelowa, A., Stronzik, M., Eckermann, F. and Hunt, A. (2003) 'Transaction Costs of the Kyoto Mechanisms', *Climate Policy*, vol 3, pp261–278

Moomaw, W. R. et al (2001) 'Technological and Economic Potential of Greenhouse Gas Emissions Reduction', in IPCC, *Climate Change 2001: Mitigation*, Cambridge, Cambridge University Press

Organisation for Economic Co-operation and Development (OECD) (2000) *Emission Baselines: Estimating the Unknown*, Paris, OECD

Rentz, H. (1998) 'Joint Implementation and the Question of Additionality – A Proposal for a Pragmatic Approach to Identify Possible Joint Implementation Projects', *Energy Policy*, vol 4, pp275–279

Sathaye, J, Bouille, D., Biswas, D., Crabbe, P., Geng, L., Hall, D., Imura, H., Jaffe, A., Michaelis, L., Peszko, G., Verbruggen, A., Worrell, E., Yamba, F., Tolmasquim, M. and Janzen, H.. (2001) 'Barriers, Opportunities, and Market Potential of Technologies and Practices', in IPCC, *Climate Change 2001: Mitigation*, pp345–398, Cambridge, Cambridge University Press

Springer, U. and Varilek, M. (2004) 'Estimating the Price of Tradable Permits for Greenhouse Gas Emissions', *Energy Policy*, vol 32, pp611–621

Srivastava, L. and Soni, P. (1998) 'Financing Options for Protecting the Climate', in Tata Energy Research Institute, *Climate Change: Post-Kyoto Perspectives from the South*, pp95–126, New Delhi

United Nations Framework Convention on Climate Change (UNFCCC) (2003) 'Indicative Simplified Baseline and Monitoring Methodologies for Selected Small-Scale CDM Project Activity Categories', Appendix B1 of the simplified modalities and procedures for small-scale CDM project activities, Annex 6, Report of the 7th meeting of the Executive Board, Bonn

Wohlgemuth, N. and Missfeldt, F. (2000) 'The Kyoto Mechanisms and the Prospects for Renewable Energy Technologies', *Solar Energy*, vol 69, pp305–314

WWF (2003) *The Gold Standard: Clean Development Mechanism Project Design Document*, Gland, Switzerland

PART FOUR

DEVELOPING COUNTRIES

10

Putting Development First:
The Role of Renewable
Energy in Achieving the
Millennium Development Goals

Susan McDade, Minoru Takada and Jem Porcaro

THE CASE FOR MILLENNIUM DEVELOPMENT
GOAL-DRIVEN NATIONAL ENERGY STRATEGIES
IN DEVELOPING COUNTRIES

Overcoming poverty, hunger, illiteracy, the marginalization of women, major disease as well as environmental degradation will require energy, and lots of it. This is the basic premise for why developing countries need significantly greater access to energy services.

The international community agreed in 2000, and has renewed its commitment in 2005, to work towards attaining a set of development goals, namely the Millennium Development Goals (MDGs), by 2015. If energy, in all its conventional and renewable forms, is to contribute towards development, national energy strategies ought to take a development-oriented approach to energy which prioritizes energy options on the basis of their contributions to the achievement of the MDGs.

Being development oriented requires an understanding of and an ability to respond to the needs of varying people. The poor, for example, are concerned with the availability and affordability of energy services – that is, the benefits that energy use offers. For households, priorities often vary by gender: women typically need fuels for cooking and mechanical power for agricultural/food processing and water pumping, whereas men prefer electricity for entertainment

(Krause and Nordstrom, 2004). At the community level the priority is having reliable electricity for health services, schools, public lighting systems and communications.

Articulating the energy needs of the poor – especially those of women – has posed a serious challenge to policy-makers. Until recently, development policies have emphasized projects designed to provide energy for growth in the formal sectors of the economy. Almost totally excluded from development policies are the activities of women in the informal and non-monetized sectors. These activities can be quite energy intensive; however, because they take place in diffuse and difficult-to-reach segments of the economy they tend to be overlooked by energy planners and government statisticians. When attention is not paid to the energy needs of the poor, and especially those women whose livelihoods are heavily dependent on access to energy, competing priorities often dictate where government and donor resources are spent.

The failure to align government and donor energy priorities along with those of the poor has led to disappointingly low levels of energy access, especially in rural areas, where the majority of the world's poor currently live. Worldwide, an estimated 2.4 billion people rely on traditional biomass fuels for cooking and heating, while at least 1.6 billion people do not have access to electricity in their homes (IEA, 2004). It is evident that more development-oriented and results-based energy policies are needed to support economic development, including poverty reduction as called for in the MDGs.

In recognition of this failure, this chapter attempts to answer the following question: *What roles should renewable energy play within an MDG-driven national energy strategy?* In trying to address this question, the chapter bases its arguments on the following two premises.

First, the most fundamental objective of any developing country's MDG-based national energy strategy should be to expand access to modern energy services to all its population. Without access to modern energy services, it is unlikely that any serious progress towards the MDGs can be made. Moving towards more sustainable energy systems is certainly important. However, it should not be forgotten that climbing up the 'energy ladder' to more sustainable energy systems is difficult where there is no energy access to begin with. Hence, in developing countries – particularly in the least developed countries – the top priority should be to ensure that the poor are first given *access* to energy services before they are asked to 'leapfrog' to more sustainable energy options.

Second, in order for energy to contribute meaningfully to the achievement of the MDGs, at least three areas of energy access need to be ensured: (1) energy for cooking; (2) mechanical power for agro/food processing, water pumping and other productive uses; and (3) electricity for public services (Modi et al, 2005).

Box 10.1 Millennium Development Goal-compatible energy services

Energy for cooking

Cooking is the major energy end-use among poor families in developing countries, accounting for 90–100 per cent of household energy consumption in most of these countries (WEC, 1999). Because most staple foods have to be processed, conserved and/or cooked, access to affordable, clean cooking fuels and equipment is among the most basic energy needs of the world's poor. Traditional biomass is the primary source of fuel for cooking in developing countries. This is especially true in rural areas of sub-Saharan Africa and Asia. In Asia, for example, over 80 per cent of the total rural population depend on fuelwood, dung and crop residues to satisfy their cooking needs (Karekezi et al, 2005). There are a number of ways by which those dependent upon biomass can benefit from switching to cleaner-burning cooking fuels. Fuel substitution helps reduce the drudgery involved in collecting fuelwood, thereby increasing the opportunities for women and children to engage in more productive activities. It also reduces harmful indoor air pollution and greenhouse gas emissions. The multitude of benefits to health, productivity, agriculture, the environment and women's welfare that modern cooking fuels provide suggests that a rapid scale-up of modern cooking fuels should be an important part of the strategy to meet the MDGs (Modi et al, 2005).

Mechanical power for agro/food processing, water pumping and other productive uses

Over the past century, technological advances have helped reduce the drudgery of human labour through the widespread use of mechanical power. Mechanical power is perhaps second only to cooking when it comes to the sorts of energy services poor people need most. Mechanical power is critical to enhancing the productive use of labour and to poverty alleviation. Enhanced productivity in agro/food processing, water pumping and irrigation boosts economic development and improves livelihoods. In areas where electricity grids are unable to reach, the availability of decentralized mechanical power is particularly important to increasing the social and economic opportunities of the poor. Ironically, despite the importance of this energy service, there exist few data on mechanical power in developing countries. This underscores the need to view energy by the services it provides and the development outcomes it supports, not merely by the source from which it is obtained.

Electricity for public services

Developing countries need electricity to power their emerging economies and serve growing populations. Currently, roughly two-thirds of people living in developing countries have access to electricity. In regions such as sub-Saharan Africa and South Asia the electrification rate is much lower: 24 per cent and 35 per cent, respectively (IEA, 2004). These regional figures, of course, hide significant differences between countries and between rural and urban areas. Electricity is important to support productive activities at the local level and at large. It supports the delivery of social services such as education and health care, and is essential to modern information and communication systems. Electricity is also indispensable for certain activities, such as lighting, refrigeration and the running of household appliances.

The suitability of any given resource, technology or system for providing MDG-compatible energy services depends on the particular service(s) in demand and on the underlying socio-economic and environmental conditions (Flavin and Aeck, 2005). Renewable energy technologies (RETs), like any energy technology, should be promoted on the basis of the services they provide and the development outcomes they support, including of course environmental benefits. Once viewed in this light, a wide range of renewable resources and technologies become available to support the MDGs. This should be of no surprise given the abundant amount of renewable energy potential around the world.

Achieving the MDGs will clearly require developing countries to work on all areas of energy access. Which energy services are given priority will depend on the context of the need and how each developing country views its development goals. In the same vein, policy options to promote renewable energy will need to be examined and prioritized on the basis of their relevance to the achievement of the MDGs. A good starting point in this regard is to examine each energy option on the basis of how well it supports the three major areas of energy access (i.e. MDG-compatible energy services).

THE POTENTIAL AND CONSTRAINTS OF RENEWABLES IN SUPPORTING MDG-COMPATIBLE ENERGY SERVICES

There are a variety of RETs capable of supporting MDG-compatible energy services. While the following is by no means an exhaustive list, it does highlight the potential and constraints of a few of the more mature and emerging RETs available to developing countries.

Energy for cooking

Bioenergy – that is, energy derived from wood, residues, waste or other plant matter – is one of the more promising forms of renewable energy capable of addressing the cooking needs of the poor. When biomass energy is 'modernized' – that is, when it is converted into higher-'quality' and more convenient forms of gases, liquids and electricity – it can be used far more efficiently than solid biomass, while also emitting far fewer toxic pollutants (UNDP, 2000).

There are a variety of technologies for converting biomass into fuels appropriate for use in cooking. Gasifiers and anaerobic digesters are a couple of the more common and commercially available types of technologies used in developing countries. The use of biomass gasifiers and anaerobic digesters has largely been confined to countries of Asia – namely China and India – where the cost of such technologies can be more easily afforded. There are just under 16 million household-scale anaerobic digesters being used in the two countries taken together (Flavin and Aeck, 2005).

Depending on the technology used, biomass can be converted into either producer gas (generated by high-temperature gasifiers) or biogas (generated by low-temperature anaerobic digesters). Although the two fuels are derived from distinctively different conversion processes, both can provide efficiency gains during cooking and substantial health benefits in terms of reduced indoor air pollution. That said, biogas does have the added advantage of being derived from a process that produces an effluent sludge that has a high fertilizer value.

Box 10.2 Case study: biogas digesters in Tanzania

Farming is the major rural activity in Tanzania, but it does not generate sufficient income. As in most African countries, fuelwood and charcoal are the primary sources of energy for Tanzania's rural population. On average each rural household spends ten hours per week searching for fuelwood. To address this problem, the Foundation for Sustainable Rural Development, in partnership with a number of international donors and supporting organizations, implemented a project to integrate livestock into farming systems and to promote the production of biogas energy. Biogas production integrated with cattle-raising and farming has provided more than 1000 households with a reliable source of cleaner cooking fuel as well as increased income and employment opportunities.

Source: UNDP (2003)

Another promising, albeit less mature, form of bioenergy is ethanol. Ethanol and ethanol-based fuels like gelfuel are produced through the fermentation and distillation of sugars or starch crops. Although virtually absent from most developing countries, gelfuels have shown promise in the few countries in which they have been used (Utria, 2004).

Modern bioenergy compares well with commercial cooking fuels like kerosene and liquefied petroleum gas (LPG). Biogas, for example, has been shown to compete with LPG in terms of life-cycle cost in a number of developing countries and is only fractionally more expensive than kerosene. Even less mature biofuels such as ethanol-based gelfuels have been shown to be less expensive than commercial fuels on occasion (Utria, 2004). Despite their relatively low life-cycle cost, modern bioenergy systems are constrained by their high up-front costs. If, however, their initial costs can be afforded, modern bioenergy systems do have the added advantage of being more readily available in rural areas, where supply and market barriers keep access to LPG and kerosene constrained.

Solar hot water (SHW) systems are another promising RET that could help provide cleaner and more convenient means of heating water. SHWs have been demonstrated to be economical in a number of developing countries, especially in those with warmer climates. It is estimated that roughly 10 million households worldwide have SWH systems (Flavin and Aeck, 2005). The systems can often be constructed using locally available materials and with just a small amount of expertise and skill.

Although much too rudimentary and unreliable a technology to meet the full range of the poor's cooking needs, solar cookers are worth mentioning if only to discuss their niche applications and limitations. Despite having been introduced in a number of developing countries – namely China and India – solar cookers have yet to take off. The total number of households with solar cookers is roughly 1 million (Flavin and Aeck, 2005). The main problem with the technology is that it can be effectively used only during hours of strong sunlight. In addition, solar cookers require new ways of cooking (i.e. cooking outdoors) that households may not be familiar with or comfortable adopting. More promising solar-based technologies include solar water pasteurizers that provide clean potable water and reduce water-borne diseases and solar dryers that lower post-harvest agricultural losses (Karekezi and Kithyoma, 2002).

Mechanical power for agro/food processing, water pumping and other productive uses

Micro-hydro is perhaps the most reliable and mature technology for providing motive/shaft power in remote rural areas, especially mountainous ones. Although the availability of data on micro-hydro is limited, it is thought that there are in the order of tens of thousands of micro-hydro plants operating in China, and significant numbers in countries such as Nepal, Sri Lanka, Pakistan, Vietnam and Peru (Khennas and Barnett, 2000).

Compared with most other RETs, micro-hydro systems have a high power capacity and thus have the advantage of being able to support multiple end uses, including high-load mechanical applications such as grinding, threshing and water pumping. These services are most often provided by the generation of electricity but can be provided by shaft power directly. Micro-hydro compares well with other renewable and decentralized energy options and exhibits the standard characteristics of all renewables, namely relatively high initial capital costs and relatively low operating cost (Khennas and Barnett, 2000).

The data also show that in certain situations micro-hydro compares quite well with fossil fuels such as diesel when it comes to providing decentralized mechanical and electrical power. In a survey of four villages in Peru, for example, micro-hydro was found to be a less expensive alternative to diesel in all four cases, even without taking into account environmental costs (Khennas and Barnett, 2000). The cost advantage that micro-hydro can have over diesel of course depends on the availability and cost of transporting diesel and the extent to which the price of diesel is subsidized.

Box 10.3 Case study: mechanical power for productive uses in Nepal

Roughly 2200 micro-hydro schemes have been installed in Nepal over the past few decades, benefiting approximately 146,000 rural households. The most common use of micro-hydro schemes is for agricultural processes. In the village of Harichaur, for example, micro-hydro power is being used to run a rice-milling plant. With the advent of their agro-processing plant, villagers can save up to two hours a day in milling. This time is now being used to participate in a local educational programme. It is estimated that about 70 per cent of the community (an increase of about 45 per cent) have become literate and the enrolment of female students has increased every year. The number of hours for which students study has also increased, by 1–1.5 hours daily.

Source: Khennas and Barnett (2000)

Small-scale non-electrical RETs, such as wind pumps, treadle pumps and ram pumps, are also worth mentioning as they have the technical potential for irrigation and water supply and thus the potential for enhancing food security through increased food production. Despite their enormous potential for increasing agricultural output and helping support small and medium-sized enterprises, wind pumps and other non-electrical pump technologies have yet to be fully exploited in most countries. In sub-Saharan Africa only a few countries, such as Namibia and South Africa, have much experience with wind pumps (Karekezi, 2002). In those that do have experience with wind pumps, the technology is mainly limited to large and medium-scale farms or other large-scale rural institutions that can afford their high capital cost.

Transport is another service derived from motive power and one that could benefit from the further development of renewable energy. Ethanol, methanol and biodiesel are three biofuels that hold promise for the future. With record high oil prices, biofuels are becoming a more attractive alternative to traditional gasoline and diesel fuels. Despite recent developments in biofuel commercialization – especially those taking place in Brazil – biofuels still remain an elusive and expensive choice of transportation fuel for the poor.

Electricity for public services

There are a number of RETs capable of supporting lighting, ICT and other social services for rural communities and urban and peri-urban areas. Micro-hydro systems, modern biomass systems, small wind electric turbines and solar PV systems are a few examples. In total, renewable energy accounts for nearly 20 per cent of electricity generation in Asia and Africa and almost 70 per cent in Latin America (IEA, 2004).

As previously mentioned, both micro-hydro and modern bioenergy systems are capable of producing electricity. As of 2000 over 50 million households and roughly 60,000 small enterprises were electrified by small-hydro village-scale and community mini-grids while roughly 10 million households were electrified by biogas (Martinot et al, 2002). Both technologies have the advantage of being independent of an intermittent power source, or at least one that is difficult to predict. Both are also capable of powering village-scale mini-grids. This makes both technologies attractive from a community perspective, as both can serve as a node from which schools, health clinics and other public facilities can draw power. One of the main challenges facing these technologies is maintenance. Because micro-hydro schemes are often located in remote and difficult-to-reach terrain, providing them with a steady stream of parts and skilled labour is often a challenge.

Small-scale wind turbines are another option for providing rural communities and urban and peri-urban areas with electricity. Unlike the trend towards large-scale grid-connected wind turbines seen in industrialized countries, the more pressing need for wind power in developing countries is for smaller decentralized and off-grid turbines. As with other intermittent energy sources, wind turbines require some form of energy storage – typically lead-acid batteries – to become a reliable source of electricity. Because of their high cost, many wind turbines in developing countries are coupled with less expensive technologies such as diesel generators in order to remain economical.

Box 10.4 CASE STUDY: POWERING CHILEAN COMMUNITIES WITH WIND–DIESEL HYBRID POWER

Inhabitants of Chile's island of Isla Tac are benefiting from a 15-kilowatt wind–diesel hybrid power system. The system is installed on a low hill in close proximity to the island's main community. The turbines provide electricity to households and help power a school and community centre. Since its arrival, islanders have invested heavily in domestic equipment to dramatically improve their quality of life. Before the turbines had been installed, islanders were paying the equivalent of approximately 2500 pesos per kilowatt-hour for small batteries, candles, small engine generators and kerosene lanterns. With the installation of the turbines, villagers now pay between 210 and 650 pesos per kilowatt-hour.

Source: Stevens (2001)

Solar photovoltaics (PV) are emerging as one of the most popular RETs in developing countries. As of 2000, an estimated 1.1 million solar home systems (SHS) and solar lanterns had been installed in developing countries (Martinot et al, 2002). PV has grown rapidly since 2000 in countries such as Bangladesh, Brazil, China, India, Kenya, Sri Lanka and Thailand (Flavin and Aeck, 2005). One of the driving forces behind PV's growth has been a dramatic decrease in production costs.

PV technology has proved successful in certain high-tech applications of communication. It is also an ideal alternative for powering vaccine refrigeration. Despite its success in these and other public service-related areas, solar PV technology has had limited success at delivering development benefits at the rural household level (Karekezi and Kithyoma, 2002). This is partly because the most common type of solar technology promoted at the rural level, the SHS, neither reduces poor people's reliance on biomass for cooking nor has the capacity to meet the high-load mechanical applications required for agricultural/food processing – two areas of energy access that are particularly important to poor people. Another reason SHSs have shown more promise in the public arena is that there is generally more capacity to maintain the systems at a community level than there is at the household level.

A PROCESS FOR TAILORING RENEWABLE ENERGY POLICIES TO NATIONAL DEVELOPMENT PRIORITIES

As a matter of principle, policy discussions should start with putting all energy options on the table in order to assess the relative pros and cons of each option. The conventional approach to renewables in developing countries and economies in transition has relied on narrowly defined energy objectives and donor-driven agendas for promoting the expanded use of renewables rather than promoting renewables within a broader and more inclusive development framework. What is needed is a policy agenda that moves beyond a singular focus on technologies to a more diverse and context-specific approach that emphasizes the goal of sustainable human development as the key driver for promoting renewables. To ensure a systematic and an unbiased approach towards expanding renewables, it is recommended that countries first contextualize the need for renewables, then identify which renewable options are available to support national development priorities, and finally prioritize areas of public support needed to expand their use.

Determine the degree to which renewables can support national development priorities

Renewables are a necessary component of any sustainable energy strategy. However, the role of renewable energy vis-à-vis other energy sources should be viewed within the context of a country's overall development strategy. Development challenges can differ significantly across countries, as can the capacities and resources available to address them. The heterogeneity of countries therefore warrants a differentiated approach to energy policy. Although policies can and should be bifurcated along a number of different lines, the most obvious starting point is one that distinguishes less developed countries (LDCs) from wealthier, newly industrialized market economies and economies in transition (EITs). This bifurcation, although stylized, provides a convenient starting point for evaluating the role of renewables in a country's overall development strategy.

For the world's LDCs, national development strategies often place a strong emphasis on macroeconomic growth, poverty reduction and social equity. In order to advance such goals, LDCs should first assess the relative merits of renewables over other energy options in helping support social and economic development. If renewables are found not to be the best option for delivering the specific energy services needed to support development, LDCs should prioritize other energy options accordingly.

Taking the above approach does not mean that environmental goals should be ignored. On the contrary, it just means that the pace at which environmental goals are pursued in LDCs could be different – more long-term in nature – from those of more economically well-off countries. This suggests that energy options need to be aligned to the full spectrum of priorities found in LDCs, not just those related to an environmental agenda. Irrespective of the particular development goal an LDC chooses to focus on, it is critical that renewables be viewed as an instrument to achieve the goal, rather than a goal itself. Making this distinction is important, as many policy-makers in the environmental arena view the 'maximization' of renewable energy as a goal in itself (Reddy, 2002). Proponents of this view often fail to see the utility of balancing global environmental concerns with issues of equity and development. This is particularly important for LDCs, as they have the dual challenges of tackling extreme poverty and an environmental problem – namely, climate change – that has, ironically, arisen from energy usage primarily in developed countries (Reddy, 2002).

Newly industrialized market economies and EITs may view renewables slightly differently as compared with LDCs. For middle-income countries with highly energy-intensive and rapidly growing economies, meeting future energy needs may not be as easily decoupled from the environmental impacts associated with increased energy use. The impetus for increased use of renewables in middle-income countries comes from the need to improve energy efficiency, enhance energy security and move towards a more sustainable energy future. This opens the door to using renewables in a number of areas. Countries such as China, for example, face enormous environmental challenges as a result of their growing demand for commercial fuels, dismally low efficiency and their largely coal-dominated resource base. Fortunately, for wealthier countries like China, taking a more sustainable approach to energy use is something they can begin to afford to do. For EITs, the impetus for expanding the use of renewables may include, in addition to environmental benefits, the economic benefits of implementing greenhouse gas reduction projects through the Kyoto Protocol's Joint Implementation mechanism.

Identify specific renewable options

Once the role of renewables is properly contextualized within a country's broader development goals, the question of what specific renewable energy options countries should prioritize arises. The answer once again depends on the context of the need for renewables.

For example, if an LDC prioritizes the provision of mechanical power to help increase income generation, then an assessment should be made to determine which renewable technologies/resources are best suited to provide motive power. If on the other hand energy for cooking is the priority, then the assessment might focus on finding clean renewable cooking fuels. In either case, options do exist. Studies from sub-Saharan Africa show that renewables such small hydro plants and wind turbines can provide sufficient mechanical power to boost the poor's income (Karekezi, 2002, p8; Karekezi and Kithyoma, 2002, p1076). Others point out that there are a number of promising biomass-based options that are capable of addressing the cooking needs of women (Utria, 2004, p110; Reddy, 2002, p77). These options have the potential of yielding substantial near-term benefits in terms of income generation and productivity, and helping reduce the negative environmental and health impacts of traditional energy use.

The renewable priorities of newly industrialized market economies and EITs can differ from those of LDCs. For some, the priority is to expand access to electricity in an environmentally conscious manner. In this case, countries could take an aggressive approach to renewable energy-based power generation, especially in areas such as wind power and solar PV. For others, the priority may be finding cleaner and more affordable fuels for transport, in which case an assessment of renewable options might point to biomass-derived fuels as an appropriate alternative to gasoline and diesel.

Prioritize areas of public-sector support

Once renewable energy priorities are established, the next question that arises is where public resources should be allocated among the policy options available to support renewable energy. Conventional wisdom would suggest that resources are needed in every policy area. This is probably true for most developing countries. However, owing to the limited budgets of developing countries it is the prioritization of policies, not merely policies themselves, that is needed.

What is required, therefore, is a process by which governments prioritize the areas of public support required to advance their various renewable energy options. Undertaking such a process requires an understanding of the specific barriers that impede renewable energy development. As before, the issues requiring attention (i.e. the barriers) will vary depending on the country in question, as will the areas of public support needed to overcome the barriers.

For example, in some LDCs the most important policy for promoting renewables may be the expansion of new and flexible financing tools. For others it may to ensure that commercial banks and other large-scale financial intermediaries are better acquainted with RETs. For newly industrialized and transition economies, the priority may be different altogether. Many middle-income countries view the technological development and market penetration of renewable energy as a key area requiring public support. Admittedly, this is just a small sample of policies developing countries ought to consider when crafting a policy agenda to support renewable energy. Other policies, such as those aimed at energy sector reform, R&D and capacity development, are all important areas for consideration.

It is clear that in order to achieve the MDGs, developing countries will require greater access to energy. Renewables certainly have a role to play, alongside fossil fuels, in helping countries secure affordable and reliable energy services for the poor. How large a role they will play is of less certainty. This will partly depend on the financial, economic, policy and environmental choices governments make over the course of the next decade. Ultimately, however, it will be the extent to which renewables provide the services or benefits that people, rich and poor alike, need that will determine their use in the future.

REFERENCES AND FURTHER READING

Flavin, C. and Aeck, M. (2005) *Energy for Development: The Potential Contribution of Renewable Energy to Meeting the Millennium Development Goals*, Washington, DC, Worldwatch Institute

IEA (2004) *World Energy Outlook 2004*, Paris, OECD/IEA

Intergovernmental Panel on Climate Change (IPCC) (2000) *Methodological and Technological Issues in Technology Transfer*, Cambridge, Cambridge University Press

International Energy Agency (IEA) (2003) *Economies in Transition, the IEA and Renewable Energy*, Background Paper, Budapest, IEA/Renewable Energy Unit

Karekezi, S. (2002) *Renewables in Africa: Meeting the Energy Needs of the Poor*, Nairobi, African Energy Policy Research Network/Foundation for Woodstove Dissemination

Karekezi, S., Khennas, S., Natu, S., Rakos, C. and Wangeci, J. (2005) *Status of Biomass Energy in Developing Countries and Prospects for International Collaboration*, Nairobi, African Energy Policy Research Network/Foundation for Woodstove Dissemination

Karekezi, S. and Kithyoma, W. (2002) 'Renewable Energy Strategies for Rural Africa: Is a PV-Led Renewable Energy Strategy the Right Approach for Providing Modern Energy to the Rural Poor of Sub-Saharan Africa?', *Energy Policy*, vol 30, pp1071–1086

Khennas, S. and Barnett, A. (2000) *Best Practices for Sustainable Development of Micro Hydro Power in Developing Countries*, Rugby, UK, Intermediate Technology Development Group

Krause, M. and Nordstrom, S. (2004) *Solar Photovoltaics in Africa: Experiences with Financing and Delivery Models*, New York, United Nations Development Programme

McDade, S. (2004) 'Fueling Development: The Role of LPG in Poverty Reduction and Growth', *Energy for Sustainable Development*, vol 8, no 3, pp74–81

Martinot, E., Chaurey, A., Lew, D., Moreira, J. and Wamukonya, N. (2002) 'Renewable Energy Markets in Developing Countries', *Annual Review of Energy and the Environment*, vol 27, pp309–348

Martinot, E., Cabraal, A. and Mathur, S. (2001) 'World Bank/GEF Solar Home Systems Projects: Experiences and Lessons Learned 1993–2000', *Renewable and Sustainable Energy Reviews*, vol 5, no 1, pp39–57

Modi, V., McDade, S., Lallement, D. and Saghir, J. (2006) *Energy and the Millennium Development Goals*, New York: Energy Sector Management Assistance Programme, United Nations Development Programme, UN Millennium Project and World Bank

Rao, K. and Ravindranath, N. (2002) 'Policies to Overcome Barriers to the Spread of Bioenergy Technologies in India', *Energy for Sustainable Development*, vol 6, no 3, pp59–73

Reddy, A. (2002) 'A Generic Southern Perspective on Renewable Energy', *Energy for Sustainable Development*, vol 6, no 3, pp74–83, Wireless Energy Chile, Ltda, Region de Los Lagos

Stevens, N. (2001) *Isla Tac Power System First Year Status Report: October 2000 through October 2001*

United Nations Development Programme (UNDP) (2000) *Bioenergy Primer: Modernised Biomass Energy for Sustainable Development*, New York, United Nations Publications

UNDP (2003) *Community Action to Address Climate Change: Case Studies Linking Sustainable Energy Use with Improved Livelihoods*, New York, UNDP

UNDP (2004) *Gender and Energy for Sustainable Development: A Toolkit and Resource Guide*, New York, UNDP

UNDP (2005a) *Achieving the Millennium Development Goals: The Role of Energy Services*, New York, UNDP

UNDP (2005b) *Energizing the Millennium Development Goals: A Guide to Energy's Role in Reducing Poverty*, New York, UNDP

UN Energy (2005) *The Energy Challenge for Achieving the Millennium Development Goals*, New York, United Nations Development Programme

UN Millennium Project (2005) *Investing in Development: A Practical Plan to Achieve the Millennium Development Goals*, New York, United Nations Development Programme

Utria, B. (2004) 'Ethanol and Gelfuel: Clean Renewable Cooking Fuels for Poverty Alleviation in Africa', *Energy for Sustainable Development*, vol 8, no 3, pp107–114

World Bank (2005) *Renewable Energy Toolkit Needs Assessment*, ESMAP Technical Paper 077, Washington, DC

World Energy Council (WEC) (1999) *The Challenge of Rural Energy Poverty in Developing Countries*, London, WEC

11

Traditional Biomass Energy: Improving Its Use and Moving to Modern Energy Use

Stephen Karekezi, Kusum Lata and Suani Teixeira Coelho

INTRODUCTION

Why is biomass energy important?

Biomass energy plays a vital role in meeting local energy demand in many regions of the developing world. Modern biomass energy is widely used in many developed countries as well as in parts of the non-industrialized world. Biomass is a primary source of energy for close to 2.4 billion people in developing countries (IEA, 2002b; Trossero, 2000; FAO, 2000). It is easily available to many of the world's poor and provides vital and affordable energy for cooking and space heating. Biomass-based industries are a significant source of enterprise development, job creation and income generation in rural areas (Karekezi et al, 2002c; Goldemberg, 2003; Trossero, 2000; Reddy et al, 1997; Barnes et al, 1994; Kammen, 1995; Larson and Williams, 1995). With proper management backed by adherence to appropriate ecological practices, modern biomass can be a sustainable source of electricity as well as liquid and gaseous fuels. Subject to its sustainable exploitation, biomass is not only a vital source of energy for many today but is likely to remain an important source of energy in the future (Yamamoto et al, 2001; Trossero, 2000; Hall and Rosillo-Calle, 1998; Kammen, 1995; Barnes et al, 1994).

Available statistics indicate that the share of biomass in global energy consumption has remained roughly the same over the past 30 years (Houck and Tiegs, 2005; ESMAP, 2005; Larson and Williams, 1995). Biomass energy[1] accounted for 14 per cent of the world's final energy consumption in 2002. The International Energy Agency (IEA, 2004c) estimates that at global level the share of biomass in total final energy consumption (Figure 11.1) is comparable to that of electricity (16 per cent) and gas (16 per cent).

At a regional level, however, the share of biomass energy in total energy consumption varies significantly (Table 11.1). Developing regions (Africa, Asia and Latin America) record high levels of reliance on biomass energy.

Biomass is the principal energy source for the poor of the developing world. This is especially true for traditional biomass energy, which is often collected as a 'free'[2] fuel (Reddy et al, 1997; Karekezi and Kithyoma, 2002b; Kgathi et al, 1997; Hall and Mao, 1994; Karekezi and Ranja, 1997). There appears to be a correlation between poverty levels and traditional biomass use in many developing countries (Figure 11.2). As a rule, the poorer the country, the greater the reliance on traditional biomass resources (IEA, 1998, 2000; UNDP, 2003; Trossero, 2000; ESMAP, 2005).

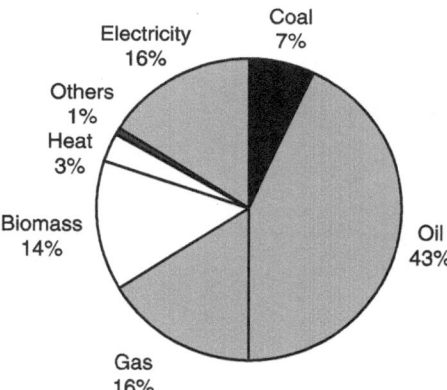

Source: IEA (2004c)

Figure 11.1 *World final energy consumption, 2002*

Table 11.1 *Biomass supply as a percentage of total primary energy supply, 1971 and 2002*

Region	1971 (%)	2002 (%)
OECD	2	3
Non-OECD Europe	4	5
Latin America	31	18
Asia	59	31
Africa	62	49

Source: IEA (2004a, b, c); ESMAP (2005)

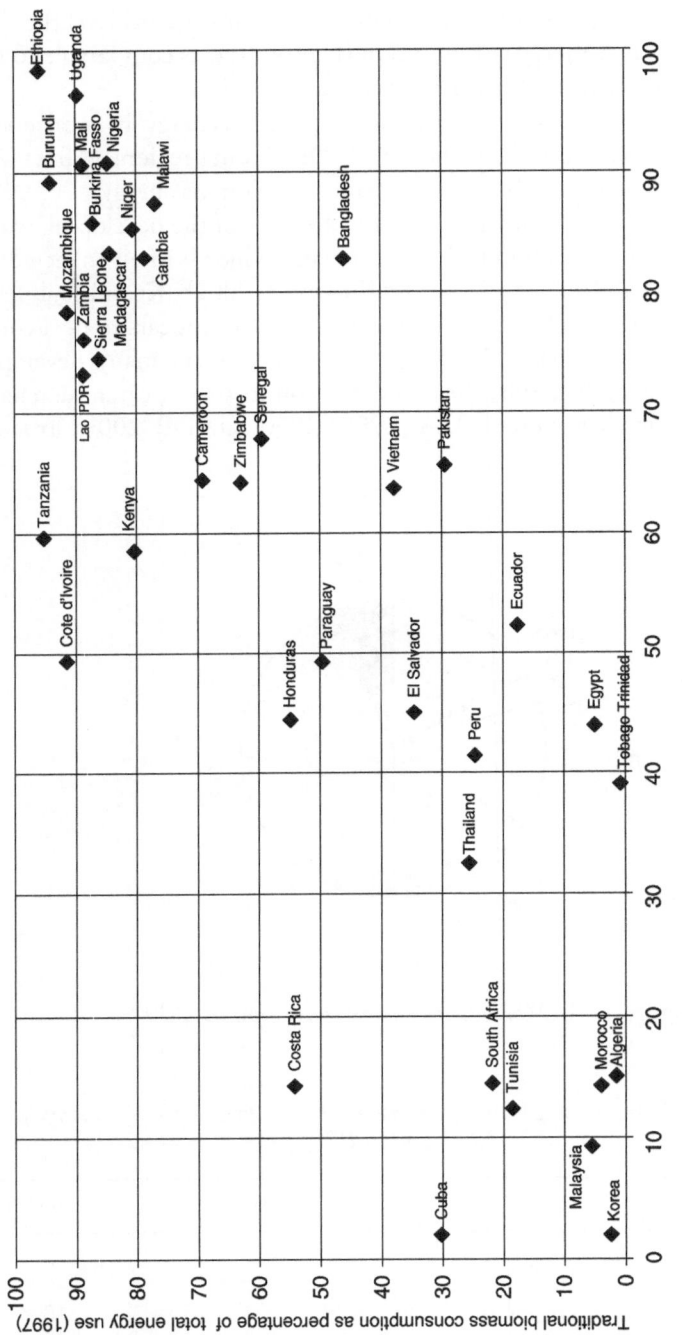

Figure 11.2 *Poverty and traditional energy use*

Source: UNDP (2003); IEA (2000)

Why simultaneously address traditional, improved and modern use of biomass energy?

The phrase 'traditional biomass energy use' as used in this chapter refers to the direct combustion of wood, charcoal, leaves, agricultural residues, animal or human waste and urban waste, for cooking, drying and charcoal production. Improved biomass technologies (IBTs) refers to improved and efficient technologies for direct combustion of biomass such as improved cooking stoves and improved biofuel kilns. 'Modern biomass energy use' refers to the conversion of biomass energy to advanced fuels, namely liquid fuels, gas and electricity (AFREPREN, 2002). Although primarily focusing on developing countries which significantly rely on traditional biomass energy as well as improved use of traditional biomass (which from now on will simply be referred to as improved biomass), this chapter also examines the use of modern biomass energy in more advanced economies, for several reasons.

First, all the three biomass energy forms largely rely on the same natural resource base. Second, many of the options aimed at addressing problems associated with traditional biomass energy use entail the deployment of improved and modern biomass energy technologies. While biomass energy, particularly traditional biomass energy use, is often perceived in a somewhat negative light, there are attractive opportunities for using biomass energy in more modern, efficient and environmentally friendly ways (Karekezi and Ranja, 1997; Hall and Rosillo-Calle, 1998; Goldemberg, 2002; Institute for Sustainable Energy, 2005). Improved and modern biomass energy technologies reduce the negative impacts of current traditional biomass energy use patterns.

Many policy-makers and researchers in the developing world (as well as interested analysts and decision-makers in the more developed parts of the world) are keen to see a progressive shift from traditional biomass use to improved use and eventually to modern biomass energy use (Karekezi et al, 2002c; Leach, 1992). Of priority interest is the need, first, to improve the current use of traditional biomass, and second, to transform biomass into high-quality low-emissions electricity, fuels and gases (Goldemberg and Coelho, 2003).

Third, existing data sets do not differentiate between traditional, improved and modern uses of biomass energy.[3] This is particularly true of aggregated global and regional data sets. Most statistical sources combine biomass energy used in both sustainable and unsustainable technologies. In spite of existing statistical uncertainties, biomass can indeed be one of the most important energy sources for developing countries (Goldemberg and Coelho, 2003).

Global scenarios

Biomass energy dominates current renewable energy statistics. About 80 per cent of current global renewable energy supply comprises biomass energy (IEA, 2004a).

Various global studies on the potential of biomass indicate that its contribution to total energy supply is expected to increase in the future (Trossero, 2000;

Table 11.2 *Global renewable energy supply, 2002*

Country/region	TPES[a]	Of which total renewables	Share of total renewables in TPES	Share of the main fuel categories in total renewables		
				Hydro	Geothermal, solar, wind, etc.	Combustible renewables and waste (CRW)[d]
	(Mtoe)	(Mtoe)	(%)	(%)	(%)	(%)
Africa	540	269	49.8	2.7	0.2	97.1
Asia[b]	1184	391	33.0	3.6	3.7	92.7
China[c]	1245	242	19.4	10.2	0.0	89.8
Latin America and Caribbean	455	129	28.4	35.9	1.6	62.5
Non-OECD Europe	100	9	9.0	44.9	0.8	57.3
OECD	5345	317	6.0	33.4	10.2	56.3
Former USSR	930	31	3.3	62.8	0.6	36.6
Middle East	431	4	0.9	48.8	20.9	30.3
World	**10,230**	**1392**	**13.6**	**16.1**	**3.6**	**80.3**

Notes:
a Total primary energy supply calculated using the physical energy content methodology.
b Asia excludes China.
c China includes People's Republic of China and Hong Kong, China.
d Combustible renewables and waste (CRW) refers to:

- *Solid biomass and animal products:* Biomass refers to any plant matter used directly as fuel or converted into other forms before combustion. Included are wood, vegetal waste (including wood waste and crops used for energy production), animal materials/wastes, sulphite lyes, also known as 'black liquor', and other solid biomass (including charcoal).
- *Gas/liquids from biomass:* Biogas is derived principally from anaerobic fermentation of biomass and solid wastes and combusted to produce heat and /or power.
- *Municipal waste:* Municipal waste consists of products that are combusted directly to produce heat and/or power and comprises wastes produced by residential, commercial and public services sectors that are collected by local authorities for disposal in a central location. Hospital waste is included in this category.
- *Industrial waste* consists of solid and liquid products (e.g. tyres) that are combusted directly, usually in specialized plants, to produce heat and/or power and that are not reported in the category solid biomass (IEA, 2002b).

Source: IEA (2004a)

Kammen, 1995). The IEA estimates that final consumption of biomass energy will increase in most regions (Table 11.3), although at a slower rate than conventional energy consumption. However, the share of biomass energy in total energy supply is expected to decrease from 14 per cent to 12 per cent (IEA, 2004c).

In Africa, available estimates indicate that up to the year 2020 biomass energy use is expected to increase roughly at the same rate as population growth rates (IEA, 1998), resulting in modest changes in the share of biomass in total final energy supply. In contrast, the share of biomass in total final energy supply in developing countries as a whole (Africa, Asia and Latin America) is expected to decrease in the same period, particularly for Asia and Latin America, which are expected to register a substantial reduction.

A study conducted jointly by the International Institute for Applied Systems Analysis (IIASA), and the World Energy Council (WEC) also projects an increase in global biomass energy use. By 2050 the biomass potential[4] is expected to increase to between 8.8 and 10.8 gigatonnes of oil equivalent (Gtoe) (Fischer and Schrattenholzer, 2001).[5] The IIASA–WEC scenario takes into account competition for land between bioenergy and food production, and the sustainable production of biomass energy. Other studies[6] also indicate growth of biomass energy in global energy supply, albeit at different rates.

In 2000, approximately 40 per cent of the world's population depended on biomass energy (IEA, 2002b; FAO, 2000). The proportion of the population relying on biomass energy is expected to increase in some developing countries. Table 11.5 shows the projected increase in the number of people depending on biomass energy. In some regions (e.g. Africa), biomass energy use will increase

Table 11.3 *Past and projected final biomass consumption in relation to total energy use, 2002 and 2020*

Country/ region	2002				2020			
	Biomass	Conventional energy	Total	Share of biomass	Biomass	Conventional energy	Total	Share of biomass
	(Mtoe)	(Mtoe)	(Mtoe)	(%)	(Mtoe)	(Mtoe)	(Mtoe)	(%)
China	216.16	608.13	824.29	26.22	203	1059	1262	16.0
Asia	350.58	501.84	852.42	41.13	404	1559	1963	20.6
Latin America	71.59	283.37	354.96	20.17	81	494	575	14.0
Africa	232.68	174.00	406.68	57.21	336	296	632	53.0
Total non-OECD	883.37	2519.65	3403.02	25.96	1037	4134	5171	20.1
OECD countries	116.35	3575.60	3691.95	3.15	173	4414	4587	4.0
World	**999.72**	**6095.25**	**7094.97**	**14.09**	**1210**	**8548**	**9758**	**12.0**

Source: IEA (2004a, c)

Table 11.4 *World biomass energy potential (Mtoe)*

	2020
Crop residue	480–499
Wood	1791–2025
Energy crops	2971–3535
Animal waste	994
Municipal waste	516
Total	**6752–7569**

Note: Fischer and Schrattenholzer use the term 'bioenergy', which refers to 'the sustainable use of biomass for energy uses', in place of 'biomass energy'.

Source: Fischer and Schrattenholzer (2001)

at the same rate as the population (IEA, 1998). South Asia and Africa are expected to have the highest growth in the number of people relying on biomass energy.

The case for differentiated regional assessment of biomass energy issue

The use of biomass energy varies significantly across the globe. Biomass energy is an important source of energy in many developing countries, especially in sub-Saharan Africa. The role of biomass energy in industrialized countries is more modest. Even in developing parts of the world there are variations in the type of biomass energy that is dominant. For example, in sub-Saharan Africa traditional biomass dominates national statistics while in Asia one sees greater use of

Table 11.5 *Number of people relying on biomass for cooking and heating in developing countries (millions)*

Country/region	2000	2030	2000–2030 (%)
China	706	645	−9
Indonesia	155	124	−20
Rest of Asia	137	145	6
India	585	632	8
Rest of South Asia	128	187	46
Latin America	96	72	−25
Africa	583	823	41
Developing countries total	**2390**	**2628**	**10**

Note: The figures for 2030 assume that biomass use per capita is constant, at some 0.3 toe per capita, over the projected period. This figure is an average across all regions and countries. Analysis indicates that average per capita biomass use varies between some 0.24 toe in South Asia to nearly 0.4 toe in many countries in East Asia (IEA, 2002b).

Source: IEA (2002a)

improved biomass energy technologies. Use of modern biomass technologies is more prevalent in Latin America (IEA, 1998; Hall and Rosillo-Calle, 1998; Bhattacharya, 2004; Kammen et al, 2001).

Biomass energy resources vary geographically, and are not uniformly distributed (IEA, 2002b; Reddy et al, 1997). Biomass energy use is dependent on various factors, such as geographical location, land use patterns, preferences, and cultural and social issues. Urban biomass energy problems are quite different from the rural problems. In rural areas biomass energy is collected from the environment with the use of women's and children's labour, which could be used much more productively (Leach, 1992; Rosillo-Calle and Hall, 1991; Arvidsond and Ellegard, 1999; McDade, 2004). In urban areas biomass energy is mostly purchased, and in some cases accounts for a large part of urban consumers' overall expenditures. Sometimes it is even more costly than competing petroleum products, once energy efficiency is taken into account. Income distribution patterns also contribute to variations in biomass energy use, with poorer regions relying on traditional forms of biomass, and industrialized regions using more modern biomass energy technologies (Leach, 1992; Rosillo-Calle and Hall, 1991; Arvidsond and Ellegard, 1999; Kammen et al, 2001). These variations point to the need for a regional assessment of biomass energy issues.

REGIONAL PERSPECTIVES

Africa

Biomass energy accounts for the bulk of Africa's total final energy supply. It is, however, important to note that data on biomass in Africa are particularly problematic. Most countries do not have reliable and up-to-date databases on energy, and especially biomass energy.[7] Available data estimates indicate that biomass constituted 58 per cent of total final energy consumption in Africa in 2002. According to the same source, in 2002 biomass accounted for 49 per cent of total primary energy supply (IEA, 2004c). Although there was a decrease from the share of biomass in total primary energy supply over a 30-year period (from 62 per cent to 49 per cent; IEA, 2004c), biomass still plays a dominant role in Africa's energy sector.

Variations within Africa exist, with biomass accounting for only 5 per cent of energy consumption in North Africa and 15 per cent in South Africa (IPCC, 2003). The heavy reliance on biomass is notably prominent in sub-Saharan Africa, where biomass accounts for 70–90 per cent of primary energy supply in some countries (UNDP, 2003; Karekezi et al, 2002c), and an estimated 86 per cent of energy consumption (IPCC, 2003; Kammen, 1995). The bulk of biomass energy used in sub-Saharan Africa is traditional biomass (UNDP, 2003).

The heavy reliance on biomass energy in sub-Saharan Africa is unlikely to change in the near future, given the stagnant (or sometimes declining) per capita national incomes as well as slow growth in conventional energy use. For Africa as whole, available estimates indicate that by 2020, biomass energy use is expected to increase roughly at the same rate as population growth rates (IEA, 1998, 2002b), resulting in modest changes in the share of biomass in total final energy supply (Table 11.6). In contrast, the share of biomass in total final energy supply in developing countries is expected to decrease in the same period. The absolute number of people relying on biomass energy in Africa is also expected to increase between the years 2000 and 2030 – from 583 million to 823 million, an increase of about 27 per cent (IEA, 2002a, b).

Asia

Biomass energy plays a significant role in Asia's energy sector. The share of biomass energy in total primary energy supply for Asia in 2002 was estimated to be 30.6 per cent. Asia also records significant variations in biomass energy consumption at regional and national levels. Biomass energy use accounts for 26 per cent of the total energy consumption in China, 26 per cent in East Asia and 53 per cent in South Asia (IEA, 2004c). The types of biomass energy used in Asia are a mixture of traditional biomass, improved biomass and modern biomass energy (FAO, 2000).

Over 80 per cent of the total rural population and about 22 per cent of the total urban population of Asia depend on biomass to satisfy their cooking energy needs. Fuel wood, dung cakes and crop residues still remain important household fuels, with their share in household energy consumption being well above 50 per cent in many Asian countries (Lefevre et al, 1997; Jingjing et al, 2001; Trossero, 2000). Non-commercial activities consume the highest proportion of solid biomass (10.4 per cent of total primary energy supply) in Asia. Figure 11.3 presents the contribution of biomass to primary energy in selected Asian countries (Sudha and Ravindranath, 1999; Lata, 2004).

Table 11.6 *Total final energy supply, including biomass energy, in Africa*

	2020		Annual growth rate, 2002–2020 (%)
	Biomass (Mtoe)	*Share of biomass in total supply (%)*	*Biomass*
Africa	367	43	1.9
Total developing countries	1127	18	1.1
World	1428	10	1.4

Source: IEA (2004c)

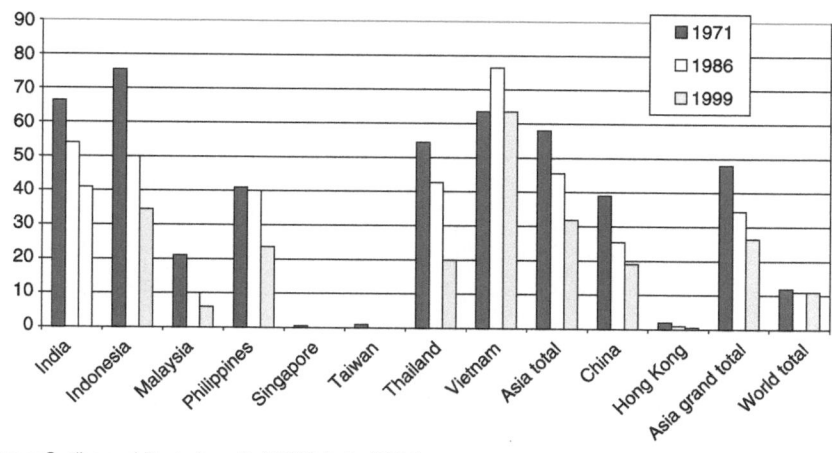

Source: Sudha and Ravindranath (1999); Lata (2004)

Figure 11.3 *Share of biomass in total energy supply of selected Asian countries (%)*

Latin America and the Caribbean

The share of biomass in the total primary energy supply of Latin America and the Caribbean (LAC) was estimated at 18 per cent in 2002 (IEA, 2004c), making it the developing region with the lowest share of biomass in energy consumption (Table 11.3). More recent estimates (IEA, 2002a) indicate that the proportion of primary energy derived from biomass (combustibles and renewable wastes, CRW) has decreased to 13.5 per cent, but the statistics vary significantly within the countries.

Biomass energy use in Latin America is more modernized – for example, alcohol production from sustainable biomass contributes to the transport sector in a number of countries (Coelho et al, 2003). A large proportion (49 per cent) of biomass energy use in Latin America is in the industrial sector (IEA, 2004c). Even in the countries where there are experiences with modern biomass – Brazil, for example – there are several parts of the country where reliance on inefficient traditional biomass energy is still prevalent.

Industrialized countries

Industrialized countries record significantly lower levels of biomass energy supply, most of which is modern biomass energy use (IEA, 2003b). The Organisation for Economic Co-operation and Development (OECD) estimated the share of biomass in total primary energy supply in industrialized countries at 3.3 per cent in 2002. This was an increase of 0.9 per cent

Table 11.7 *Biomass electricity production in selected industrialized countries, 2001*

Country	Biomass electricity (TWh)	Percentage of total electricity
US	70.60	1.80
Japan	12.83	1.24
Germany	12.97	2.23
Finland	8.69	11.67
Brazil	9.40	2.87
UK	5.08	1.32
Canada	7.21	1.23
Netherlands	3.46	3.70
Australia	1.36	3.00
Sweden	3.52	2.18

Source: IEA (2003b)

since 1971 (IEA, 2004b). The bulk of biomass energy use in industrialized countries is through the deployment of modern biomass energy technologies (IEA, 2002b). Biomass contributed about 2 per cent of fuels used for electricity generation in industrialized countries in 2002 (IEA, 2004b). Table 11.7 presents the contribution of biomass energy to electricity generation in selected industrialized countries.

Biomass energy use in industrialized countries is expected to increase in the future, although its contribution to final energy consumption will not grow substantially (IEA, 1998). According to the IEA, the share of biomass energy in electricity generation in industrialized countries is expected to increase from 2 per cent in 2002 to 4 per cent in 2020 (IEA, 2004b, c).

Categorization of biomass energy: traditional, improved and modern

As mentioned earlier, biomass energy use can broadly be categorized into the following three clusters:

1 traditional biomass energy technologies (TBEs);
2 improved biomass energy technologies (IBTs);
3 modern biomass energy technologies (MBTs).

The goal is to move from TBEs to IBTs and eventually to MBTs. The following section discusses the benefits and challenges of each of these categories of biomass energy, and the prospects for their development in the developing world.

TRADITIONAL BIOMASS ENERGY TECHNOLOGIES

Traditional biomass energy is a local energy source, and one that is readily available to meet the energy needs of a significant proportion of the population – particularly the poor in rural areas of the developing world. Traditional biomass is low cost and does not require processing before use (Hall and Mao, 1994). Fuelwood and charcoal are extremely important sources of employment and income for many rural households in developing countries (Trossero, 2000).

Traditional biomass use, however, has significant drawbacks. The indoor air pollution from unvented biofuel cooking stoves is linked to respiratory diseases in many highland areas of developing countries[8] (Karekezi and Ranja, 1997; Karekezi et al, 2002a; Ezzati and Kammen, 2002; Smith, 1991, 1994). Rural and poor women and children in many developing countries spend a significant portion of their time gathering and collecting fuelwood, crop residues and animal dung for use as cooking and space heating fuels (Trossero, 2000). Traditional biomass energy use has direct negative impacts on women and children,[9] who are the most vulnerable group in terms of biomass energy scarcity and adverse indoor air pollution impacts[10] (Tuck and Uku, 2002). Studies in some regions such as India, however, show that young boys and men have a higher prevalence of respiratory infections compared to young girls and women. These findings indicate the need for further research on the differentiated impacts of biomass energy on men and women (Schirnding, 2001).

Reliance on traditional biomass (especially in the form of charcoal) contributes to land degradation (Scully, 2002) and deforestation in countries where charcoal (sourced from natural forests and not planted forests) is widely used (Table 11.8). The unreliability of biomass energy data complicates attempts to

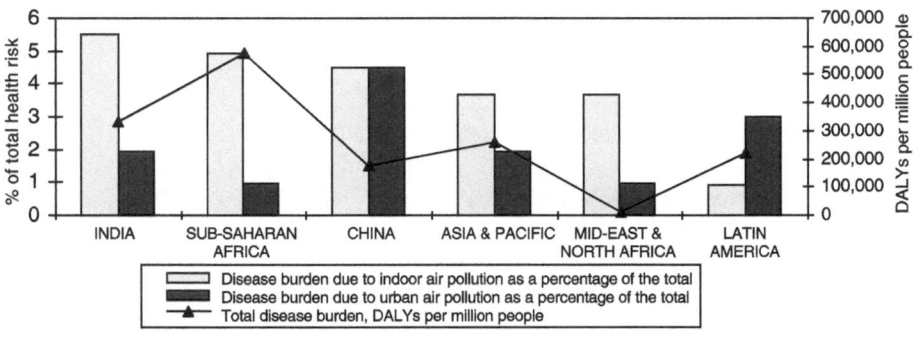

Note: DALY = death and disability adjusted life-years

Source: Schirnding (2001)

Figure 11.4 *Comparison of total disease burden and disease burden arising from indoor and urban air pollution*

link deforestation to biomass use, but the consensus among leading biomass energy experts is that inefficient charcoal production from natural forests and woodlands contributes to deforestation[11] (FAO and ADB, 1995; Ellegard and Nordstrom, 2003; Manila Times, 2005; Kammen et al, 2001).

In some areas (for example, around major cities such as Lusaka, Zambia; Nairobi, Kenya; and Dar es Salaam, Tanzania), charcoal demand appears to contribute to degradation of the surrounding woodlands and forests (Scully, 2002). Traditional charcoal production relies on the traditional and rudimentary earth kiln and is considered to be a major contributor to deforestation and land degradation in many peri-urban and rural regions of the developing world.[12] Traditional charcoal production is a particularly inefficient process, resulting in significant loss of energy in the conversion of fuelwood to charcoal (Karekezi and Ranja, 1997; IEA, 1998; Rosillo-Calle et al, 1996; Manila Times, 2005; ESDA, undated). In addition, charcoal production often leads to uncontrolled fires, which destroy biodiversity and contribute to regional air pollution.

Table 11.8 *Losses in charcoal production in developing countries*

	1995	2010	2020
East Asia	5%	7%	8%
Share of charcoal in final biomass use (%)	5.6	7.8	9.2
Charcoal production/use (Mtoe)	16.5	21.7	25.1
Wood input in charcoal production (Mtoe)	**10.8**	**14.0**	**15.9**
Energy losses in charcoal transformation (Mtoe)			
South Asia	2%	3%	4%
Share of charcoal in final biomass use (%)	3.5	7.9	11.1
Charcoal production/use (Mtoe)	12.6	28.2	39.5
Wood input in charcoal production (Mtoe)	**9.1**	**20.3**	**28.4**
Energy losses in charcoal transformation (Mtoe)			
Latin America	9%	9%	9%
Share of charcoal in final biomass use (%)	6.4	7.0	7.2
Charcoal production/use (Mtoe)	13.2	14.5	14.9
Wood input in charcoal production (Mtoe)	**6.8**	**7.5**	**7.7**
Energy losses in charcoal transformation (Mtoe)			
Africa	3%	6%	8%
Share of charcoal in final biomass use (%)	6.8	19.1	30.8
Charcoal production/use (Mtoe)	27.0	72.1	112.1
Wood input in charcoal production (Mtoe)	**20.3**	**53.0**	**81.3**
Energy losses in charcoal transformation (Mtoe)			
Total developing countries	3%	4%	5%
Share of charcoal in final biomass use (%)	22.3	41.8	58.3
Charcoal production/use (Mtoe)	69.3	136.5	191.6
Wood input in charcoal production (Mtoe)	**47.0**	**94.8**	**133.3**
Energy losses in charcoal transformation (Mtoe)			

Source: IEA (1998)

The ownership of traditional biomass resources presents an additional problem. Forests are often public property (communal), and the entire community harvests products from the forest (e.g. wood and timber). However, few people are willing to pay for the resource recovery through protection and reforestation (Scully, 2002). Often termed 'the crisis of the commons', the question of ownership of traditional biomass resources bedevils both researchers and policy-makers and has yet to be satisfactorily resolved. This is often compounded by the intricate relationship between control over biomass energy resources and prevailing land tenure practices, policies and regulatory frameworks (FAO and ADB, 1995; Ellegard and Nordstrom, 2003).

Key challenges facing many countries that rely heavily on traditional use of biomass include, first, how to ensure that the biomass used is sourced from sustainable biomass resources (e.g. wood plantation, sustainable management of native forests); second, how to disseminate IBTs widely; and finally, how to promote MBTs that use a wide range of biomass resources (fuelwood, agro-industrial residues, rural and urban residues) to generate high-quality fuels, gases and electricity (Hall and Rosillo-Calle, 1998; Masera et al, 2000).

IMPROVED BIOMASS ENERGY TECHNOLOGIES

Benefits and challenges

Improved biomass technologies (IBTs) contribute to more efficient and environmentally sound use of biomass energy. Improved cooking stoves, for instance, are designed to reduce heat loss, decrease indoor air pollution, increase combustion efficiency and attain a higher heat transfer (Karekezi and Ranja, 1997; Masera et al, 2000). This results in savings in the amount of fuel used, which translates to direct cash savings (Table 11.9).

There are several advantages of using improved biomass technologies such as more efficient cooking stoves, charcoal kilns and dryers. The provision of more efficient stoves can reduce respiratory health problems associated with smoke

Table 11.9 *Savings from improved stoves in Africa*

	Average daily charcoal consumption (kg per person per day)		Yearly savings per family (kg)	Value of savings (US$)	GNP per capita (US$)
	Traditional stove	Improved stove			
Kenya	0.67	0.39	64.70	613	360
Rwanda	0.51	0.33	84.10	394	210

Source: Karekezi and Ranja (1997); World Bank (2004)

emission from biofuel stoves (Khennas et al, 1999; Karekezi and Kithyoma, 2002b) and reduce the negative deforestation impact of, for example, traditional charcoal production. Improved use of biomass in households, institutions and industries leads to reduced fuel consumption, faster processing, improved product quality, products with better shelf life, and a better home and working environment for rural housewives and cooks/operators in institutions, respectively (Schirnding, 2001; Karekezi and Ranja, 1997; Karekezi et al, 2002c).

Other benefits include the alleviation of the burden placed on women and children in fuel collection, freeing up more time for women to engage in other activities, especially income-generating activities. Reduced fuel collection times can also translate into increased time for education of rural children, especially girls (Karekezi et al, 2002a, c). The production and dissemination of improved biomass energy technologies provides employment and job opportunities for a significant proportion of the population, particularly women.

IBTs provide an attractive option for small and medium-sized enterprises. IBTs improve the efficiency of biomass use in traditional energy-intensive rural productive activities such as charcoal production, crop drying, fish drying and beer brewing (Reddy et al, 1997; Karekezi and Kithyoma, 2002b).

Initiatives to disseminate IBTs have delivered significant benefits to both the urban and the rural poor in developing parts of the world. Urban improved stove initiatives deliver several benefits to both the urban and rural poor: first, in terms of jobs created in improved-stoves programmes; and second, in terms of reduced charcoal consumption through the use of improved charcoal stoves (Khennas et al, 1999; Karekezi and Kithyoma, 2002b). The informal sector, which provides employment to the urban poor, is the principal source of improved stoves (see the following case studies).

Box 11.1 Case study i: the Kenya Ceramic Jiko (improved cooking stove)

The Kenya Ceramic Jiko (KCJ) is one of the most successful stove projects in Africa. The KCJ is suitable for cooking and space heating (KENGO, 1991; Karekezi and Kithyoma, 2002b; Houck and Tiegs, 2005; ITDG, 2005; Kammen, 1995). Its manufacture is a relatively mature cottage industry. The level of specialization in the manufacture of the stove has increased, as has the level of mechanization. There is now a discernible labour division. Shauri Moyo is the principal artisanal production centre in Nairobi, where there are artisans whose occupation is to purchase clay liners and metal claddings and to assemble complete stoves and retail them to customers. There are two types of stove producers in Nairobi: mechanized manufacturers and semi-mechanized producers. It is estimated that mechanized producers are manufacturing close to 3200 liners a month. Semi-mechanized producers are now producing an estimated 10,600 liners per month (Karekezi and Ranja, 1997; Karekezi et al, 2002c).

Box 11.2 Case study 2: the Maendeleo/Upesi stove in Kenya

The Women Energy project of the Ministry of Energy in Kenya initially spearheaded the production and dissemination of the Upesi stove. The German Technical Cooperation (GTZ)-funded project had the overall objective of improving the living conditions of Kenya's rural population by reducing fuelwood requirements (Muriithi, 1995). The Intermediate Technology Development Group (ITDG) actively participated in the second phase of this programme and promoted the commercial production of the Upesi stove in western Kenya. ITDG focused on benefits to the producers and the development of a commercial market.

Women were the main implementers of the project by ITDG, and 19 women's groups were trained in the manufacture of the stove. To date, a total of 10 women's groups are recognized as producers of the stove. The annual production is over 12,000 Upesi stoves and 2500 liners for the Kenya Ceramic Jiko. The total profit generated by the production of stoves is estimated to be between Ksh 217,500 (US$2788) and Ksh 397,500 (US$5096) (Khennas et al, 1995). The project provided the opportunity for women to engage in income-generating activities, and has undoubtedly improved their livelihood and welfare (Khennas et al, 1999).

Prospects

Given the relatively low levels of dissemination of IBTs in developing countries (especially sub-Saharan Africa), and the projected increase in the number of people relying on biomass, the potential for IBTs is vast. For example, almost every country in developing regions has put in place a programme for the dissemination of improved cooking stoves, and this provides a good basis for significant increases in the dissemination of these technologies. Increased use of improved cooking stoves is likely to result in significant energy savings, efficiency improvements (Table 11.10) and health benefits. There is also significant potential for increased use of other improved biomass technologies in the developing countries.

Table 11.10 *Potential energy savings in developing countries from improved cooking stoves*

	Rural household bioenergy use (Mtoe)	Efficiency improvements (%)	Energy savings (Mtoe)	Maximum fuelwood savings[a] (million tonnes)
China	198	20–30	40–59	180
India	168	20–35	34–59	178
Latin America	28	10–40	3–12	36
Africa	116	30–40	35–46	141

Note:
a Using the conversion factor 1 tonne of firewood = 0.33 toe.

Source: IEA (2001)

Although improved biofuel stoves are often perceived as low-tech transitional options for very low-income developing countries, there are examples of very successful development of improved biofuel stoves and heaters in industrialized countries – a notable example is Austria (Box 11.3).

Box 11.3 CASE STUDY 3: AUSTRIA: SUCCESSFUL DISSEMINATION OF IMPROVED BIOFUEL STOVES AND HEATERS

Austria has a century-old tradition of using heavy brick stoves (also known as ceramic stoves, tile stoves, masonry stoves, Russian stoves or Kachelofen), which are heated by burning logwood in a strong fire for about an hour. When the fire is finished the stove is closed and the heat stored in the bricks slowly heats the room for up to 24 hours. About 500,000 brick stoves are in use in Austria (which has a population of 8 million).

By the beginning of the 1990s, wood chip heating was an established business in Austria. The market was rather modest, with about 2000 units sold per year. Typical clients for wood chip heating systems were farmers. In 1994, wood pellets were first introduced into the market: 4–10 millimetre granules made from compressed sawdust. The result was a completely unexpected dynamic growth of markets and the expansion of heating with wood fuels from farmers to single family house owners. For several years, sales of wood pellet boilers nearly doubled every year. In some provinces, pellet boilers have exceeded a market share of 50 per cent in newly built houses.

Source: Rakos (2005)

MODERN BIOMASS ENERGY TECHNOLOGIES

Benefits and challenges

Modern biomass technologies (MBTs) have the potential to provide improved rural energy services based on available biomass resources and agricultural residues.[13] The availability of low-cost biomass power in rural areas could help provide cleaner, more efficient energy services to support local development, promote environmental protection, provide improved domestic fuels and improve rural livelihoods. Bioenergy technologies based on sustainable biomass supply are carbon neutral and lead to net CO_2 emission reduction if used to substitute fossil fuels (IPCC, 2003; Coelho and Walter, 2003; Fischer and Schrattenholzer, 2001).

In addition, MBTs can contribute to better bio-waste management. For example, landfill gas can assist urban waste management, while bagasse-based co-generation solves the problem of safe disposal of bagasse at sugar plantations (Veragoo, 2003; Deepchand, 2002). Another way of modernizing biomass energy use is through the use of ethanol gel to meet household energy needs. Ethanol gel can substitute for wood, charcoal, gas and kerosene for domestic cooking in developing countries with ethanol production potential. In Zimbabwe, ethanol gel is used for camping, starting barbeques and fires, and in the army. Ethanol gel is ranked third among the household energy sources in Zimbabwe (Biomass Technology Group, 2003).

Another advantage of modern biomass energy is its job generation potential – a very important attraction for many developing countries faced with chronic levels of unemployment or underemployment. Existing studies (Goldemberg, 2003; FAO, 2000) indicate that, in comparison to other primary energy sources, the job generation potential of modern biomass is among the highest (Table 11.11). For example, in Brazil the annual production of 14 billion litres of ethanol from sugarcane is responsible for the creation of 462,000 direct and 1,386,000 indirect jobs in the country, corresponding to a rate of 263,000 annual jobs per megatonne of oil equivalent generated (Goldemberg, 2003).

One of the main challenges facing modern biomass use is the extent to which it can compete on cost and reliability with conventional fossil fuel options – both for transportation and for electricity supply. There is, however, a growing body of assessments of national implementation programmes demonstrating that modern large-scale biomass energy systems can be competitive on both economic and technical grounds. Examples include co-generation using a wide range of agro-residues (wood residues, sugarcane bagasse, rice husks, etc.) in many agro-industries, use of biofuels in a transport fuel blend and the well-known large-scale biofuel programme in Brazil (IEI, 2001; Winrock International Brazil, 2002).

On the other hand, smaller-scale applications of modern biomass energy technologies still face numerous challenges, particularly at the level of cost-competitiveness (although many argue that this is due to an absence of a level playing field) (IEI, 2002; Coelho and Walter, 2003). Small-scale modern biomass-based systems have registered encouraging levels of success in India, South-East Asia and parts of Latin America.

Table 11.11 *Comparison of job creation: biomass and conventional energy forms*

Sector	Jobs-person-years (terawatt-hour)
Petroleum	260
Offshore oil	265
Natural gas	250
Coal	370
Nuclear	75
Wood energy	**1000**
Ethanol (from sugarcane)	**4000**

Source: Goldemberg (2003)

Box 11.4 CASE STUDY 4: THE TRANSPORTATION SECTOR IN LATIN AMERICA

The Brazilian alcohol programme was established in 1975 for the purpose of reducing oil imports by producing ethanol from sugar cane. Ethanol is used in cars as an octane enhancer and oxygenated additive to gasoline (blended in a proportion of 20–26 per cent anhydrous ethanol in a mixture called gasohol) or in dedicated hydrated ethanol engines. In 1999 the Brazilian government eliminated controls on prices, and hydrated ethanol is sold for 60–70 per cent of the price of gasohol at the filling station, owing to significant reductions in production costs. These results show the long-term economic competitiveness of ethanol fuel when compared to gasoline (Goldemberg et al, 2004b).

The world leader in alcohol production continues to be Brazil, where alcohol prices are competitive and the development of the new flexible fuel cars promotes greater ethanol use by providing flexibility to consumers. Ethanol has made a valuable contribution to the development of the country's agro-industry. Moreover, the addition of alcohol appears to have contributed to the reduction of air pollution in mega-cities such as São Paulo (Coelho et al, 2003).

In Colombia there is a Federal law that requires the addition of 10 per cent of ethanol in gasoline. By 2006 the seven largest cities in Colombia are expected to have switched to gasohol. The gasohol fuel will be introduced in other cities of the country in line with the development of sugar-alcohol agro-industry. About 700 million litres of ethanol will be required per year, corresponding to 150 thousand hectares of sugar cane crops (Campuzano, 2003). According to the Bariloche Foundation there are four biodiesel plants in Argentina using sunflower, cotton and soybean as feedstock (www.bariloche.com.ar/fb).

Box 11.5 CASE STUDY 5: GASIFICATION IN INDIA

India has developed and successfully commercialized small-scale biomass gasifier-IC engine systems for remote electricity generation. More than 28 MWe of such technology is currently installed (Kartha and Larson, 2000).

The growing of biomass energy resources can also present several challenges. First, high-input mono cropping results in the loss of biodiversity, soil fertility and land degradation, and is often accompanied by the use of fertilizers and pesticides, which could lead to pollution of underground and surface water sources (Johansson et al, 2004; Masera et al, 2000). Second, it could lead to competition for land between food production and biomass resources (Masera et al, 2000; UNDP, 2000).

In many countries there is justified concern as to whether bioenergy diverts land from food production and other essential needs (Johansson et al, 1992; Kartha and Larson, 2000; Larson and Kartha, 2000). Owing to the land-intensive nature of bioenergy, some experts believe that bioenergy development could compete for the limited amounts of available biomass and land (Gielen et al, 2001) – that land requirements for future energy crop and forest plantations could compete with land used for traditional food production (Sims, 2002). For example, pulpwood plantations in many developing countries have led to the displacement of farmers and a decline in production of agriculture and other forest products. In Brazil, as a result of the ethanol programme, large sugar cane plantations were established in regions where previously small farms existed. Consequently, the subsistence crops of small farms (corn, vegetables, black beans, etc.) were eliminated, leading to the importing of food from distant regions (Kartha and Larson, 2000).

BOX 11.6 LAND AVAILABILITY FOR FOOD AND FUEL

The availability of land for the production of biomass in developing countries is determined by the demand on land for food production. With increasing population, food production and consumption in developing regions is expected to increase (FAO and ADB, 1995). Estimates by the Response Strategies Working Group of the Intergovernmental Panel on Climate Change indicate that the use of land for food production in developing regions (Asia, Africa and Latin America) will increase by 50 per cent by 2025 (IPCC, 1996). In addition, the demand for biomass energy is also expected to increase with population increase. Estimates by the World Energy Council indicate that by 2100 about 1700 million hectares of additional land will be needed for agriculture, while about 690–1350 million hectares of additional land would be needed to support biomass energy requirements (UNDP, 2000). The challenge, therefore, is ensuring sustainable biomass supply to meet growing energy demand, without taking up land required for food production. Some of the options for avoiding the competition for land between food and fuel are (1) increasing food production on current agricultural lands; (2) the establishment of large tree plantations; and (3) the use of modern forestry practices (IPCC, 1996).

Sources: Sudha and Ravindranath (1999)

The impact of modern biomass energy technologies on the poor is not well understood. It can complicate and compound existing competition over available biomass resources and land (Masera et al, 2003). Without appropriate, sensitive and equitable management, large-scale modern biomass energy development can lead to further marginalization of the rural poor. It is, however, possible that the growth and development of these technologies could lead to increased incomes for the poor (e.g. smallholder sugar farmers) if a well-designed revenue-sharing scheme were established. Mauritius provides a model case example of where the share of the benefits from large-scale co-generation plants that flow to low-income farmers has increased over time through direct policy interventions and an innovative revenue-sharing mechanism (Deepchand, 2002; Karekezi et al, 2002c).

The development of modern biomass energy often requires significant capital investments and technical expertise, which may not be readily available in many developing countries. In addition, there are cases where the legal and regulatory framework in place does not support the development of modern biomass energy technologies (AFREPREN, 2001). With the exception of Mauritius (Box 11.7), this has been a major barrier, for example, in the co-generation of electricity for sale to the national grid by sugar companies in many countries of sub-Saharan Africa (AFREPREN, 2003).

Box 11.7 CASE STUDY 6: CO-GENERATION IN MAURITIUS

The Mauritian experience in co-generation is one of the success stories in sub-Saharan Africa. As a result of extensive use of co-generation in Mauritius, the country's sugar industry is self-sufficient in electricity and sells excess power to the national grid. In 1998 close to 26 per cent of the country's electricity was generated from the sugar industry, largely using bagasse, a by-product of the sugar industry. By 2002, electricity generation from sugar estates stood at 40 per cent (half of it from bagasse) of the total electricity demand in the country.

Government support and involvement has been instrumental in the development of a co-generation programme in Mauritius. First the Sugar Sector Package Deal Act 1985 was enacted to encourage the production of bagasse for the generation of electricity. The Sugar Industry Efficiency Act 1988 provided tax incentives for investments in the generation of electricity and encouraged small planters to provide bagasse for electricity generation. Three years later the Bagasse Energy Development Programme (BEDP) for the sugar industry was initiated. In 1994 the Mauritian government abolished the sugar export duty, an additional incentive to the industry. A year later, foreign exchange controls were removed and the centralization of the sugar industry was accelerated. These measures have resulted in the steady growth of bagasse-based electricity in the country's electricity sector.

Bagasse-based co-generation development in Mauritius has delivered a number of benefits, including reduced dependence on imported oil, diversification in electricity generation and improved efficiency in the power sector in general. Using a wide variety of innovative revenue-sharing measures, the co-generation industry has worked closely with the government of Mauritius to ensure that substantial benefits flow to all key stakeholders of the sugar economy, including poor smallholder sugar farmers. The equitable revenue-sharing policies that are in place in Mauritius provide a model for emulation in ongoing and planned large-scale modern biomass energy projects in Africa.

Sources: Veragoo (2003); Deepchand (2001)

Prospects

Although modern biomass energy technologies have not been widely disseminated in many parts of the developing world, the International Energy Agency has attempted to assess the prospects of biomass-based power generation in different developing regions of the world (Table 11.12). More comprehensive assessments that examine a wide range of modern biomass energy options (electricity, gas and fuels) are hampered by the poor quality of biomass energy data that are available.

Table 11.12 *Biomass-based power generation in developing countries*

	2002	2010	2020
China			
Biomass-based power generation (TWh)	2	31	58
% of total electricity generation	–	1	1
Biomass used in power generation (Mtoe)	1	14	33
East Asia			
Biomass-based power generation (TWh)	8	11	11
% of total electricity generation	1	1	1
Biomass used in power generation (Mtoe)	2	4	4
South Asia			
Biomass-based power generation (TWh)	2	10	10
% of total electricity generation	–	1	1
Biomass used in power generation (Mtoe)	1	3	3
Latin America			
Biomass-based power generation (TWh)	20	31	45
% of total electricity generation	2	3	3
Biomass used in power generation (Mtoe)	5	8	12
Africa			
Biomass-based power generation (TWh)	3	3	3
% of total electricity generation	1	–	–
Biomass used in power generation (Mtoe)	2	2	2
Total developing countries			
Biomass-based power generation (TWh)	35	86	127
% of total electricity generation	1	1	1
Biomass used in power generation (Mtoe)	11	31	54

Source: IEA (2004c)

THE WAY FORWARD: POLICY OPTIONS

As this chapter is largely aimed at developing countries, the policy implications and recommendations will be restricted to developing countries and associated development partners.

In spite of the importance of biomass energy in developing regions, biomass energy policy planning in most developing countries is often undertaken in an ad hoc fashion. This is in part due to the absence of a single focal institution responsible for biomass energy. In many countries a wide range of institutions have some jurisdiction on biomass energy issues ranging from the ministries of energy, the environment and forestry to a plethora of national and sub-national agencies responsible for land tenure policy and rural development. Designing and establishing an appropriate and effective institutional and associated legal and regulatory framework for biomass energy is a key challenge that decision-makers and analysts need to address urgently (Karekezi et al, 2002c; Karekezi and Ranja, 1997).

Data and information on biomass energy use in most countries are outdated and often unreliable, which makes it difficult to plan. In comparison to the conventional energy sector, which has comprehensive five- to ten-year plans, planning for biomass energy in most developing countries is often incoherent, sporadic and starved of the necessary budgetary allocation. The mobilization of additional financial and technical resources to support data collection and associated biomass energy planning is of primary importance (IEA, 2003a, b).

One of the key challenges facing many developing countries as well as respective development partners is the level of effort and resources that should be expended on the previously mentioned three clusters of biomass options, namely:

1   traditional energy options;
2   improved biomass energy options;
3   modern biomass energy options.

Traditional biomass energy

Initiatives pertaining to inefficient and environmentally unsound traditional energy options should primarily be aimed at research and analysis as well as data collection to provide the basis for developing effective strategies for reducing reliance on traditional energy options. As mentioned earlier, many poor developing countries do not have reliable databases on traditional biomass energy use. This makes it difficult to formulate appropriate policy and field-oriented interventions. Mechanisms for collection and documentation of data on traditional biomass supply and consumption, data that are regularly updated and validated, need to be instituted (IEA, 2003a, b). This is primarily a national activity. Developing-country governments should be encouraged to integrate biomass energy assessments into regular surveys of their national statistical authorities. Regional commissions and agencies could assist in developing common methodologies that would facilitate exchange of data and information.

Above all, planning for biomass energy development should have a decentralized component and should involve end-users. Special attention should be devoted to involving women, because they bear the burden of traditional energy systems and are likely to be the greatest beneficiaries of improved systems. Decentralization of rural energy planning is wise because these systems are based primarily on traditional biomass. Consequently, an assessment of the demand and supply flows and of desirable interventions must also occur on the same geographic scale. Through their superior knowledge of the local situation, local people – women in particular – can be integral parts of the solution (World Bank, 2003a, b; Karekezi and Kithyoma, 2002b).

Improved biomass energy

While there is no full consensus among policy analysts and researchers, there is a growing body of evidence indicating that for low-income developing countries with large and very poor rural populations the accent should be on the promotion and disseminating of improved biomass energy options. This approach is likely to yield large near-term developmental benefits in terms of job generation and increased incomes, and to assist in reversing the negative environmental impacts of traditional energy use (Masera et al, 2003).

Many policy analysts stress the need for aggressive dissemination of improved biomass technologies (IBTs) in developing regions, to mitigate the negative effects of traditional biomass energy use – particularly indoor air pollution, which is linked to respiratory diseases, one of the main causes of death for children under the age of 5. Governments should put in place policies that support the development and dissemination of IBTs (ESMAP, 2005; Karekezi et al, 2002a, 2002c). The private sector, NGOs, CBOs (Community Based Organization) and donor organizations should implement projects aimed at ensuring the rapid dissemination of IBTs. In particular, efforts to reduce the cost of widely used IBTs such as improved cooking stoves should be accelerated, so that they come within the reach of even the poorest of the poor in Africa (Smith, 1991, 1994; Ezatti and Kammen, 2002). National institutions in developing countries would be expected to play a lead role in the promotion of IBTs, but international development institutions and partnerships could be instrumental in piloting effective approaches to bringing down the cost of widely used IBTs.

Given the harmful environmental impacts of charcoal production in the region, there is a need to regulate the production of charcoal (Scully, 2002). Afforestation and reforestation projects should be established as part of all charcoal production programmes. The use of improved and efficient charcoal kilns should be promoted (Karekezi and Ranja, 1997).

It is important for improved biomass energy system development and dissemination programmes to recognize the gender- and income-differentiated impacts of biomass energy use. In particular, improved biomass energy technologies that alleviate the burden and negative health effects of traditional biomass energy on the rural poor (a section of the population with a high proportion of women and children) should be promoted and given prominence in government policies.

Although consensus from decision-makers on policy measures for accelerating access to IBTs is still to be realized, there are a number of options that have been analysed by leading biomass energy experts and could provide an embryonic base for broad national, regional and global IBT initiatives. Notable options that could be considered for implementation by policy-makers in developing countries and respective partners include:

- Setting targets, which include the identifying and setting of goals for the incremental contribution of improved biomass energy options to total energy supply. The targets should preferably include financial commitments by governments and development partners. Poverty Reduction Strategy

Papers (PRSPs) would be instrumental in setting targets, although more specific targets would need to be stipulated in national energy plans first.

- Introduction of new and innovative financing mechanisms, such as allocating a proportion of available energy subsidies (e.g. levies on electricity and petroleum) to the adaptation and wide-scale dissemination of improved biomass energy technologies as well as encouraging the active involvement of local small credit institutions.

Modern biomass

For developing countries with lower levels of poverty and higher levels of industrialization, the emphasis should probably be best placed on the encouragement of modern biomass energy technologies that can be used as levers to develop agro-industries and a basis for leapfrogging to cleaner biomass-based advanced fuels, electricity and gases. International development organizations with specialized expertise and experience in the agro-industry would be ideal anchors for a global initiative for the promotion of modern biomass energy technologies in the sector.

The development of modern biomass energy technologies will require supportive legal and regulatory frameworks that attract investment in modern biomass energy systems. Owing to the substantial amount of resources required to develop these technologies, it is important that a clear legal and regulatory framework is put in place. In addition, new and innovative ways of financing modern biomass energy projects should be pursued (Goldemberg et al, 2004a; Karekezi and Ranja, 1997).

In the case of ethanol production, collaboration within the sugar industry would facilitate rapid improvement of agricultural practice (to increase productivity and reduce adverse environmental impacts) and allow the capture of substantial scale benefits associated with larger and more efficient plant. Ethanol producers could fully utilize economies of scale if some form of collaboration at an international level were initiated. At the moment, international trade in ethanol is constrained by various trade- and non-trade-related constraints. Increased trade in ethanol could provide an important impetus to the further development of the biofuel industry (Berg, 2001).

Long-term energy training programmes designed to develop the critical element of a locally trained workforce with the requisite technical, economic and socio-cultural skills are needed. Many of the engineering and technical courses that are currently taught at universities and colleges in developing countries provide little exposure to advanced biomass energy technologies. Both capacity and demand for local analytical expertise to provide comprehensive evaluations of available biomass energy resources and options for utilizing them are needed. Non-partisan groups, such as NGOs and independent research institutes and networks, are well placed to carry out such studies (IEI, 2001; Karekezi and Ranja, 1997).

As in the case of IBTs, there is no general consensus on what policy options would accelerate the use of modern biomass technologies but the following options could provide an initial menu for action:

- Ensuring a level playing field for modern biomass and conventional energy forms.
- Enacting a legal and regulatory framework that allows for the development of modern biomass energy and provides, among other incentives, access to the grid and transport fuel market. An international organization or partnership with expertise in legal and regulatory issues could play an instrumental role in identifying opportunities for promoting modern biomass energy technologies in a reforming energy industry.
- Setting targets, which includes identifying and setting goals for the incremental contribution of modern biomass energy options to total energy supply. The use of tradable renewable energy certificates could assist in further promotion of modern biomass energy technologies.
- Establishing regional and international funds for financing large-scale biomass energy technologies. Regional funds would be placed in regional development banks. International funds would be managed by international multilateral development agencies and banks.
- Taking advantage of the growing number of climate-related financing opportunities (Global Environment Facility, the Clean Development Mechanism (CDM), etc.) to support wider dissemination of advanced modern biomass energy technologies.

NOTES

The following are reviewers of and contributors to this chapter: Dirk Aßmann, Gustavo Best, Akanksha Chaurey, Elizabeth Cecelski, Suani Teixeira Coelho, Jose Goldemberg, Patricia Guardabassi, Anita Khuller, Waeni Kithyoma, Kusum Lata, Oswaldo Lucon, Ezekiel Manyara, Maxwell Mapako, Omar Masera, Abel Mbewe, Derrick Okello, Christian Rakos, Jamal Saghir, Jonathan Scurlock, Margaret Skutch, Miguel Trossero and Jennifer Wangeci.

An earlier version of this chapter was presented at the GFSE (Global Forum for Sustainable Energy) Event with the kind support of the Government of Austria.

1 'Biomass energy' as used in this chapter refers to firewood, agricultural residue, animal wastes, charcoal and other derived fuels (IEA, 1998).
2 The real price of traditional biomass has always been underestimated in the energy markets because of its perceived low economic value (RWEDP, 2001).
3 Most developing countries do not have reliable databases on traditional energy consumption and use. Data on traditional biomass energy in these regions should, therefore, be treated with caution (IEA, 2002).

4 Fischer and Schrattenholzer (2001) use the term 'bioenergy', which refers to 'the sustainable use of biomass for energy uses'.

5 Estimates by Fischer and Schrattenholzer are based on the units of primary energy leaving out unspecified forms of biomass whether converted to alcohol, electricity or any other form of final energy.

6 Fischer and Schrattenholzer compare the IIASA–WEC bioenergy potential to various bioenergy potentials reported in other studies (Dessus et al, 1992; Greenpeace, 1993; Woods and Hall, 1994; Kusumikawa and Mori, 1998; Johannson et al, 1993; Leemans et al, 1996; Lashof and Tirpak, 1990; Shell, 1996; Yamamoto et al, 2001). Although, strictly, comparing the various potentials is not possible, the general trend indicates an increase in future bioenergy potential (Fischer and Schrattenholzer, 2001).

7 The IEA (2003a) rates the data on Africa's biomass energy sector as being of low quality. There is an urgent need to redouble efforts to improve biomass energy databases in this region.

8 In many highland areas of rural parts of developing countries, biomass is used for both cooking and space heating in poorly ventilated homes, which aggravates indoor air pollution. The need for space heating is less acute in low-lying areas, thus reducing exposure to indoor air pollution from biomass-fuelled cooking stoves.

9 High indoor pollution is linked to acute respiratory infection (ARI), particularly pneumonia. ARI is the leading health hazard to children in developing countries and results in an estimated 4.3 million deaths per year (Kammen, 1995).

10 An estimated 500,000 women and children die in India each year as a result of indoor air pollution-related diseases, because of the use of traditional biomass fuels (Tuck and Uku, 2002).

11 Deforestation in countries where charcoal is not widely used (or is sourced from planted forests) is often linked to clearing for agriculture.

12 For instance, in central Zambia charcoal production can destroy up to 50 per cent of total woody biomass of a typical miombo woodland area, which takes 10–15 years of growth to re-establish itself (Hibajane et al, 1993; Hosier, 1993). Before the introduction of LPG, improved stoves and charcoal production control measures, Senegal, with about 70 per cent forest cover, depleted her forests at a rate of approximately 1.2 per cent per annum, in part due to charcoal production. This translated to about 165,000 ha/year (Ribot, 1993). In China, three consecutive forest inventories (1981, 1988 and 1993) reveal that in the seven provinces of northwest China the percentage of fuelwood forestland has decreased. Energy analysts attribute this trend in part to increased fuelwood consumption in rural areas (Zheng, 2000)

13 Transforming biomass into advanced and modern heat, liquid fuels, gas and electricity services.

REFERENCES

African Energy Policy Research Network (AFREPREN) (2001) *Power Sector Reform in Africa: Proceedings of a Regional Policy Seminar*, Occasional Paper 5, Nairobi, AFREPREN

AFREPREN (2002) *African Energy Data Handbook*, Occasional Paper 13, Nairobi, AFREPREN

AFREPREN (2003) African Energy Database, Nairobi, AFREPREN

Arvidsond, A. and Ellegard, A. (1999) 'Charcoal Potential for Southern Africa', *Renewable Energy for Development*, vol 11, no 1, p11

Barnes, D. F., Openshaw, K., Smith, K. R. and van der Plas, R. (1994) *What Makes People Cook with Improved Biomass Stoves? A Comparative International Review of Stove Programs*, Washington, DC, World Bank

Berg, C. (2001) 'World Ethanol Production 2001', www.fo-licht.com

Bhattacharya, S. (2004) 'Fuel for Thought', www.earthscan.co.uk/news/, London, Earthscan

Biomass Technology Group (2003) *Ethanol Gel as Domestic Fuel*, Twente, the Netherlands, University of Twente

Campuzano, H. (2003) 'Se despeja el camino para los alcoholes carburantes en Colombia', www.corpodib.com

Chidumayo, E. N. (1993) 'Zambian Charcoal Production: Miombo Woodland Recovery', *Energy Policy*, vol 21, no 5, pp586–597, London, Butterworth-Heinemann Ltd

Coelho, S. T., Guardabassi, P. and Lucon, O. (2003) Continental Paper: 'Latin America and the Caribbean (LAC)', second draft

Coelho, S. T. and Walter, A. S. (2003) 'Brazil – Country Study for Sustainable Development', in *Indigenous Technologies for Sustainable Development*, Study under development funded by the International Atomic Energy Agency (chairman: J. Goldemberg)

Deepchand, K. (2001) 'Bagasse-Based Cogeneration in Mauritius – A Model for Eastern and Southern Africa', AFREPREN Occasional Paper 2, Nairobi, African Energy Policy Research Network

Deepchand, K. (2002) 'Promoting Equity in the Large-Scale Renewable Energy Developments: The Case of Mauritius', *Energy Policy*, vol 30, no 11/12

Dessus, B., Devin, B. and Pharabod, F. (1992) *World Potential of Renewable Energies*, Paris

Ellegard, A. and Nordstrom, M. (2003) 'Deforestation for the Poor?', *Renewable Energy for Development*, vol 16, no 2, pp4–6

Energy for Sustainable Development Africa (ESDA) (undated) 'Biomass Conversion Technologies', www.esda.energyprojects.net/links/biomass.htm, Nairobi, ESDA

Energy Sector Management Assistance Programme (ESMAP) (2005) *Household Energy Use in Developing Countries*, Washington, DC, ESMAP, World Bank

Ezzati, M. and Kammen, D. M. (2002) 'Household Energy, Indoor Air Pollution and Health in Developing Countries: Knowledge Base for Effective Interventions', *Annual Review of Energy and the Environment*, vol 27, Palo Alto, CA, Annual Reviews

Food and Agriculture Organization (FAO) (2000) *Environment and Natural Resource Working Paper: Rural Energy Supply*, Rome, FAO

FAO and African Development Bank (ADB) (1995) *Future Energy Requirements for Africa's Agriculture*, Rome, FAO and ADB

Fisher, G. and Schrattenholzer, L. (2001) 'Global Bioenergy Potentials through 2050', *Biomass and Bioenergy*, vol 20, no 3, pp151–159

Gielen, D. J., de Feber, M. A. P. C., Bos, A. J. M. and Gerlagh, T. (2001) 'Biomass for

Energy or Materials? A Western Europe Systems Engineering Perspective', *Energy Policy*, vol 29, no 4, pp291–302]

Goldemberg, J. (2002) 'Brazilian Initiative', *Renewable Energy for Development*, vol 15, no 1/2, p9

Goldemberg, J. (2003) 'The case for Energy Renewables', Thematic Background, the Secretariat for the International Conference for Renewable Energies, 1–4 June 2004, Bonn, Germany

Goldemberg, J. and Coelho, S. T. (2003) 'Renewable Energy – Traditional Biomass vs. Modern Biomass', *Energy Policy*, vol 32, no 6, pp711–714

Goldemberg, J., Coelho, S. T. and Lucon, O. (2004a) 'How Adequate Policies Can Push Renewables', *Energy Policy*, vol 32, no 9, pp1141–1146

Goldemberg, J., Coelho, S. T., Nastari, P. M. and Lucon, O. (2004b) 'Ethanol Learning Curve – The Brazilian Experience', *Biomass and Bioenergy*, vol 26, no 3, pp301–304

Greenpeace (1993) *Towards a Fossil Free Future – The Technical and Economic Feasibility of Phasing Out Global Fossil Fuel Use*, London, Greenpeace International

Hall, D. O. and Mao, Y. S. (1994) *Biomass and Coal in Africa*, London, Zed Books

Hall, D. O. and Rosillo-Calle, F. (1998) 'Evaluating Environmental Effects and Carbon Sources and Sinks Resulting from Biomass Production and Use in Developing Countries', in *Biomass: Data Analysis and Trends*, Paris, Organisation for Economic Co-operation and Development/International Energy Agency

Hibajane, S. H., Chidumayo, E. N. and Ellegard, A. (1993) *Summary of the Zambia Charcoal Industry Workshop on Policy and Management Challenges for the Future, May 10–14, 1993 Siavonga, Zambia*, Stockholm, Stockholm Environment Institute

Houck, D. J. and Tiegs, P. (2005) *Cookstoves in Developing Nations*, Washington, DC, World Bank

Institute for Sustainable Energy (2005) 'Distributed Generation', www.easternct.edu/depts/sustainenergy/energy/distributedGeneration.htm, Willimantic, CT, Eastern Connecticut State University

Intergovernmental Panel on Climate Change (IPCC) (1996) 'Technologies, Policies and Measures for Mitigating Climate Change', Technical Paper 1, Geneva, IPCC

IPCC (2003) www.grida.no/climate/ippc-tar/

Intermediate Technology Development Group (ITDG) (2005) 'Weighing Up the Cost of Smoke Alleviation', www.itdg.org

International Energy Agency (IEA) (1998) *World Energy Outlook, 1998*, Paris, IEA

IEA (2000) *World Energy Outlook, 2000*, Paris, IEA

IEA (2001) *World Energy Outlook, 2001*, Paris, IEA

IEA (2002a) *Energy Balances of Non-OECD Countries 1999–2000*, Paris, IEA

IEA (2002b) *World Energy Outlook, 2002*, Paris, IEA

IEA (2003a) *Energy Balances of Non-OECD Countries 2000–2001*, Paris, IEA

IEA (2003b) *Energy Balances of OECD Countries 2000–2001*, Paris, IEA

IEA (2004a) *Energy Balances of Non-OECD Countries 2001–2002*, Paris, IEA

IEA (2004b) *Energy Balances of OECD Countries 2001–2002*, Paris, IEA

IEA (2004c) *World Energy Outlook, 2004*, Paris, IEA)

International Energy Initiative (IEI) (2001) *Energy for Sustainable Development*, vol 5, no 1, Bangalore, IEI

IEI (2002) *Energy for Sustainable Development*, vol 6, no 2, Bangalore, IEI

Jingjing, L., Xing, Z., DeLaquil, P. and Larson, E. D. (2001) 'Biomass Energy in China and Its Potential', *Energy for Sustainable Development*, vol 5, no 4, pp66–80

Johansson, T. B., Kelly, H., Reddy, A. K. N. and Williams, R. H. (eds) (1992) *Renewable Energy: Sources for Fuels and Electricity*, Washington, DC, Island Press

Johansson, T. B., Kelly, H., Reddy, A. K. N. and William, R. (eds) (1993) *Renewable Fuels and Electricity for a Growing World Economy: Defining and Achieving the Potential, Renewables for Fuels and Electricity*, Washington DC, Island Press

Johansson, T. B., McCormick, K., Neij, L. and Turkenburg, W. (2004) 'The Potentials of Renewable Energy', Thematic Background Paper for the 2004 International Conference for Renewable Energies, Bonn, Secretariat for the International Conference for Renewable Energies, 1–4 June 2004

Kammen, D. M. (1995) 'From Energy Efficiency to Social Utility: Lessons from Cookstove Design, Dissemination, and Use', in Goldemberg, J. and Johansson, T. B. (eds) *Energy as an Instrument for Socio-economic Development*, New York, United Nations Development Programme

Kammen, D. M., Bailis, R. and Herzog, A. V. (2001) *Clean Energy for Development and Economic Growth: Biomass and Other Renewable Energy Options to Meet Energy and Development Needs in Poor Nations*, Berkeley, University of California

Karekezi, S. and Ranja, T. (1997) *Renewable Energy Technologies in Africa*, London, Zed Books

Karekezi, S., Banda, K. B. and Kithyoma, W. (2002a) 'Improving Energy Services for the Urban Poor in Africa – A Gender Perspective', *ENERGIA News*, vol 5, no 4, pp16–19

Karekezi, S. and Kithyoma, W. (2002b) 'Renewable Energy Strategies for Rural Africa: Is a PV-Led Renewable Energy Strategy the Right Approach for Providing Modern Energy to the Rural Poor of Sub-Saharan Africa?', *Energy Policy*, vol 30, no 11/12, pp1071–1086

Karekezi, S., Teferra, M. and Mapako, M. (eds) (2002c) Special Issue: Africa: Improving Modern Energy Services for the Poor, *Energy Policy*, vol 30, no 11/12, pp1059–1086

Kartha, S. and Larson, E. D. (2000) *Bioenergy Primer: Modernised Biomass Energy for Sustainable Development*, New York, United Nations Development Programme

KENGO (1991) *How to Make and Use the KCJ*, Nairobi, KENGO/Regional Wood Energy Programme for Africa (RWEPA)

Kgathi, D. L., Hall, D. O., Hategeka, A., Mlotshwa, C. V. and Sekhwela, M. B. M. (eds) (1997) *Biomass Energy Policy in Africa: Selected Case Studies*, London, Zed Books

Khennas, S., Anderson, T., Doig, A. and Rees, D. (1999) *Rural Energy Services: A Handbook for Sustainable Energy Development*, London, Intermediate Technology Publications

Khennas, S., Walubengo, D. and Weyman, A. (1995) *Rural Stoves West Kenya Evaluation*, Intermediate Technology Development Group (ITDG) internal document, Nairobi

Kusumikawa, J. and Mori, S. (1998) Development and Analysis of a World Energy – Economic-Environment Model Including the Forestation Option, Tokyo, University of Tokyo

Larson, E. D. and Kartha, S. (2000) 'Expanding Roles for Modernized Biomass Energy', *Energy for Sustainable Development*, vol 4, no 3, pp15–25

Larson, E. D. and Williams, R. H. (1995) 'Biomass Plantation Energy Systems and Sustainable Development', in Goldemberg, J. and Johansson, T. B. (eds) *Energy as an Instrument for Socio-economic Development*, New York, United Nations Development Programme

Lata, K. (2004) 'Biomass in Asia', Continental Paper, New Delhi, The Energy Resource Institute (TERI)

Lashof, D. A. and Tirpak, D. A. (1990) Policy Options for Stabilizing Global Climate: Draft Report to Congress (unpublished), US Environmental Policy Agency, Washington DC

Leach, G. (1992) 'The Energy Transition', *Energy Policy*, vol 20, no 2, pp116–123

Leemans, R., van Amstel, A., Battjes, C., Kreilman, E. and Toet, A. (1996) 'The Land Cover and Carbon Cycle Consequences of Large-Scale Utilizations of Biomass as an Energy Source', *Global Environmental Change 1996*, vol 6, no 6, pp335–58

Lefevre, T., Todoc, J. L. and Timilsina, G. R. (1997) 'Status of Wood Energy Data in Asia', paper presented at the International Energy Agency's First Biomass Workshop, February, Paris

McDade, S. (2004) 'Fuelling Development: The Role of LPG in Poverty Reduction and Growth', *Energy for Sustainable Development*, vol 8, no 3, pp74–81

Manila Times (2005) 'Charcoal, yes; wood, no', www.manilatimes.net/national/2005/apr/12/, Manila, Manila Times Publishing Corporation

Masera, O. R., Drigo, R. and Trossero, M. A. (2003) 'Woodfuel Integrated Supply/Demand Overview Mapping (WISDOM): A Methodological Approach for Assessing Wood Fuel Sustainability and Supporting Wood Energy Planning', FAO Report TCD/D/Y 47 19E/ 6.03/1000, Wood Energy Program, Forest Product Division, FAO, Rome, March

Masera, O. R., Saatkamp, B. D. and Kammen, D. M. (2000) 'Energy and Health Transactions in Development: Fuel Use, Stove Technology, and Morbidity in Jaracuaro, Mexico', *Energy for Sustainable Development*, vol 4, no 2, pp7–16

Muriithi, J. (1995) 'Women and Energy Project – Kenya: An Impact Study', *Boiling Point*, no 35, pp7–8

Rakos, C. (2005) 'Austrian Experiences with Small-Scale Biomass Systems – Opportunities for Development Cooperation', Nairobi, AFREPREN/FWD

Reddy, K. N., Williams, R. H. and Johansson, T. B. (1997) *Energy after Rio: Prospects and Challenges*, New York, United Nations Development Programme

Regional Wood Energy Development Programme (RWEDP) (2001) *Wood Energy News*, vol 16, no 1, pp1–32

Ribot, J. C. (1993) 'Forestry Policy and Charcoal Production in Senegal', *Energy Policy*, vol 21, no 5, pp559–585

Rosillo-Calle, F., Furtado, P., Rezende, M. E. A. and Hall, D. O. (1996) *The Charcoal Dilemma: Finding Sustainable Solutions for Brazilian Industry*, London, Intermediate Technology Publications

Rosillo-Calle, F. and Hall, D. O. (1992) 'Biomass Energy, Forests and Global Warming', *Energy Policy*, vol 20, no 2, pp124–136, Butterworth-Heinemann Ltd

Schirnding, Y. (2001) 'Gender Differences in the Impact of Biomass Fuel on Health', *ENERGIA News*, vol 4, no 4, pp10–11

Scully, J. (2002) *Ecoforum*, vol 25, no 4

Sims, R. (2002) 'Biomass, Bioenergy and Barriers', *Renewable Energy World*, vol 5, no 4, pp118–131

Smith, K. R. (1991) 'Improved Biomass Cookstove Programs: A Global Evaluation', *The Health Effect of Biomass Smoke: A Brief Survey of Current Knowledge*, no 4, Hawaii, Environment and Policy Institute

Smith, K. R. (1994) 'Health, Energy, and Greenhouse-Gas in Household Stoves', *Energy for Sustainable Development*, vol 1, no 4, pp23–29

Sudha, P. and Ravindranath, N. H. (1999) 'Land Availability and Biomass Production Potential in India', *Biomass and Bioenergy*, vol 16, no 3, pp207–221

Trossero, M. A. (2000) 'Wood Energy: The Way Ahead', www.fao.org/docrep/005/y4450e/y4450e02.htm, Rome, Food and Agriculture Organization

Tuck, M. and Uku, R. (2002) 'Making Infrastructure Work for the Poor', www.worldbank.org/infrastructure, Washington, DC, World Bank

United Nations Development Programme (UNDP) (2000) *Energy and the Challenge of Sustainability*, New York, UNDP

UNDP (2003) *Human Development Report: 2003*, New York, UNDP

Veragoo, D. (2003) 'Cogeneration: The Promotion of Renewable Energy and Efficiency in Mauritius', paper presented to the Regional Consultative Meeting on the Renewable Energy and Energy Efficiency Partnership in East Africa (REEEP), 9–10 June, Nairobi, African Energy Policy Research Network/Foundation for Woodstove Dissemination

Winrock International Brazil (2002) *Trade Guide on Renewable Energy in Brazil*, October, Salvador Bahia, Winrock International Brazil

Woods, J. and Hall, D. O. (1994) 'Bioenergy for Development: Technical and Environmental Dimensions' *FAO Environment and Energy*, Paper 13, Rome, Food and Agricultural Organisation of the United States (FAO)

World Bank (2003a) *African Development Indicators, 2003*, Washington, DC, World Bank

World Bank (2003b) *World Development Report, 2003*, Washington, DC, World Bank

World Bank (2004) *African Development Indicators, 2004*, Washington, DC, World Bank

Yamamoto, Y., Fujino, J. and Yamaji, K. (2001) 'Evaluation of Bioenergy Potential with a Multi-regional Global-Land-Use-and-Energy Model', *Biomass and Bioenergy*, vol 21, pp185–203

Zheng, Y. (2000) 'Fuelwood Production in Northern China: A Case Study', *Wood and Energy News*, vol 15, no 1, pp8–10

12

Gender Equity and Renewable Energies

Joy Clancy, Sheila Oparaocha and Ulrike Roehr

RENEWABLE ENERGY AND GENDER

The aim of this chapter is to explore the gender dimension of different aspects of renewable energy in the North and South, in particular how renewable energies can contribute to gender equity so that both women and men benefit fairly from access to energy services. The term 'gender' is used here deliberately rather than 'sex'. The term 'gender' refers to socially constructed roles of women and men rather than biologically determined differences. These gender roles of men and women, with their accompanying responsibilities, constraints, opportunities and needs, are defined by a particular society. Gender roles are learned by children as part of their socialization process. The roles change over time and vary widely within and across cultures. Men and women play different roles in both Southern and Northern societies, although at first glance the roles in the North may not be as clearly visible as they are in the South.

The reason that a gender perspective on energy production and use patterns is important is not merely that men and women often use, are impacted by, or benefit from energy services differently, but more importantly because the activities of one may impact the opportunities of the other. The same energy service may indeed impact on men and women differently, with different social or economic outcomes. For example, men may choose to locate a light outside the house for security reasons (such as protecting livestock from theft) while women may choose to locate the light in the kitchen. An evaluation of the distribution of the benefits of a micro-hydro scheme in rural Sri Lanka found that in connected households, men and women obtained equal (but different) benefits from the energy services provided by electricity (mainly television and lighting) (Dhanapala (1995) quoted in Barnett (2000)). However, it was the unconnected households where there was unequal distribution of benefits. These households benefited by access to TV (by visiting neighbours with sets) and the possibility of hiring lights for special occasions. It was the men who had greater access to TV because they had greater freedom of movement, particularly at night.[1]

This chapter first explores the reasons why gender analysis can help those people trying to increase the dissemination of renewable energy technologies. We give a brief description of the gender aspects of household energy and how different renewable energy technologies can contribute to drudgery reduction and time-saving, particularly for women. The role of women in renewable energy is analysed. The chapter concludes with an analysis of lessons learned. However, the point has to be stressed that there is now only beginning to emerge information about gender and energy in the South and there is very little information about the situation in the North. The data that are available are primarily in the form of case studies, mainly related to stove programmes and rural electricity grid extension. The systematic collection of gender-disaggregated statistical data by energy ministries is non-existent and it is very rare to find energy project evaluation that uses gender analysis.

RENEWABLE ENERGY ENABLING DEVELOPMENT

Modern forms of energy are considered a necessary (but not sole) input for economic development and the elimination of poverty (Barnett, 1999). There is a particular need to provide substitutes for human energy to remove drudgery and increase well-being. This has been one of the gains that increased energy provision by fossil fuels and electricity has brought in the North.[2] In the South, similar improvements in the quality of life can be enjoyed by a small percentage of the population, the so-called urban elites, who have sufficient income to purchase modern forms of energy and the appropriate conversion technologies. In development terms it is the other part of the population – the substantial majority living in rural and urban areas often referred to as 'the poor' – that needs assistance to gain an improved quality of life and more sustainable livelihoods.[3] This is the context in which renewable energy technologies have to work.

However, development practitioners do not see poverty as purely a problem requiring a technical solution of improved access to particular commodities, such as energy, but as one requiring a much more holistic approach.[4] This will come as a surprise to many energy professionals who tend to focus on more technical issues of technology selection,[5] efficiency and management of production and use. This means that if the renewable energy sector wishes to contribute to development, it also needs to take a more holistic view, and a good starting point is to use gender analysis to gain an understanding of energy needs and the context in which the technology is to operate. Gender analysis is not about looking at women alone, nor is it about complaining that women suffer more than men, but rather gender is about reaching a better understanding of how communities work from the perspective of relationships between men and women. Gender interests are not always obvious, neither are potential impacts of energy interventions. Sometimes, inappropriate interventions are made because they are made on the basis of assumptions. For example, the emphasis in energy planning for the benefit of women has long concentrated around cooking, with firewood collection being seen as the central problem to be tackled. A proper analysis of women's workload often reveals quite different priorities. For example, a study by Mehretu and Mutambira (1992) in Zimbabwe measured the time and energy used in transport connected with regular household activities by different family members. They found that the most significant energy intervention may be for water collection and market trips, because the women's own metabolic energy costs are higher for these activities than for interventions for collecting firewood and cooking. The outcomes of gender analysis can then be used in developing solutions with communities themselves, and by taking gender into account and thereby achieving more sustainable solutions.

Box 12.1 A GENDER APPROACH LEADS TO GREATER PROJECT EFFICIENCY: CASE STUDY: PHOTOVOLTAIC SOLAR HOMES IN GUATEMALA

Fondación Solar, operating a photovoltaic (PV) project in Guatemala, found that men mostly attended the training sessions on maintenance of the equipment, and those women who did attend merely stood by and watched while their husbands got involved in hands-on activities, such as changing the batteries. As a consequence, when the PV system needed maintenance, such as topping up the batteries, and the men were not at home, the women did not have the skills or confidence to take the appropriate action, which was not positive for the long-term durability of the system. Fondación Solar found much better results in overall system care (and hence in project perform-ance) when they took specific action to train the women in system maintenance at home, but when the men of the household were out. By taking this approach to training, the NGO created an environment in which the women were not afraid to make mistakes or ask questions.

Source: Wides (1998) cited in Skutsch (1998)

It is somewhat unfortunate that gender analysis has not been extensively applied in the energy sector, unlike in other sectors such as water and agriculture. How-ever, taking a gender approach in project planning and implementation can

benefit both the project management (*project efficiency*) and the intended beneficiaries (*women's welfare* or *gender equity* or *women's empowerment*). In the
former, a project manager is responsible for meeting the goals of a project
within time and to budget. Awareness that men and women have different
perspectives and needs and constraints can lead to a better fit of project interventions with the intended beneficiaries and thus create greater management
efficiency in terms of delivery (Skutsch, 1998). Box 12.1 gives an example of a
PV project in Guatemala which found that taking a gender approach meant
that the overall performance of the project improved significantly. Box 12.2
gives an example of an energy project that has contributed to women's
empowerment.

Box 12.2 Energy project contributes to women's empowerment in Kenya

Thirteen women's groups (200 people) have been trained in making stoves in the Rural
Stoves West Kenya project, and many have also benefited from business management
training. Annual production is estimated at 11,000 stoves annually; the profit generated by
the stoves is comparable to wages in rural areas. As a result, the women potters have
gained in status, self-confidence and financial independence.

Source: ITDG (1998)

Energy in poor households: women's burden

In terms of household energy in the South and in some war-ravaged rural areas
in the North (e.g. Kosovo and Chechnya), women are usually responsible for
energy provision within the household and for tasks that ensure the survival of
the family. Women should be the direct beneficiaries in any improvements in
availability or diversification in choice of energy carrier or conversion technology. Discussions on household energy tend to focus on women, since they are
usually responsible for household energy provision; however, men can be indirect
beneficiaries of access to modern energy forms (for example, meals may be
prepared faster). Men can also play an important role because they are very
influential in determining the outcomes of any energy interventions and where
and by whom the benefits to the household will be felt.

The South

Women carry a physical and metaphorical burden in energy provision. In rural
areas it can mean spending several hours a day collecting fuelwood loads of
20 kilograms or more. In urban areas it can mean juggling with tight household

incomes to buy charcoal or kerosene. Many of these tasks are demanding of both human energy and time, and they have a disproportionate effect on women's health compared to men's. For example, the higher levels of lung and eye diseases suffered by women as compared to men are attributed to the longer hours of exposure to smoke in kitchens (Smith, 1999). Fuel collection also reduces the time women have available for contributing to other aspects of livelihood strategies.

Women in the South are also responsible for a number of other survival tasks needed to sustain the household, such as water collection and food processing. Again, many of these tasks are demanding of both human energy and time. Energy interventions, many using renewable energy, are available that would do much to reduce the drudgery involved in these daily household activities. However, a number of the tasks would easily be served by diesel engines run on biofuels, such as biogas and vegetable oils. For example, the preparation of many staple root crops takes an hour of vigorous pounding, which can be simply substituted by milling.

The North

Clancy and Roehr (2003) have argued that there are gender and energy issues in Northern households linked to poverty and poor-quality housing. In the North, women with low incomes are disproportionately found as heads of households either as single-parent families or, because of their greater average longevity as compared to men, living alone at pensionable age. For example, the United Kingdom's Office for National Statistics recently released data that showed that 40 per cent of divorcees aged over 65 have such low incomes that they are able to qualify for state aid. This compares with 1 per cent of married women and 23 per cent of divorced men. Another report from the United Kingdom claims that 25 per cent of single women over 60 are living below the poverty line (Guardian Weekly, 2004).

Young children and older people have special requirements for heating to reduce their vulnerability to illness. Heating/cooling and cooking for people on low incomes can be problematic if they have a restricted choice of energy forms: electricity is expensive; and solid fuel produces smoke, which has negative health impacts. People on low incomes often live in housing with poor insulation and frequently use second-hand or old equipment with poor energy efficiency. Also, they often have to pay for their electricity and gas with pre-payment systems which can be charged on a higher unit cost basis than for households with monthly billing systems. A worrying trend of energy market liberalization has been the abandonment of utility programmes that support the purchase of energy-efficient equipment. All these factors contribute to the high energy costs of people on low incomes, and these negative effects fall disproportionately on women who have restricted options.

ENERGY FOR POOR HOUSEHOLDS:
LIGHTENING WOMEN'S BURDEN

Energy can be an enabler for moving people out of poverty by increasing disposable income, by improving the energy efficiency of micro-enterprises or by opening up new entrepreneurial opportunities. In other words, sustainable energy is at the heart of the transition out of poverty. In addition, access to modern forms of energy, in part enabled by increased income-generating opportunities, contributes to a sense of well-being, a primary objective of development. However, it is important to match the energy services people want, in particular reduction of drudgery and cooking, with the form of energy supplied while at the same time recognizing the particular challenges women face compared to men: lack of access to credit and limited knowledge about (and hence confidence with) new technologies.

As was indicated above, women in the South need sustainable energy services that address two recurrent problems in their lives: cooking and drudgery. There are a number of choices of modern energy carriers for cooking, each with advantages and disadvantages. Renewable energy technologies have to compete either with biomass collected at zero financial cost or with other petroleum-based cooking fuels (kerosene and LPG) which have the advantage that they are available through well-established commercial distribution channels. Petroleum fuels provide controllable heat, which is popular with cooks, and it is possible to purchase kerosene in small quantities that match household cash flows.

Electricity

Much of the focus in the energy sector is currently on electricity, in part linked to the liberalization of energy markets and the promotion of solar home systems. Unfortunately, electricity is not the cheapest option for cooking many basic foods, although cooks do appreciate the cleanliness of the energy form. There are examples of successful projects, for example in Nepal, involving electricity for cooking based on mini- or micro-hydro power schemes (Anderson et al, 1999). It should be kept in mind that it is not enough to supply the electricity: there must also be appropriate stoves available, as well as the need to work with cooks to help them adapt their cooking techniques. Solar home systems cannot be used for cooking since their output is too low. Rather, their strength lies in the provision of good-quality light and home entertainment.

The World Bank ENPOGEN study in Sri Lanka, China and Indonesia found that rural people regarded electricity as a basic necessity of daily life. The most significant benefits ascribed to electricity are that it makes home life more convenient and housework easier (Ramani and Heijndermans, 2003). In Sri Lanka the major benefit of electricity was found to be the time that women save. Eighty per cent of the interviewees reported saving between one and two hours through avoided journeys (taking batteries to be recharged, and going to the city to buy kerosene, medication and vaccinations) and on household activities (such as firewood collection, cooking, ironing, boiling water, house cleaning and chimney cleaning) (Massé and Samaranayake, 2003). It is interesting to note how women made use of their extra 'free time'. Twenty-nine per cent of the female household members said that the time they saved was spent on extra housework, while less than 5 per cent reported using it for productive activities.

It is also possible that electricity can contribute to improvements in gender equity with regard to household tasks if the views of one man from an electrified household quoted in the ENPOGEN Sri Lanka study become the norm: 'I am now prepared to do ironing and assist my wife in her work: ironing, boiling of water, cooking.'

Solar cookers

There are many enthusiastic promoters and users of solar cooker technology. However, there are still a number of challenges to overcome with this technology, including variations in sizing to accommodate the cooking needs of different household sizes. One person's advantage (portability for moving around the household) soon becomes another's disadvantage: easy to steal! The technology requires changes in cooking practices and these can be difficult to achieve and sustain. A major drawback of this technology is that it has often been taken up by enthusiastic well-meaning amateurs, and cookers with poor aesthetic design are produced. While this might work in emergency situations, such as refugee camps, when people are asked to part with hard-earned cash they do not want to buy something that looks second-rate.[6]

Biogas

Cooks who use biogas respond enthusiastically to its controllability and cleanliness. However, the cost of a digester and the number of animals required to produce sufficient gas for the household's daily cooking needs is usually beyond low-income households. The collection of the water needed as an input adds considerably to women's burden (in other words, the savings of metabolic energy and time spent on fuel collection is switched to water collection). There can be gender differences in perceptions of the benefits of biogas. An evaluation of a biogas programme in rural India found that men and women both valued the time-saving element but for different reasons (Dutta et al, 1997). The women mainly spoke of time saved in fuelwood collection and cooking (allowing more time to be spent with their families), while the men appreciated faster cooking and more timely meals.

Improved biomass stoves

Biomass will remain the fuel option for many households for the foreseeable future. Therefore, there is a need to produce wood and charcoal stoves that are more efficient and pay attention to safety issues (smoke with wood, and carbon monoxide with charcoal). A programme promoting fuel-efficient stoves in Madagascar is reported as bringing annual fuel savings equivalent to the minimum monthly salary (approximately US$24) to households that adopt the stoves (Bazile, 2002). This level of savings should have a significant impact in low-income households and may be of sufficient order for households to begin to accumulate assets. Supply-side issues of sustainable biomass also need to be addressed.

Each energy carrier has its enthusiastic supporters and detractors. However, solutions for moving people out of poverty should not focus on arguments about *which technology* is best for an end-user, but should be about enabling women to *choose* which option meets their needs and fits their circumstances. The issue is not only that technologies are available to choose from, but also that women can afford to pay for them. The question then becomes: what role can energy play in increasing women's purchasing power, which in turn enables them to buy more energy-efficient equipment to meet their needs?

ENERGY FOR MICRO-ENTERPRISES

Access to income is a keystone in strategies to move people out of poverty. Women already have income-generating activities, many of which are small-scale, often run from the household.

The role of energy as an input factor in small business sustainability is a particularly under-researched area. What constraining role does energy play in the sustainability of small-scale enterprises? The use of stand-alone renewable energy systems is being increasingly promoted in response to the high delivery costs of grid electricity and petroleum energy products in rural areas.[7] The focus in the 1990s was on the provision of electricity, usually for consumptive rather than productive uses (van Campen et al, 2000). There is no doubt that electric light extends the possibility for working in the evening hours. There is quite a debate about whether or not access to electricity adds to the burden of a woman's working day (see, for example, Cecelski 1995; Clancy et al, 2003). Unfortunately, there are insufficient empirical data on what use is actually made of the lighting for a definitive answer to be possible and it would be sensible not to pre-judge.

Certainly electric light can benefit small enterprises, enabling greater efficiency and quality of work, a better working environment and a more attractive and secure environment for customers (security is particularly important for women) (Barua, 1998, quoted in Cecelski, 2002). However, there is a need to promote more productive uses of electricity, thereby increasing household income, which would help to offset the high up-front costs associated with renewable energy technologies.

Women's enterprises use heat and light provided by purchased fuels, such as biomass and kerosene, and are highly energy-intensive. Hence, their viability and costs are affected by prices and availability. Alternatively, women's income generation can also involve significant inputs of their own energy (e.g. oil seed processing). Electricity or biomass-fuelled diesel engines could potentially contribute to this type of need.

Unfortunately, the role of energy in the sustainability of women's enterprises is not well understood. In food processing it has been estimated that energy costs are 20–25 per cent of the total inputs, which would indicate that technological inputs could possibly reduce costs. Women entrepreneurs do want technologies that improve their incomes and the viability of their businesses. For example, women's groups in Uganda use solar dryers for fruit preservation, giving them a better-quality product than air drying, which enables them to export 50 tonnes annually, with significant positive impacts on their income (Okalebo and Hankins, 1997).

WOMEN IN THE ENERGY SECTOR

Women as energy entrepreneurs

The liberalization of energy markets is opening up new opportunities for the provision of energy services. Rural energy service companies (RESCOs) are springing up, offering the potential of good incomes. Women should not be excluded from these opportunities – particularly when their exclusion is based on prejudices that women are not interested in technical matters. Women are already energy entrepreneurs, as a number of stove programmes would testify (see, for example, Khatami-Njenga, 2001). There can be no logical reason why women, given some appropriate training, cannot fill the role of energy entrepreneur just as adequately as men. Box 12.3 gives some details about a women's energy entrepreneur project for solar PV home lighting systems and solar lamps in rural Bangladesh, and shows that with appropriate attention to where women need help to establish themselves as entrepreneurs, success follows. This project also has a possible best-practice model of selling renewable energy technologies in rural communities. The women energy entrepreneurs have set up cooperatives, which allows them to pool resources and offer micro-credit to purchase the lamps the cooperatives produce (Khan, 2001). Box 12.4, which gives details of a community-owned micro-hydro system operated by village women in the western Solomon Islands, goes a step further, demonstrating that an approach that enables women to be entrepreneurs also had the advantage of

---

Box 12.3 BATTERY-OPERATED LAMPS PRODUCED BY RURAL WOMEN IN BANGLADESH

A project funded by the World Bank Energy Sector Management Programme (ESMAP) has been running on the remote island of Char Montaz in the south of Bangladesh since 1999. The project aims to improve the lighting and indoor air quality of rural households by replacing the traditional kerosene lamps with modern fluorescent battery-operated lamps. The fluorescent lamps are produced and marketed by a women's micro-enterprise, and so far about a thousand households are using these lamps. The long-term potential is good, with grid extension within the next 20 years a highly remote possibility and a market of 20,000 households.

The lamp business represents an important opportunity for the women to earn a relatively good wage. If a woman is able to construct and sell two lamps a day, she is able to earn a wage equivalent to that of a skilled labourer, a significant opportunity that both benefits her family and improves her social status.

The remote community also benefits from the lamps, which have a high efficiency and low energy consumption. Their advantage over kerosene lamps is the reduced risk of fire, as well as the elimination of smoke and other emissions with their negative health impacts.

The project from the start has recognized the knowledge of rural women about local conditions and has used major inputs by rural women in the design of the energy service mechanisms. Recognizing that women had gaps in their knowledge about electronic components and lack of skills with the tools needed to work with the components, the project gave appropriate training to ensure that reliable lamps were produced. Training was also given in accounting and bookkeeping. Male family members have also been encouraged to act as advisers to the women, especially on marketing and sales, and in operating battery-charging services, a new activity that has developed from the original project.

What are the indicators of success for the project? Taking gender issues into account; using women's existing knowledge in the project design; providing compensatory training for gaps in technical and business knowledge; gaining male family members' support; providing income-generating opportunities; providing a service the community wants.

Source: Khan (2001)

---

improving women's standing and confidence in society – a change also appreciated by their menfolk.

Women are good candidates to be successful energy entrepreneurs (Batliwala and Reddy, 1996). Women who live in rural areas know local circumstances and understand local needs. A woman may be able to sell more effectively to other women, and access to potential female clients is not hindered by social constraints.

As well as women being entrepreneurs, renewable energy companies can also use women as operation and maintenance technicians based at the local level. The water sector long ago began to train women in the operation and maintenance of handpumps. Women have proved to be more effective in regular and preventive maintenance than men (Cecelski, 2000).

BOX 12.4 COMMUNITY-OWNED MICRO-HYDRO SYSTEM OPERATED BY VILLAGE WOMEN IN THE WESTERN SOLOMON ISLANDS

The women in Bulelavata, a small, remote village in the western Solomon Islands, used to live a subsistence lifestyle typical of women in tens of thousands of other Pacific island villages. In 1998 the community chose to begin the process of establishing an energy-for-development project. In 2001 the community-owned micro-hydro system was officially opened. The system produces 24 kW and has 1.5 km of high-voltage transmission line, enabling the community to sell power to the provincial secondary school.

For the women of Bulelavata the energy project has had some significant and profound impacts, ranging from the practical, quantifiable advantages of lighting and community income to qualitative outcomes such as solidarity and empowerment. The project design used a women's participatory action agenda, exploiting 'action learning', or learning by doing. They had the decided advantage of a context in which a relevant project was happening in their lives, in which workshops could be grounded. The facets encompassed project policy support, female project management, female role modelling at varying levels, specific women's awareness and training workshops (although community ones were also held in which men too participated), visits by women to other villages, management committee positions for women, a new village institution for women, technical team leadership by women, and logistical project support teams being given equal status to that of technical project teams. This affirmative agenda was designed to encourage and facilitate active and meaningful opportunities for participation by the village women, and operated within existing Melanesian cultural and village religious mores while at the same time challenging the boundaries of perceived gender roles through the medium of the new technology.

The Bulelavata village men say that the electricity project has changed their women; that they are now more confident and outspoken and participate more in community development activities. The men think this is a good outcome in terms of the whole project, and rate it second only (by consensus) to the community's understanding of 'planning for tomorrow'.

Source: Bryce and Ching Soo (2004)

Women working in the renewable energy sector

There are no separate statistics giving gender-disaggregated data for the renewable energy industry, but the figures for the energy sector in the North as a whole are depressing. In Germany the share of female technical staff in the energy industry is around 6 per cent, in decision-making positions 4 per cent and in the top management less than 1 per cent (Hoppenstedt-Analyse, 2000). Women in the energy industry work mainly in administration, sales, finance, catering and personnel. However, there are examples of women taking the initiative. For example, women in Germany established their own cooperative to generate electricity for the grid using wind energy (Delfs, 2000).

There is also no gender-disaggregated data available concerning women's participation in the renewable energy sector in the South, although it would appear that women's involvement has not been significant as employees in the formal sector.

So why are women not making headway in the field of renewable energy? There is a lack of empirical evidence to answer this question. Some researchers have suggested that the energy sector has a highly masculine image that deters women (Clancy et al, 2001). There are perceptions that the work involves heavy labour, which is regarded as a significant barrier to female participation. However, in recent years automation has reduced the need for significant inputs of muscular labour. Women are certainly interested in technical subjects – for example, environmental technology – but not courses involving energy. Is it because women miss the ecological and social dimensions in energy courses?

Does it matter if women are not employed in the renewable energy industry? Social justice arguments of fairness or equality would imply that all women and men have the right to the equal enjoyment of socially valued goods, opportunities, resources and rewards, and equal participation in decision-making about those goods (Clancy and Roehr, 2003). Energy is a commodity that both provides services and offers job opportunities. Therefore, women and men should, on the basis of equality principles, have equal opportunities not only to make use of energy but also to be able to participate in all aspects related to energy, including employment. Women should be able to *choose* whether or not to work in the renewable energy sector and not be excluded by artificial barriers. From an economic perspective, can the renewable energy industry afford to miss out on a pool of (female) talent?

LESSONS LEARNED TO DATE: IMPLICATIONS FOR POLICY

Renewable energy can play an important role in increasing access to modern energy services and thereby free woman's time from domestic tasks, permit home study and reading, enable access to educational media and communications in schools and at home, mitigate the impacts of indoor air pollution on women, allow access to better medical facilities for maternal care, including refrigeration and sterilization, and permit income-generating activities. All these activities contribute to improving gender equity. By taking a gender approach, renewable energy suppliers can increase the potential client base and the sustainability in use of their technologies. However, this cannot take place in a vacuum and needs a supporting environment. This section looks at some of the elements of that environment.

Policy can create an enabling environment that ensures access to more sustainable energy services. There is a need for more gender-sensitive energy policies that address women's and men's energy needs equally. One way of ensuring this engendered policy is through creating awareness in policy-makers. This can be done through advocacy activities at all levels. A good example of this type of initiative is a regional meeting in Africa linked to the Pan-African Energy

Ministers Conference, which asked ministers to explore ways of using policy to empower and promote safer and healthier technologies for women and produced the Durban declaration, which committed ministers to these objectives (US Department of Energy, 2000).

More women as decision-makers in the energy sector would be helpful, but men also have a role to play, and they need to be more gender sensitive. The same arguments apply at the implementation level. A critical mass of women and gender-sensitive men is needed in the energy sector to ensure that issues are raised in a concerted and consistent fashion as well as to overcome institutional resistance to gender and energy issues. They also have to have knowledge about renewable energy technologies, the benefits they bring and also their limitations.[8] Training and networking can play a valuable role here. An example of best practice in training women in renewable energy technologies is the programme by the United Nations Environment Programme (UNEP) for training women leaders on the uptake of renewable energy technology (ACRE, 2001).

Networking plays an important role in sharing knowledge and experiences and in advocacy on gender and energy. While there is an active network on advocacy for gender and sustainable energy (ENERGIA[9]), in terms of networking for women working in the renewable energy sector there is no professional association comparable to those found in the other energy sectors – for example, WIN, which is for women working in the nuclear industry.[10]

There is a need for more gender-disaggregated data in the energy sector, not only quantitative data about energy use but also qualitative data about the sort of choices women and men want to make regarding the energy sources they would like to see used. Decision-making should reflect equally men's and women's preferences. The provision of gender-disaggregated data could be done relatively easily, at least within Europe. The European Commission surveys European citizens biennially on their attitudes to a key range of policy issues, including energy. At present the data are not published on a gender-disaggregated basis. A simple adjustment could provide a wealth of information about how men and women within the European Union view energy policy and choices (Clancy and Roehr, 2003).

All too often, the starting point for an intervention to remove people from poverty is the promotion of a particular technology, including energy, without references to context. However, communities in different parts of the South differ in many ways, so sustainable solutions are more likely to be achieved if the process begins with a needs assessment. Drawing on local knowledge about customs and practices, for example harvesting, can be crucial in identifying energy demand. Remembering women's key role in household energy provision is essential for including them in the design, implementation and evaluation of energy projects. Gender analysis can play an important role in ensuring project sustainability and better marketing opportunities, while promoting gender equity.

Reaching the Millennium Development Goals (MDGs) cannot be achieved without the input of energy, and renewable energy has an important role to play here, benefiting both women and men. Women can potentially benefit more than men from increased energy inputs in terms of reduced drudgery and of time-saving. If renewable energy is to contribute to those benefits it has to match the form that energy end-users want. There is overemphasis on the technologies that generate electricity, whereas the major uses of energy both in the household and in small enterprises are for process heat.

Evidence shows that energy projects must have some aspect of income generation for them to be taken up with enthusiasm and any hope of sustainability. The prospects for income generation have to be well mapped out in advance to ensure long-term financial viability. Women entrepreneurs tend to be more risk-averse than men. This means that they need more support to adopt new technologies or set up businesses. The nature of the support has to match the reality of women's lives: their assets and the control over those assets as well as their work patterns. Much has been written about the need for micro-financing and credit schemes, especially for women, and many innovative schemes exist. However, projects aimed at enabling women's access to technologies can still go wrong if they do not also take into account the reality of women's lives. A project in Uganda that set out to encourage women entrepreneurs to purchase solar systems by offering credit through a women's bank failed to reach the target group because interest rates were set well above levels women could meet, repayment schedules were too short, and collateral requirements did not match women's resources (Sengendo, 2001). An example of best practice in micro-credit is the ENSIGN project (Ramani, 2002) of the Asia/Pacific Development Centre and the United Nations Development Programme (UNDP), which combined micro-credit loans for energy services and for corresponding income-generating activities for the poor, including women. The services are co-financed by a revolving fund and national financing institutions, such as the Self-Employed Women's Association (SEWA) Bank in India. An average growth of 124 per cent in income was found in the participating households.

Evidence has been presented that energy is not gender-neutral in the North but gender-blind. There is a lack of empirical evidence from which broad conclusions can be drawn. There is concern, from both a social justice perspective and the economic perspective of loss of quality human resources, that women are not gaining access to the renewable energy sector for technical employment. This lack of women's involvement would also seem to be the case in the South, where women's contact with renewable energies tends to be as passive beneficiaries. An example of best practice in encouraging women into renewable energy as technical expertise is the ENEQO project[11] within the electricity industry in Europe, which aims to advance equal opportunities by promoting the positive benefits that employing women brings to the working environment (Clancy et al, 2001).

NOTES

1 The role of television in women's empowerment should not be underestimated. For example, in Tunisia, watching television enabled women to become more aware of political events and to have a greater knowledge of world events than their husbands. Through this knowledge they have gained confidence to speak out and take up leadership roles (Chaieb and Ounalli, 2001).
2 Although some might argue that new forms of drudgery and new health concerns have appeared with computerization.
3 Using a Northern measure of poverty – that is, defining poverty purely in economic terms as consuming less than US$1/day of goods and services – the number of people living in poverty in 1993 was estimated as around 1.2 billion, which is around one-fifth of the world's population (World Bank, 2001). There is little evidence to suggest that this figure has decreased significantly, at least in absolute terms.
4 In fact, many development practitioners do not recognize energy poverty (defined by Reddy (2000) as the absence of choice in access to adequate, affordable, reliable, high-quality and environmentally benign energy services) as a concept.
5 It has to be acknowledged that this includes advocating the benefits of renewable energy technologies over fossil and nuclear fuels without reference to context.
6 This applies to any technology.
7 See, for example, the Village Power Conferences: www.rsvp.nrel.gov/vpconference/vpconference.html; Turkenburg (2000).
8 There are a lot of disappointed solar home owners who cannot use their electric irons.
9 www.energia.org.
10 www.win-global.org.
11 www.sydkraft.se/eneqo/home.htm.

REFERENCES

ACRE (2001) *Report on the Global Seminar for Women Leaders on the Uptake of Renewable Energy Technology*, 27 June to 4 July, Murdoch University, Perth, Australia, http://acre.murdoch.edu.au/unep/

Anderson, T., Doig, A., Rees, D. and Khennas, S. (1999) *Rural Energy Services*, London, ITPublications

Barnett, A. (1999) *Increasing Access to Sustainable Energy Sources: A Summary of Recommendations Made to the Sustainable Energy Programme of the Shell Foundation*, 20 September

Barnett, A. (2000) *Energy and the Fight against Poverty*, Paper given as part of series of Economic Research Seminars at Institute of Social Studies, 29th June 2000, The Hague

Batliwala, S. and Reddy, A. (1996) 'Energy for Women and Women for Energy: Empowering Women through Energy Entrepreneurship', *ENERGIA News*, vol 1, no 1, pp11–13

Bazile, D. (2002) 'Improved Cookstoves as a Means of Poverty Alleviation', *Boiling Point*, no 48, pp20–22

Bryce, D. and Chin Ching Soo (2004) 'Bulelavata Women Speak', *ENERGIA News*, vol 6, no 2, pp7–9

Cecelski, E. (1995) 'From Rio to Beijing: Engendering the Energy Debate', *Energy Policy*, vol 23, no 6, pp561–575

Cecelski, E. (2000) *The Role of Women in Sustainable Energy Development*, NREL/SR-550–26889, National Renewable Energy Laboratory, Golden, CO

Cecelski, E. (2002) 'Enabling Equitable Access to Rural Electrification: Current Thinking on Energy, Poverty and Gender', Briefing Paper for the Asia Alternative Energy Policy and Project Development Support: *Emphasis on Poverty Alleviation and Women*, Asia Alternative Energy Unit, World Bank, Washington, DC

Chaieb, S. and Ounalli, A. (2001) 'Rural Electrification Benefits Women's Health, Income and Status in Tunisia', *ENERGIA News*, vol 4, no 4, pp18–20

Clancy, J. S. and Roehr, U. (2003) 'Gender and Energy: Is There a Northern Perspective?', *Energy for Sustainable Development*, vol vii, no 3, pp44–50

Clancy, J. S., Skutsch, M. M. and Batchelor, S. (2003) 'The Gender–Energy–Poverty Nexus: Finding the Energy to Address Gender Concerns in Development', paper prepared for the UK Department for International Development (DFID), London, DFID Project CNTR998521, www.utwente.nl/tdg

Clancy, J. S., Stienstra, M., Gregory, J. and Cornland, D. (2001) 'Gender and Energy – Women's Concerns in Energy: Background and State of the Art', Working Paper 1, study carried out for the European Commission Research Directorate

Delfs, C. (2000) 'Case Study: Women Producing Electricity from Renewables: From "Project" to "Market": The "Windfang" Project in Germany', *ENERGIA News*, vol 3, no 1, pp9–11

Dhanapala, K. (1995) *Report on the Gender Related Impact of Micro Hydro Technology at the Village Level*, Intermediate Technology, Study Report 2

Dutta, S., Ibrahim, H., Malhotra, P. and Venkata Ramana, P. (1997) *Biogas: The Indian NGO Experience, AFPRO-CHF Network Programme*, New Delhi, Tata Energy Research Institute

Guardian Weekly (2004) 'Divorced women face poor old age', *Guardian Weekly*, 5–11 February, p11

Hoppenstedt-Analyse (2000) *Frauen im Management 1999: Branchenanalyse 1* (Women in Management 1999; Analysis of Industrial Branches – Large and Medium-Scale Industries), Hoppenstedt-Verlag, Darmstadt

International Programmes: Focus on Intermediate Technology Development Group, *ENERGIA News*, vol 2, no 2, pp4–5, 2 May 1998

Khan, H. (2001) 'Battery-Operated Lamps Produced by Rural Women', *Generating Opportunities: Case Studies on Energy and Women*, New York, United Nations Development Programme

Khatami-Njenga, B. (2001) 'Upesi Rural Stoves Project', in Misana, S. and Karlsson, G., United Nations Development Programme, *Generating Opportunities: Case Studies on Energy and Women*, New York, United Nations Development Programme

Massé, R. and Samaranayake, M. R. (2003) 'EnPoGen Study in Sri Lanka', *ENERGIA News*, vol 5, no 3, pp14–16

Mehretu, A. and Mutambira, C. (1992) 'Gender Differences in Time and Energy Costs of Distance for Regular Domestic Chores in Rural Zimbabwe: A Case Study of the Chiduku Communal Area', *World Development*, vol 20, no 11, pp1675–1683

Okalebo, J. and Hankins, M. (1997) 'Why Women Adopt Solar Dryers', *ENERGIA News*, vol 1, no 3, pp6–7

Ramani, K. V. (2002) 'Energy as an Instrument of Women's Economic Empowerment', *ENERGIA News*, vol 5, no1, pp8–10

Ramani, K. V. and Heijndermans, E. (2003) *Energy, Poverty, and Gender: A Synthesis*, Washington, DC, World Bank

Reddy, A. K. N. (2000) 'Energy and Social Issues', *World Energy Assessment*, New York, United Nations Development Programme

Sengendo, M. C. (2001) 'Photovoltaic Project for Rural Electrification', in S. Misana and G. Karlsson (eds) *Generating Opportunities: Case Studies on Energy and Women*, New York, United Nations Development Programme

Skutsch, M. M. (1998) 'The Gender Issue in Energy Project Planning: Welfare, Empowerment or Efficiency?', *Energy Policy*, vol 26, no 12, pp945–955

Smith, K. R. (1999) 'Indoor Air Pollution', *Pollution Management in Focus*, Discussion Note 4, August, Washington, DC, World Bank

Turkenburg, W. C. (2000) 'Renewable Energy Technologies', *World Energy Assessment*, New York, United Nations Development Programme

US Department of Energy (2000) *Proceedings of the Women in Energy Ministers' Conference*, Durban, 11–15 December, www.worldbank.org/html/fpd/esmap/pdfs/wew_Proceedings_eng.pdf

van Campen, B., Guidi, D. and Best, G. (2000) 'Solar Photovoltaics for Sustainable Agriculture and Rural Development', Environment and Natural Resources Working Paper 2, Rome, Food and Agriculture Organization

Wides, L. (1998) 'Gender Lessons from Community Based PV Projects in Rural Guatemalan Villages', *ENERGIA News*, vol 2, no 1, p9

World Bank (2001) *World Development Report 2000/2001: Attacking Poverty*, Washington, DC, World Bank/Oxford, Oxford University Press

PART FIVE

CAPACITIES

13

Research and Development: The Basis for Widespread Employment of Renewable Energies

Joachim Luther and Carsten Agert

INTRODUCTION

Research and development (R&D) is a prerequisite for the evolution of global civilizations towards sustainable schemes. The development of the energy sector is especially relevant as it is inevitably linked to many aspects of sustainability, including protection of the natural life-support systems, the eradication of energy poverty in developing countries and the prevention of geopolitical conflicts. This evolutionary process may to a certain extent already be based on successful R&D and industrial activities in many countries. However, in order to set up a global sustainable energy system, much more R&D is needed in a wide area of fields, for example:

- on sustainable technologies for energy conversion, storage, transport and use;
- on economic, political and institutional schemes for the transformation of energy systems on all levels (local, regional, global);
- on appropriate implementation of new energy technologies into societies;
- on methods of financing the evolution of the energy system on a global scale;
- on management structures to implement, organize and monitor global transformation.

The basis for R&D in these areas must be set up or reinforced globally. The chances and prospects of international collaboration and task-sharing must be taken into account. Further, the significant time lag between the application-oriented basic research, the development of strategies and technologies, and their market implementation must be considered in R&D strategies.

Apart from a renewable energy focus, many other aspects related to a sustainable evolution of the global energy system require strong R&D efforts, such as energy efficiency in all sectors and on all levels, and sustainable transport concepts. It is important to realize that three major aspects are integral parts of the transformation strategy towards a sustainable energy system:

- renewable energies;
- efficiency;
- less carbon-intensive fossil energy conversion.

It is argued that the subjects for non-technological R&D described below apply equally well to these three major aspects. The focus in this chapter, however, will be on the R&D on renewable sources of energy. The other elements of a transformation strategy towards a sustainable energy system were, in most cases, beyond the scope of the Renewables 2004 conference.

In the field of renewable energies, non-technological R&D comprises a wide variety of subjects (economic, sociological, political, etc.). For example, it is of major importance to develop well-adapted dissemination strategies in both industrialized and developing countries. In this context, the effects of liberalization and globalization on the application of renewable energies should be further analysed. There is a pressing need to develop strategies for enhanced capacity-building, to perform studies on acceptance and on raising awareness. Stronger R&D on improved financing schemes and the minimizing of overall costs is of utmost importance.

SUSTAINABLE POTENTIALS AND LONG-TERM R&D

According to estimates made using the direct equivalent method (Roehrl and Riahi, 2000), the world's primary energy consumption was approximately 420×10^{18} joules in 2000, of which 85 per cent came from fossil fuels. In the context of increasing the share of renewable energies against the background of a rising overall energy demand, the potential of various renewable energies to contribute to a sustainable global energy supply is obviously of major importance. Frequently the related technical potential of renewable energies is much higher than the actual sustainable potential in which other aspects of sustainability are considered, such as land use for biomass energy. In 2003 the German Advisory Council on Global Change (WBGU) identified the sustainable potential of different renewable energy sources: for example, 100 exajoules (EJ) per year for modern biomass use, 140 EJ per year for wind power and 15 EJ per year in the long term for hydro energy (WBGU, 2004). Only the sustainable potential of solar energy can be termed quasi-unlimited against the background of all projections of future human energy demand.

Renewable energies can be classified into two groups of energy sources: (1) already competitive, but limited overall potential (e.g. wind, water, biomass, photovoltaics for off-grid applications); and (2) still expensive, but quasi-unlimited overall potential (e.g. solar electricity by solar thermal power plants or photovoltaics). It is possible for the contributions to the global energy supply from the first group to be significantly increased over a short time period. Bulk solar electricity, however, will require a few decades to make large contributions on a global scale. It is absolutely essential to continue and enhance R&D work and efforts towards market introduction for all forms of renewable energy. Only the technologies with quasi-unlimited potential can cover human energy demand in the long run (Figure 13.1).

The 'renewable energies' curve assumes high dissemination rates up to the maximum sustainable potential of the related sources, while the 'solar electricity' curve assumes a tenfold capacity increase per decade. All renewable energies together (= the sum of these two curves) are not expected to fully cover the primary energy consumption in the foreseeable future, so that the (decreasing) difference will have to be delivered by conventional sources. As a result, efficiency improvements and less carbon-intensive fossil energy conversion are also of high importance. All the curves shown are normalized to the current global primary energy consumption. The figure underlines the long-term aspects of R&D in renewable energies and illustrates some major issues, as follows.

Ambitious efficiency improvements have to be an integral part of a transition towards a sustainable energy system.

An ambitious, but exclusive, expansion of the mostly *already competitive*

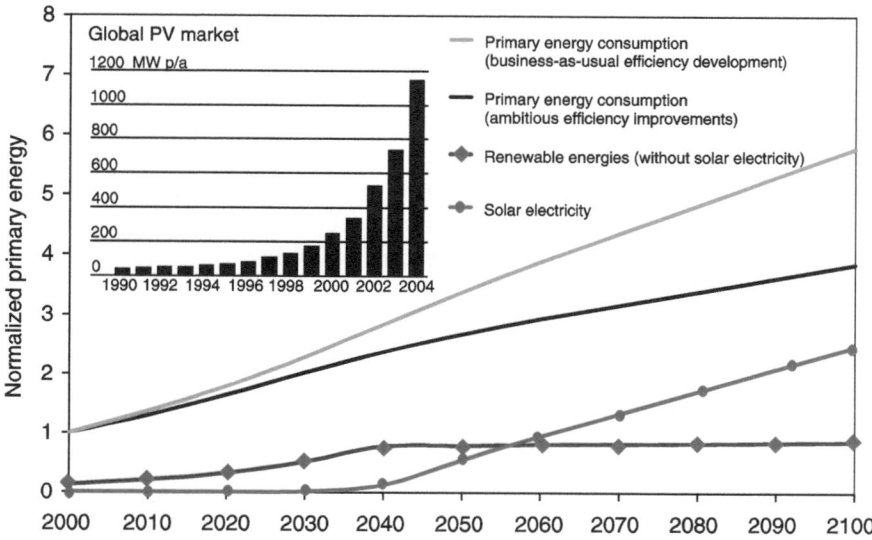

Figure 13.1 *Global primary energy consumption in a fictitious high-demand scenario*

renewable energy sources with limited overall potential (wind, hydro, biomass, etc.) is a necessary but not a sufficient condition for a future sustainable energy system, as the limit of the related sustainable potentials might be reached rather soon. Only the further expansion of the nowadays *still expensive renewable energy sources with quasi-unlimited overall potential* (i.e. mainly solar electricity and probably solar hydrogen) can lead to a fully sustainable energy system in the long term (WBGU, 2004).

The 'solar electricity' curve represents a tenfold increase per decade of the solar electricity capacity from the very beginning of the century. However, even this ambitious development will not lead to significant contributions to global primary energy consumption until the middle of the century. Exponential growth starting from a very low level needs a long time to achieve high values and must be sustained continuously to meet the related targets. The inset shows the observed exponential growth of global photovoltaics shipments.

General goals of R&D

In addition to mere energy conversion, it is vital to develop novel supply structures: distributed generation and grid optimization, transregional energy transport and eventually a global link for electricity, energy storage, load management, and so on. It is also essential to improve the solutions for end-users: domestic co-generation of electricity, heat and cold based on renewable energy sources, biogenic bottled gas as decentralized energy carrier, and cars powered by fuel cells. In all these cases, emphasis should be put on ecologically benign technologies (including recycling) and the cost-effective generation, transport and utilization of energy. In this context, continuous R&D efforts are essential for economies of scale, as exemplarily shown in Figure 13.2 for photovoltaics.

So-called price experience curves (or 'learning curves') are an established economic tool with which to express economies of scale and R&D-based techno-logical improvements. They are based on the empirical observation of a constant percentage at which the unit cost generally decreases with every doubling of cumulative production. Learning curves can therefore be used to assess future price levels of a particular product.

Photovoltaics, for example, has shown a price reduction of approximately 20 per cent with every doubling of the cumulative production over many years. This development must be sustained if a competitive price level is to be achieved in the future. The arrows given in the figure illustrate that, while enhanced R&D will contribute to reduce prices, a reduction of R&D will bend the price curve into the direction of slower price reductions. R&D is therefore an essential prerequisite for competitive renewable energy prices in the future.

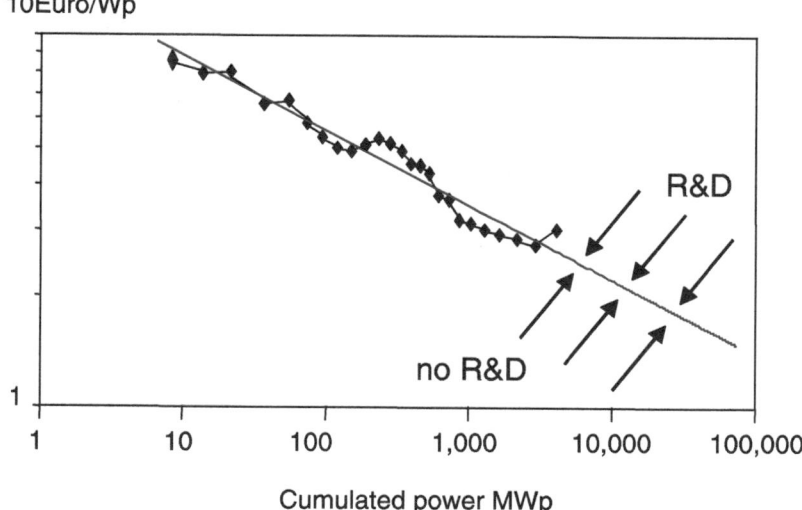

10Euro/Wp

R&D

no R&D

1

1   10   100   1,000   10,000   100,000

Cumulated power MWp

Figure 13.2 *Exemplary price experience curve: photovoltaic modules (crystalline silicon)*

R&D should address today's technologies (upper left arrow), technologies on the horizon as well as 'beyond-the-horizon' ideas of energy conversion (upper right arrow). The given price experience curve for photovoltaics displays a very favourable gradient. To a certain extent other renewable energy conversion techniques show lower price reductions. The slight price increase indicated by the last data point is caused by a silicon shortage due to the strong market growth in 2007/2008. New raw material will soon be available on the market and prices are then expected to follow the learning curve again.

Two main approaches must be followed. First, new technologies have to be developed in some areas, including biogenic bottled gas as decentralized sustainable energy carrier, low-cost energy-efficient houses, and additional storage schemes for high-quality energy. Second, cost reductions for existing renewable energy technologies have to materialize. The latter include higher efficiencies of energy conversion, longer technical lifetimes for components, less maintenance and less material consumption.

Parallel to this, R&D indirectly linked to the energy sector can foster the development of renewable energy technologies to a considerable extent. Developments in the fields of, for example, stand-alone electronic devices, off-grid telecommunication systems, and space applications may trigger additional developments and stimulate a synergetic market growth. Related applications may therefore increase the market volume of certain renewable technologies and thus help drive down the price experience curves.

STATE OF AFFAIRS

Public global expenditure on R&D in the energy sector stems primarily from a small number of industrialized countries (WBGU, 2004). For example, in 1995, 98 per cent of all International Energy Agency (IEA) member country energy R&D was carried out by only ten countries (Margolis and Kammen, 1999). Figure 13.3 shows the public RD&D (research, development and demonstration) budgets of 23 IEA member countries for selected fields of energy-related research. About 70 per cent of the expenditure shown was put into nuclear fission and fusion. Only 10 per cent was in the field of renewable energies. The overall energy R&D expenditure peaked in 1980. Since then it has been continuously declining to less than half its maximum level. At the same time, private R&D expenditure declined significantly in many industrialized countries, with the energy sector showing a particularly low R&D intensity – that is, R&D as a percentage of net sales. This is especially worrying since evidence shows that energy R&D expenditure and patents are highly correlated – that is, investment and innovation are closely linked. It is obvious that the global trend in energy-related R&D does not reflect the increasing strategic importance of the energy sector's evolution.

Figure 13.4 shows the public RD&D budgets for renewable energy research in 23 IEA member countries. Similar to the pattern shown in Figure 13.3 for overall energy expenditure, the expenditure on renewable energies peaked in

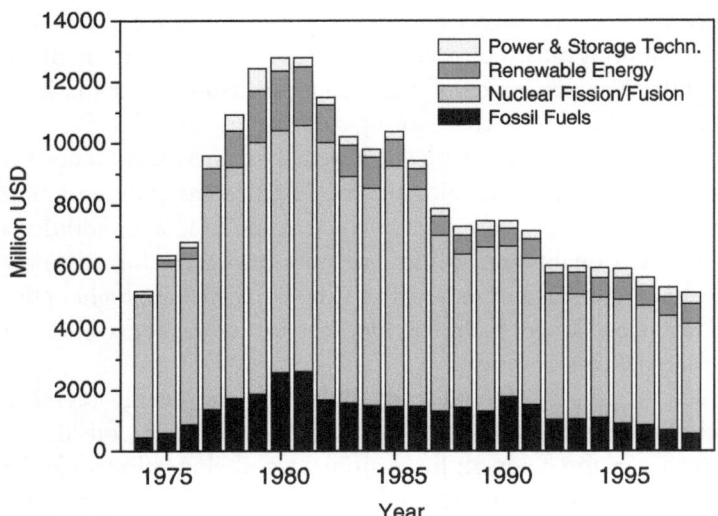

Source: IEA Energy Technology R&D Statistics Service

Figure 13.3 *Public RD&D budgets of 23 IEA member countries for selected fields of energy-related research*

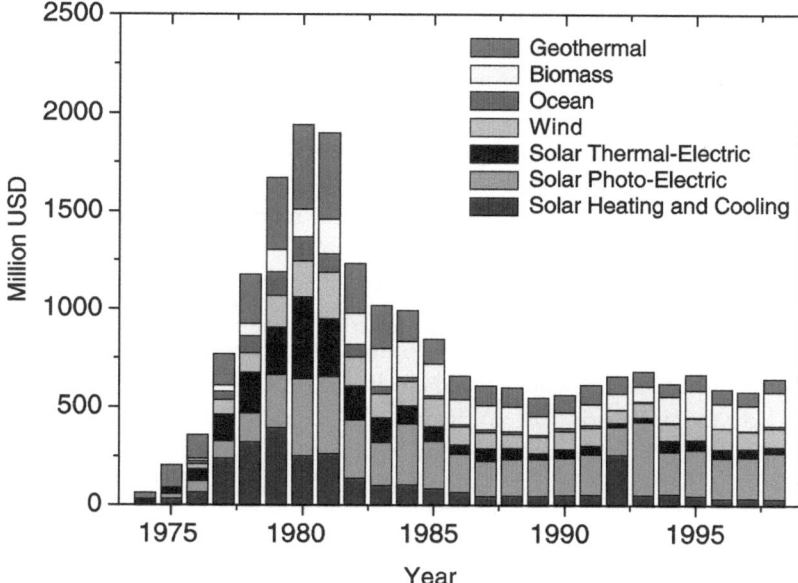

Source: IEA Energy Technology R&D Statistics Service

Figure 13.4 *RD&D budgets of 23 IEA member countries for research on renewable energies*

1980 and has since declined to about one-third of its maximum level. Because less investment means less innovation, this trend is also in clear contrast to the ever-increasing importance assigned to renewable energies.

Two major developments may be identified when analysing the relative proportions of the different renewable energies within the overall renewable energy RD&D budget: while biomass and photovoltaics (solar photoelectric conversion) show a trend to rising proportions, the other sectors remain on a constant or slightly decreasing relative level.

Various conclusions can be drawn from the above analysis concerning industrialized countries:

- R&D on energy technologies is concentrated in a small number of countries.
- Overall expenditure on energy R&D as well as the related expenditure on renewable energies peaked in 1980 and has significantly declined since then.
- Expenditure on nuclear technologies (fission and fusion) is many times higher than expenditure on renewable energies.

Information on R&D activities in developing countries is quite limited and not available in an analysed format, but it is evident, as for the OECD countries, that only a small number of the larger countries have real R&D programmes on renewable energy technologies. Countries such as China, India, Brazil, South Africa, Egypt and a few others do have major programmes dedicated to related R&D with significant developments. There are only very limited or no co-ordinated national programmes on R&D in most of the smaller and poorer countries, although there have been some attempts to create small-scale programmes in universities. However, these have often received extremely limited funding, and laboratories sometimes become a random selection of donated equipment with little associated work.

As already mentioned, the energy sector is inevitably linked to many aspects of sustainability, such as protection of natural life-support systems, the eradication of energy poverty in developing countries and the prevention of geopolitical conflicts. The above trends in R&D expenditure in both industrialized and developing countries go against the related significance of R&D on renewable energies and therefore do not meet the challenges of sustainability.

R&D requirements

R&D measures must be adapted to the overall strategy for the evolution of energy systems. Certain technological solutions and non-technological approaches have to be ready for implementation at particular points in time (Figure 13.1), and hence the significant time lag between R&D and market launch must be borne in mind. R&D on renewable energies is therefore a strategic field of research and industry policy which is inadequately steered and supported at present (Figures 13.3 and 13.4). The following subsections provide an overview of important R&D aspects for the future. The list clearly has an exemplary character and cannot be complete. It shows the wide span of interrelated R&D challenges. There is no 'golden road' based on just a small number of approaches. However, it is possible to define certain regional focal points.

R&D on non-technological aspects (e.g. economic, sociological, political)

- *R&D to develop proper sustainability indicators.* The 'sustainability level' of a certain development can be measured in part. It also depends, however, on background and cultural upbringing. There are no ideal independent indicators. Nevertheless, a society needs guidelines to decide which developments seem more attractive or less dangerous than others. R&D to determine realistic but flexible sustainability indicators is vital for proper policy choices.
- *R&D on the way innovation processes work.* The way innovation processes work should be analysed in more detail in relation to renewable energies.

The best ways to stimulate innovation in the development of renewable energy systems should be found (Jacobsson and Johnson, 2000). Barriers to a speedy transformation of the energy sector should be identified and overcome.

- *R&D on worldwide price developments.* More reliable long-range price assessment techniques should be established for the technologies within the renewable energy sector. With these, better scenarios for R&D investments can be developed.
- *R&D on model projects.* Five large-scale model demonstration projects have been proposed (WBGU, 2004) that would need intensive accompanying R&D, especially in the areas of monitoring and analysis of such undertakings:
    — a strategic energy partnership between the European Union and North Africa, integrating into European power supply the potential of solar energy use in a manner profitable for both sides;
    — developing the infrastructure needed to substitute traditional biomass use by biogenic bottled gas;
    — low-cost energy-efficient buildings in developing countries;
    — improving the power quality in weak electrical grids in rural regions of developing countries via the integration of novel distributed power supply technologies (renewable energies, storage systems, suitable power electronics, etc.);
    — a 'one million huts electrification programme' for developing countries, generating the necessary internal dynamics for rural electrification.
- *R&D on economic foundations.* The worldwide economic foundations of energy system development schemes should be further analysed. Research on well-adapted dissemination strategies in industrialized and developing countries should be enhanced. Improved financing schemes for the effective evolution of today's energy system should be developed.
- *R&D on the best use of international instruments such as the Clean Development Mechanism (CDM) and Joint Implementation (JI).* CDM and JI should be further analysed with regard to promoting renewable energies in the countries involved. The lessons learned from past experiences should be shared. Options for improvement should be considered in the further development of the Kyoto Protocol.
- *R&D on external costs of renewable, fossil and nuclear energy supply systems.* The external costs of renewable, fossil and nuclear energy supply systems should be analysed in more detail. Strategies to internalize external costs and thus level the playing field for renewable energies should be developed.
- *R&D on the effects of liberalization and globalization in the energy sector.*
- *R&D on improved capacity-building.*
- *Best practice analyses.*
- *R&D on acceptance and awareness-raising.*

- *R&D on data and statistics.* Sound planning decisions require a reliable data background, such as resource assessments of renewable energies in developing countries, and data on market penetration of renewable energy technologies.
- *R&D on the effects of the energy system on human health.* R&D on energy-related health problems should be intensified. Related solutions based on renewable energy should be developed.
- *R&D on income-generating end-user energy services.* Renewable energies have the potential to make income-generating end-user energy services available in currently non-electrified regions. Related research, including regional analyses of well-suited options to cover the resulting energy demand, should be enhanced.

R&D on renewable energies for electricity production

- *Wind energy.* Development of offshore potentials up to water depths of approximately 100 metres, environmental impacts, adaptation of the existing onshore technology to the particular requirements of applications in developing countries and extreme climates, hybrid applications (e.g. water pumping and desalination), further cost reductions through improved technologies, better grid integration and better operation management.
- *Photovoltaics.* Cost reductions for silicon solar cells, thin-film cells, cells for optical concentration, novel physical approaches, life-cycle analyses and recycling, power stations, system technology, building integration.
- *Solar thermal power plants.* Plant automation, thermal energy storage, novel materials, direct evaporation and novel concentration concepts (e.g. Fresnel concentrators), solar towers; implementation of gas turbines, solar–fossil and solar–biomass hybrid systems.
- *Hydro power.* Further improvement of micro-hydro technologies, advanced environmental risk assessments and full greenhouse gas accounting for dam projects.
- *Biomass energy.* Interface technologies to thermodynamic energy converters (e.g. co-generation plants, micro-gas turbines, Stirling engines), integration of related systems into (micro-)grids, optimization of the agricultural food–energy link, infrastructure/logistics for collecting resources, interface technologies to hydrogen technologies (syngas, fuel cells).
- *Geothermal energy.* Exploration methods, stimulation techniques, efficient thermodynamic energy converters for low-temperature input, assessment of the overall systems efficiency, environmental impacts (waste heat, minerals, etc.).
- *Maritime sources.* Technologies to exploit tidal and marine currents as well as wave energy, ocean thermal energy conversion (OTEC).

R&D on renewable energies for the production of heating and cooling energy

- *Solar water heating.* Long-term low temperature thermal storage, efficient collectors for process heat and sea water desalination, improved operation management.
- *Solar cooling.* Development of electric/thermal hybrid systems, materials science (e.g. sorbents), system technology, dehumidification.
- *Biomass energy.* Adaptation of conventional heating plants to biomass-based fuels, combustion with reduced emissions and less corrosion.
- *Geothermal energy.* Improved thermally driven heat pumps, seasonal heat energy storage, exploration methods, stimulation techniques.

R&D on solar and energy-optimized buildings for domestic and non-domestic use (including solar heating and cooling)

- *Solar optimized windows.* Improved heat insulation at high solar transmission, switchable glazings, daylighting concepts (e.g. microstructures).
- *Wall insulation.* Active solar opaque façade for heat storage, vacuum insulation.
- *Heat/cold storage.* Micro-encapsulated phase change materials as wall coatings, air-to-ground heat exchangers, nocturnal ventilation concepts.
- *Novel HVAC (heating, ventilation, air conditioning) concepts.* Compact ventilation and heating units, solar thermal cooling, heat pumps.

R&D on renewable energies for fuel production

- *Biogenic fuels.* Feedstock suitability, cost reductions for mature technologies (e.g. biodiesel), production of syngas and synthetic fuels, gas purification for fuel cells.
- *Hydrogen from various renewable sources.* Advanced electrolysis, utilization of solar thermal energy in methane reforming, direct water splitting.
- *Solar chemistry.* Photochemical synthesis of energy carriers, photobiological and photochemical hydrogen production.

R&D on comprehensive technological aspects

- *Distributed generation and grid optimization.* Integration of fluctuating and dispersed energy sources into electricity grids, co-generation of heat/cold/electricity, advanced demand-side management, power electronics, related information technology, demand forecasting.
- *System technology for off-grid applications.* Advanced maintenance and diagnostics concepts, electricity storage systems, theft protection.

- *Transregional electricity and gas transport, global link.* High-voltage direct-current transmission, high-temperature superconductors.
- *Heat storage.* Phase change materials, sorbents, long-term storage concepts.
- *Storage of high-exergy energy.* Hydrogen and fuel cell technology, new electrochemical storages (batteries), storage based on superconductors.
- *Energy meteorology.* Spatially resolved prognosis of renewable energy potentials, local forecast of energy fluxes (e.g. wind speeds).
- *R&D on the sustainability effects of the large-scale application of renewable energies.* Possible recycling of wind turbine blades, minimization or reuse of toxic materials in photovoltaic module production, reduction of emissions during biomass conversion, reduction of energy payback times of renewable systems.
- *R&D on standardization of renewable energy products.*
- *R&D on solar city planning.*

Some of the above-mentioned technologies are already in application; some are visible on the horizon. Technologies that are still beyond the horizon are broadly discussed in academia. It is important to identify a time-dependent, optimized mix of the different technologies and beyond-the-horizon-approaches (FVS, 2003).

Research requirements have to be well analysed not only in respect of time but also in respect of place. For some countries it might be advantageous to develop novel high technologies (e.g. in photovoltaics); for others it might be beneficial to work primarily on the adaptation of existing technologies to local needs. Concepts that appear well suited to a certain region might be inappropriate in other cultural or climatic environments. In addition, the acceptable costs of renewable energy applications vary widely between different regions. Country-tailored approaches are therefore an essential element of research planning.

It is important to perceive R&D and the application of its results in the energy system as a continuous interrelated process that leads to a sustainable energy system in the long term. In particular, R&D strategies have to be reviewed and updated on a regular basis.

POLICY IMPLICATIONS AND RECOMMENDATIONS

Both non-technological and technological R&D on renewable energies are essential for the evolution of the energy sector towards sustainable schemes. R&D is a vital prerequisite for the development of well-adapted dissemination strategies, schemes for enhanced capacity-building and for raising awareness, price reductions, the optimization of existing conversion technologies, the exploration of novel potentials as well as the development of new supply structures and of more user-friendly applications. Against this background it is highly paradoxical that R&D expenditure on renewable energies peaked in 1980 and thereafter has declined to about one-third of its maximum level. In order to reverse this trend it is recommended that the average direct state expenditure for R&D in the renewable energy sector in industrialized countries be increased at least tenfold up to 2020. At the same time, significant international support must also be directed to R&D in developing countries.

R&D structures and institutions

In order to reach the ambitious targets, educational and scientific institutions must adapt to meet the challenges. Synergies should be created and parallel work avoided. Networks of universities and even virtual R&D institutions could serve as powerful and efficient elements of future structures. In this context, R&D on renewable energies should be integrated into existing R&D structures in order to use mutually advantageous conjunctions (e.g. in the areas of material sciences and nanotechnologies). Partnerships should be based on a win–win philosophy. The issue of intellectual property rights on technologies has to be addressed. Particular support should be given to R&D cooperation between developing countries. Perhaps initial catalytic support could come from well-established R&D institutions, for example involving technology transfer and capacity-building.

R&D on renewable energies can be improved at international academic institutions in a variety of ways:

- Related curricula could be fostered and coordinated internationally (see the note later on a proposed 'World Energy Research Coordination Programme').
- Academic degrees that are approved internationally should be established.
- Exchange programmes for scientists and lecturers should be intensified.
- Stimulating rewards should be created for outstanding young scientists and lecturers.
- New research groups should be founded, particularly in places with little activity at present.
- Materials sciences should receive distinctive support, as they lay foundations for novel technological solutions in a wide range of fields.
- Appropriate internet discussion forums for scientists should be established on subjects related to renewable energy R&D.
- Information campaigns aimed at decision-makers and the general public should be mounted in order to raise awareness concerning the importance of R&D of renewable energies.

In order to foster and coordinate global R&D in an optimal way, the following issues should be addressed at central institutions on a global basis – for example, within the United Nations:

- *analysis and evaluation* of the global energy system with an emphasis on sustainability, identification of options for actions, and R&D;
- global and regional *consulting and coordination of R&D*, especially in scientific and technological areas of paramount importance;
- *financing* of R&D activities and related institutions in order to realize the global development and implementation of renewable energy technologies with due speed.

These issues are detailed in the following:

- *Analysis and evaluation.* It is important that the political implementation of a sustainable energy system receives continuous support through independent scientific input. This is currently the case in climate protection policy through the Intergovernmental Panel on Climate Change (IPCC). Thus, the German Advisory Council on Global Change proposed an Inter-governmental Panel on Sustainable Energy (IPSE) charged with analysing and evaluating global energy trends and identifying options for action and R&D (WBGU, 2004).
- *Consulting and coordination.* A 'World Energy Research Coordination Pro-gramme' (WERCP; WBGU, 2004) could be established, perhaps within the UN system. Here, the various strands of national-level energy research activities could be drawn together in analogy to the World Climate Research Programme. For example, shortcomings could be identified, novel approaches evaluated, advice given to local and regional scientific organiza-tions, and scientist exchange programmes organized. Task-sharing should be stimulated; that is, global research strategies could be broken down into country-tailored national activities that meet the local demands (development of novel technologies, adaptation of existing technologies, etc.). The overall focus of WERCP should be on consulting and coordinating.
- *Financing.* International financing organizations (e.g. the World Bank, the Global Environment Facility) currently focus on the implementation of renewable energies but not on related R&D. This is a particular problem for countries with smaller R&D capability that want to carry out research on the adaptation of existing technologies to local needs, say. A strategic global fund for R&D on renewable energies should therefore be established. The fund could be managed by the WERCP. In order to mobilize financial resources for the fund, innovative approaches should be taken into consideration. Apart from potential mechanisms related to the further development of the

.   Kyoto Protocol, raising charges for the use of global common goods, such as international airspace, deserves examination.

Last but not least, it should not be forgotten that strong R&D is carried out in industry. If government policy ensures proper markets (such as is currently the case for photovoltaics and wind power in several countries), then industries see the market potential and perform their own production-oriented research, complementing the work of public research institutions. A proper balance has to be found between publicly sponsored R&D and industry-financed R&D. This can be done through forming R&D consortia, where industry, R&D institutions and government work closely together.

Key issues in R&D

An exemplary list of R&D requirements has already been given in the relevant section of this chapter. The following paragraphs summarize the main issues.

Recommendations concerning non-technological R&D
Identifying the individual and institutional barriers that hinder the widespread use of renewable energies is a vital prerequisite for the future development of the energy system. Strategies to overcome the identified barriers should be developed. Further R&D aiming at well-adapted dissemination strategies in industrialized and developing countries is necessary. At the same time, research on the external costs of all energy technologies should be intensified, and strategies to internalize these costs should be developed. The following points are also of primary importance:

- national market development strategies;
- development of strategies for enhanced capacity-building, awareness-raising and information on renewable energies and their benefits;
- research on optimum financing instruments that are well adapted to regional characteristics (for example, feed-in laws and quota or certificate schemes for grid-connected renewable energies, fee-for-service schemes and micro-credits for off-grid electricity generation);
- research on the CDM and JI in the context of renewable energy promotion.

Recommendations concerning technological R&D
A large number of technologies can and must contribute to the evolution of a sustainable energy system. There is no 'golden road' based on just a small number of approaches. R&D should cover the whole range of approaches listed under 'R&D requirements'. A medium-term focus, however, should be given to:

- technologies that can be applied worldwide today in a cost-effective manner (such as solar and energy-efficient buildings, photovoltaic electricity for off-grid applications, many biomass technologies);
- technologies that need only minor development steps in order to enter completely new and large markets (for example, solar thermal power plants, wind energy in developing countries, biomass-based synfuel);
- technologies that are absolutely essential for long-term development (e.g. photovoltaic power plants);

NOTES

The final harmonization of the chapter was carried out by the energy executive board of the Global Research Alliance (GRA).

The following scientists and institutions have delivered further contributions to this thematic background paper: Dr John Christensen, Prof. Dr Dieter Holm and Dr Erik H. Lysen.

REFERENCES

Forschungsverbund Sonnenenergie (FVS) (2003) *Eckpunktepapier für ein neues Energieforschungsprogramm der Bundesregierung*, Berlin, www.fv-sonnenenergie.de/publikationen/03.06.26.eckpunkte_01.pdf

Jacobsson, S. and Johnson, A. (2000) 'The Diffusion of Renewable Energy Technology: An Analytical Framework and Key Issues for Research', *Energy Policy*, vol 28, pp625–640

Margolis, R. M. and Kammen, D. M. (1999) 'Underinvestment: The Energy Technology and R&D Policy Challenge', *Science*, vol 285, pp690–692

Roehrl, R. A. and Riahi, K. (2000) 'Technology Dynamics and Greenhouse Gas Emissions Mitigation: A Cost Assessment', *Technological Forecasting and Social Change*, vol 63, pp231–261

WBGU (2004) *World in Transition – Towards Sustainable Energy Systems*, London, Earthscan

14

Capacity Development for Renewable Energies: Know-how, Information, Awareness, Education and Training

John M. Christensen

INTRODUCTION AND RATIONALE

Capacity development, education and training are generally recognized in scientific literature, practical experience and political reality as cross-cutting issues crucial not only to promote renewable energy activities but also to move towards more sustainable development paths. The quotation from Agenda 21 in Box 14.1 illustrates the general needs perception and also describes some of the generic elements of capacity development (CD). While this chapter will not discuss various approaches to CD in theoretical detail, it is important to keep in mind that there has been significant development in both the theoretical and the practical understanding of the CD concept over the past decades. Vast experience with CD exists in most sectors in both industrialized and developing countries. The key issue is therefore to understand what the specific needs for CD are in relation to the energy sector and, as a subset of that, how to ensure that the specific conditions and requirements for CD for renewable energy technologies (RETs) are understood and addressed.

In the preparation process for the World Summit on Sustainable Development (WSSD), five areas (Water, Energy, Health, Agriculture and Biodiversity) were identified as key for sustainable development, the so-called WEHAB agenda. The Framework for Action on Energy, developed as part of the WEHAB agenda, outlines 17 potential action areas, including RETs, efficiency

---

Box 14.1   Capacity in Agenda 21

A country's ability to develop more sustainably depends on the capacity of its people and institutions to understand complex environment and development issues so that they can make the right development choices.

People need to have the expertise to understand the potential and the limits of the environment. They will face difficult policy choices when dealing with such complex problems as global climate change and protecting biodiversity. This will require scientific, technological, organizational, institutional and other skills.

Source: Agenda 21, Chapter 37: 'Creating Capacity for Sustainable Development'

---

and access. It is significant that for almost all potential areas, some elements of CD are identified as important cross-cutting action areas, indicating the crucial role of CD in energy sector development.

Similarly, the G8 Renewable Energy Task Force in its report (G8, 2001) assesses the barriers and recommends action to encourage the use of RETs in developing countries. It concludes that, as one of four key barriers, the capacity barriers are as set out in Box 14.2.

Box 14.2   Capacity barriers

To secure sustainable commercial success, renewables must overcome a number of key barriers, including:

- insufficient human and institutional infrastructure;
- limited capacity to support projects and markets, owing to a lack of experience and investment.

Source: G8 Task Force Report

These references to Agenda 21, WEHAB and G8 are all implicitly linked to CD in relation to developing countries. This is logical in the sense that all these initiatives are aimed at stimulating development, and the general capacity to undertake and implement development action is generally much more limited in developing than in industrialized countries.

This chapter will also focus predominantly on CD in relation to developing countries, but it must be underlined that in many industrialized countries the capacity to develop and implement both RE policies and technologies is still quite limited and often a real barrier to increased utilization of RETs. There is a strong interlinkage between CD and the maturity of markets – and it is obvious that emerging markets need information, R&D and professional formation for RETs.

CHANGING CONDITIONS FOR CAPACITY DEVELOPMENT IN THE ENERGY SECTOR

When assessing CD activities undertaken in the past in the energy sector, and specifically related to promoting the application of RETs, as is done in the next section, it must be realized that several parallel developments affecting both CD approaches and the energy sector have been and are still taking place. These developments are discussed in the following subsections, along with their implications for CD in the energy sector.

Changing the capacity development paradigms

The understanding of capacity development has changed especially over the past decade from a narrow focus on capacity-building (assuming that none were present), closely linked with institution-building, towards much broader approaches where capacity development is seen as an interrelated set of activities taking place principally at three levels:

- *Individual* – enabling individuals to embark on a continuous process of learning: building on existing knowledge and skills, and extending these in new directions as fresh opportunities appear.
- *Institutional* – this also involves building on existing capacities. Rather than trying to construct new institutions, existing initiatives are identified.
- *Societal* – this involves capacities in the society as a whole, or a transformation for development. This creates the opportunities, whether in the public or the private sector, that enable people to use and expand their capacities to the fullest.

Implied in these new CD paradigms is the requirement that they must pay attention to the impact of the context in which capacity development programmes occur, such as political, social, cultural, economic. Effectiveness is influenced as much by forces external to the organization as by internal forces.

Indigenous commitment and control are critical for the success of capacity development programmes, which depend on local knowledge and energy for their effectiveness. Participation by beneficiaries is thus essential at all levels of identification, planning, implementation and evaluation of an activity.

Capacity problems, solutions and constraints need to be analysed from a systems perspective. This implies a focus on the dynamics and interrelationships among the various actors, components and levels, and a sense of the interrelationships between projects, programmes and other groups and organizations within the wider context. Attention must be given to the limiting factors that constrain system performance.

It is probable that no single CD activity can address all aspects adequately. Programmes therefore need to be developed as interlinked and mutually reinforcing activities with a temporal dimension where more immediate needs should be addressed in the short term, while ensuring with longer-term activities that an overall framework is developed which will help sustain the short-term achievements.

Structural changes of the energy sector

A second development to consider when designing CD activities is related to the fact that in the past decade, macro-level structural reform programmes – and specifically reforms in the power sector – in many countries have changed the institutional, legal and regulatory frameworks for energy sector development. This has led to a number of new challenges, since the number of stakeholders has increased substantially, along with the complexity of their interrelations. At the same time the structural changes have also altered the functions and tools of existing institutions. One example is typical traditional policy-making institutions, which in the new and more market-oriented structures have to adapt and apply new types of policy instruments that in most cases require targeted CD in the form of training and education for RETs.

Parallel evolution of RETs

The assessment and recommendations for capacity development for RETs evidently needs to be linked with the realistic contribution from the technologies both technically and economically. Many early examples of RET programmes probably failed partly because the expectations for performance and costs were not fully realistic.

The World Energy Assessment (UNDP, 2000) states that many renewable technologies, although often commercially available, are still at an early stage of development and that not all are technically mature. They demand continuing research, development and demonstration efforts. Only a limited number of renewable energy technologies can compete with conventional fuels on costs, except in some niche markets. But substantial cost reductions have been and can still be achieved for most renewables, closing cost gaps and making them more competitive.

In recent years, significant progress has, for example, been made in the development of wind turbines for electricity production. Positive developments, although at smaller scale, have also been experienced with photovoltaic (PV) solar technologies and modern bioenergy. The whole renewable energy industry turnover is now in the multi-billion-dollar range, and global installed capacity is projected to more than double over the next ten years, even under present political conditions. If national efforts to support RETs market competitiveness are enhanced, this projected development can be significantly increased.

Implications for RET programmes and capacity development

Martinot et al (2002) have exemplified what this transition in the energy sector could mean for activities related to RETs moving away from what they call the old paradigm, which was generally characterized by donor-driven technology demonstration projects, as is discussed in the next section, towards more user-oriented approaches focusing on building business environments and more sustainable markets for RE technologies and services.

In order to design CD programmes reflecting the new and more complex sector structure, it is necessary to have a clear picture of all the relevant stakeholders. Bouille and McDade (2002) have identified 14 key groups of stakeholders, as presented in Table 14.1. Bouille and McDade group the stakeholders broadly into government (1 to 7), private sector (7 to 10) and into a mixed group with NGOs and energy experts/researchers. The last of these groups may both be the target of CD activities and, at the same time, act as providers of CD for the other stakeholders. This grouping is useful in designing CD activities, and presents an elaborate analysis of the potential CD needs of the different types of stakeholders.

Old paradigm		New paradigm
Technology assessment	⇒	Market assessment
Equipment supply focus	⇒	Application, value added and user focus
Economic viability	⇒	Policy, financing, institutional and social needs and solutions
Technical demonstrations	⇒	Demonstrations of business, financing, institutional and social models
Donor gifts of equipment	⇒	Donors sharing the risks and costs of building sustainable markets
Programmes and intentions	⇒	Experience, results and lessons

Source: Martinot et al (2002)

Figure 14.1 *Renewable energy markets in developing countries*

It should be emphasized that any capacity development programme has to be seen as a process rather than a product delivery activity. This implies that CD is open-ended, but for programme delivery and assessing effectiveness, evidently targets and milestones should be established, in the realization that it is not always possible to make quantitative and objective assessments of the CD outcomes.

EXPERIENCES WITH CAPACITY DEVELOPMENT IN THE ENERGY SECTOR

Capacity and institutional development in the energy sector related to RETs have gone through different stages, and a quick and qualitative historical assessment is presented in the following section, reflecting that very little systematic assessment of RET capacity development has been undertaken apart from at the individual project or programme level.

The institutions-building phase

The oil price crises in the mid-1970s and early 1980s generally put energy higher on the political agenda, with a focus on security of supply, development of domestic and renewable energy resources and increasing efficiency in both supply and demand. The typical reaction was creation of institutional frameworks either to develop relevant policy and/or to initiate new but limited efforts at R&D of RETs.

Examples are institutions such as the New and Renewable Energy Agency (NREA) in Egypt, the Department of Non-conventional Energy Sources/Ministry of Non-conventional Energy Sources (DNES/MNES) in India, the National Renewable Energy Laboratory (NREL) in the US and several similar institutions, mainly in OECD countries and the larger developing countries. These institutions were largely created through national acts and initially developed with national funding. Not all institutions have survived, but several have managed to develop into key national actors in the energy policy or RET area.

It is virtually impossible in retrospect to assess whether these institutional developments were the most appropriate. Several national institutions have, as mentioned, developed into important national entities promoting energy for sustainable development (see example in Box 14.3) while others had a short lifetime and never made an impact. In the Egyptian example the institution has grown into an important national actor and has no doubt played an important role in establishing the strong RET programmes in the country. A non-institutional example of the opposite situation is the stop–go nature of production tax credits in the US, where private-sector developments and investments

Table 14.1 *Stakeholders in energy for sustainable development*

Stakeholder	Functions/activities
1. Legislative authorities/ elected officials	Set national political priorities; social, economic and environmental goals; legal framework conditions
2. Government macroeconomic and development planners	Define development goals and macro policy; general economic policies; cross-cutting issues; subsidies and trade policy; sustainable development goals; and frameworks
3. Government energy authority or ministry	Set sectoral goals; technology priorities; policymaking and standard-setting functions; legal and regulatory framework; incentive systems; Federal, state and social level jurisdiction
4. Energy regulatory bodies	Have monitoring and oversight functions; implement the regulatory framework; administer fees and incentives
5. Market coordination agencies	Dispatch entities; have operational coordination functions; interface with industry investors; act as information brokers
6. Non-energy governmental authorities/ministries	Deal with sector policies; cross-cutting issues; inter-relation with energy policies, public sector energy consumers; require energy inputs for social services provision
7. Energy supply industry	Comprise private companies and public utilities; manage energy supply, electricity generation; fuels management and transport; finance some R&D
8. Entrepreneurs and productive industries	Involved in business development; economic value added; employment generation; private sector energy consumers
9. Energy equipment and end-use equipment manufacturers	Supply equipment for the energy industry and other industries, including vehicles and appliances; impact energy end-use efficiency; adapt/disseminate technology; finance some R&D
10. Credit institutions	Financing options for large and small-scale energy generation; capital provision for energy using enterprises; financing options for household energy consumers
11. Civil society/non-governmental organizations	Promote consumer participation and awareness; oversight and monitoring; provide environmental and social advocacy; equity considerations
12. Energy specialists and consultants firms	Offer strategic advice, problem definition and analysis; systems development; specialist services delivery; options analysis; information sharing
13. Academia and research organizations	Provide R&D, knowledge generation and sharing; formal and informal education; technical training; technology adaption, application and innovation
14. Media	Involved in awareness raising, advocacy; information sharing; journalistic inquiry, watchdog functions; monitoring, public transparency

Source: Bouille and McDade (2002)

Box 14.3 The New and Renewable Energy Agency in Egypt

The New and Renewable Energy Agency (NREA) was established in 1986 as a government R&D body working under the Ministry of Electricity. The primary objectives in establishing the NREA were:

- to contribute to Egypt's need for increasing the share of renewables in the power supply mix;
- to conduct various research projects on issues pertaining to renewable energy technologies;
- to act as a renewable energy focal point and counterpart agency for all international organizations interested in the renewables sector in Egypt;
- to advise the Ministry of Electricity on renewable energy technologies and their applications in the country;
- to collaborate with other government and non-government institutions in various research projects.

In 2003, the NREA was staffed with 122 scientists and engineers, 109 technicians and 63 administrative staff. Its budget is provided by the Ministry of Electricity.

---

have sequentially been stimulated and undermined, resulting in several industrial collapses, especially in the 1990s.

In parallel with these dedicated national CD initiatives, energy issues gradually began to receive priority in development assistance, leading to a large number of institution-building activities typically assisting developing countries to establish similar types of energy policy institutions and technological development and training centres.

RET demonstration phase

Especially in the 1980s, there was particular donor interest in what may broadly be termed RET demonstration projects. Martinot et al (2002) estimate that from 1980 to 2000 about US$3 billion of official development assistance went into RETs, and by far the greatest part of these funds was spent on technical demonstration activities or projects that may have been self-sustained but not replicable.

Many projects were considered failures owing to the absence of CD support, institutional and commercial viability and poor suitability to the user needs. National policies and institutions in addition showed limited interest in these small-scale and often decentralized RET activities, as they were not high on the political agendas, and to a large extent donor rather than needs driven. There were, however, also some successes, like the small-scale biogas projects in China – where the link between CD and successful implementation is evident.

The generally poor results led to a downturn for RET support activities in the early 1990s, but this has been reversed in recent years with impetus coming partly from the increased focus on sustainability and especially climate change concerns, combined with gradual maturing of individual RETs in terms of technical and economic performance, and compounded by recent increases in oil and gas prices.

RET developments in OECD countries in the 1980s were also limited to relatively small demonstration programmes and generally had limited success. One of the exceptions was the Danish experience with wind energy development, which has resulted in a continued development of the turbine technology into the current large units that are cost-competitive with power generated from fossil fuels. Initially, this development was, however, not driven by national policy or dedicated CD or market development programmes, but predominantly by cooperations, NGOs and small-scale industries, which then later accelerated as a result of dedicated government support.

This very simplified historic assessment does not intend to do full justice to the many good and valuable RETs efforts that were also undertaken in the 1980s and 1990s, but presents a broad-brush picture of the overriding trend of activities.

Market orientation and scaling up

During the past decade the different transitions described earlier have started to impact on activities related to implementation of RETs. With individual technologies becoming more cost-competitive compared with traditional fossil options and with industries gradually maturing, new types of CD activities have evolved, focusing on business development and finance.

Several examples of such programmes could be mentioned but here reference will only be made to one, owing to the limited size of the chapter.

A major multi-donor activity in China – the 'China Capacity Building for Rapid Commercialization of Renewable Energy' – has strengthened the already existing programmes for RET promotion in China by supporting capacity development in a number of critical areas such as:

- business development;
- establishment of standards and certification;
- strengthening resource data and collection;
- project development and entrepreneur training.

In addition, the project has involved policy-makers and industry leaders in order to ensure that the commercial potential in RETs is better understood and reflected in policies and decision-making.

Project results are in the process of being mid-term assessed, but so far results show promise, including the development of businesses for village-scale power components and industrial-scale biogas components. Programmes have been established for national testing and certification for the solar water heater industry and for national wind resource measurement. The China Renewable Energy Industries Association has been established and has already facilitated a number of business development activities, including large-scale wind development.

It must, however, be emphasized (see Chapter 4 by Janet Sawin and Christopher Flavin) that, along with the market-oriented approaches to implement RETs, there is in many cases still a need to combine these approaches with sustained government policies facilitating a 'levelling of the market playing field', including some form of government support to RET activities in order to ensure a sustained stimulation of market development and increase in the scale of the industries involved.

Environmental policy push

Even if the discussion of capacity development in this chapter mainly focuses on donor-supported activities in developing countries, it is, as mentioned in the introduction, important to realize that many of the points are also valid for industrialized countries.

Only limited systematic assessment of capacity development efforts related to RETs in OECD countries is available, maybe partly because capacity has emerged as a result of gradual processes and partly because the issue has not been given enough attention.

However, in recent years the issue of appropriate institutional structures has re-emerged as a result of the increased focus on environmental and especially climate change concerns in relation to the energy sector. This is for obvious reasons mainly a trend in the industrialized countries, reflecting the agreements in the Framework Convention on Climate Change and the Kyoto Protocol, but similar developments are likely to take place in developing countries, as is the case in relation to implementation of the Clean Development Mechanism (CDM).

An interesting analysis of the institutional tasks and structures has been presented in the Energy Review (Performance and Innovation Unit, 2002; Hartley, 2003) undertaken as input to UK policy development. This analysis of policy tasks related to implementing a more sustainable energy strategy notes that since countries differ in policies and organization, there can be no single 'right' institutional answer, and established structures do have the ability to adapt to new priorities and tasks. But in order to ensure concerted action and implementation, the review concludes that responsibilities in the United Kingdom could most effectively be concentrated in a 'sustainable energy policy unit'. Hartley sees the Sustainable Energy Ireland (the Irish energy authority) as a possible model to be considered for the United Kingdom.

Sustainable Energy Ireland was established under the Sustainable Energy Act 2002 and has the mandate to promote and assist environmentally and economically sustainable production, supply and use of energy, in support of government policy, across all sectors of the economy. Its remit relates mainly to improving energy efficiency, advancing the development and competitive deployment of renewable sources of energy and combined heat and power, and reducing the environmental impact of energy production and use, particularly in respect to greenhouse gas emissions. The Austrian Energy Agency (EVA) has similar mandates but the institutional arrangement is somewhat different.

SYNTHESIZING KNOWLEDGE INTO ACTION PROPOSALS

Taking into account the ongoing described transitions related to CD paradigms and to energy sector structures, there are a number of priority areas to focus on in future efforts related to CD for promoting RETs, as part of a broader transition towards more sustainable energy systems.

Broad awareness-raising and strengthened educational efforts at all levels will be essential for expanding RET applications. This applies both for decision-makers in public and private entities (authorities, agencies, small and medium-sized enterprises (SMEs), industry and finance) and for the end-users. RE education can contribute effectively in disseminating the information and providing appropriate training for renewable energy developers, engineers, technicians and end-users, as well as decision-makers, business people and industrialists.

Getting the policy framework right will be crucial for expanding beyond the present limited niche areas for RETs. This means that ministries involved in policy development need to acquire the necessary skills to deal with both the new, more market-oriented supply structures and the political demand so as to be able to address increasingly social and environmental aspects. This will require new analytical capabilities, understanding of market-oriented policy tools and an ability to implement such tools in practice.

A specific sub-set of the framework conditions will be to target the upcoming regulatory institutions and regulators, which will be in a central position to facilitate appropriate market structures for increased introduction of RETs. This is especially important in view of the fact that the majority of studies indicate that most national power sector reforms in their structuring process have paid very limited attention to environmental and social issues.

Private-sector involvement in terms of both large-scale industries for the commercial markets and SMEs for the rural and peri-urban markets, will be required for increased RET development and deployment. Companies need to have a wide range of technical, business, management and regulatory skills to enter new renewable energy markets. These skills need to be developed locally to promote technology development and transfer as well as the flow of international capital. Access to information and capacity to assess it are also essential for technology development. For SMEs, new types of intermediaries are required that can provide them with both appropriate forms of financing (e.g. seed capital) and enterprise development support.

Financing and lending institutions will play a crucial role, especially because of the nature of RETs, with generally high up-front investment costs combined with low running and delivery costs. This cost structure generally makes RETs inaccessible without financial institutions providing the investment capital and arranging the flow of payments. As with many other new areas of intervention, financial institutions are relatively conservative in order to protect the interests of their investors. In common with all new process development, financiers must overcome a learning curve when taking their first investments in a new sector such as RETs. Without stronger incentives, loan officers may pay only limited attention to renewable energy investments, focusing instead on the traditional benchmarks of rapid loan disbursement with minimal risk. Changing the way a financial institution considers new RE investments therefore requires both better information and new mandates to combine social and environmental factors – both risks and returns – as integral measures of economic performance.

R&D institutions in many developing countries are extremely resource constrained and CD efforts therefore need to be focused on supporting activities that have high local policy relevance. At the same time, it is important to emphasize that policy institutions should involve and utilize local expertise as far as possible instead of often preferring to rely on international expertise. One way of strengthening local research groups is by establishing regional and international networks to share information and undertake joint activities.

The listing of priority areas for CD is in no way exhaustive and should be expanded. At the same time, it is relevant to stress that there is an urgent need to make a better and more thorough assessment of CD experiences in the energy sector, and specifically the activities undertaken to support RETs. At present there is virtually no comprehensive analysis available, which means that CD activities are often designed in a piecemeal and fractured way, which is exactly the opposite approach of what is recommended in this chapter.

Specific recommendations

The overall recommendation is for a concerted international programme on capacity development, which needs to be seen as a core component of a larger effort to promote the increased development and implementation of RETs. Such a programme would need to be based on the recognition that *integrated and sustained action* is the key to successful expansion of RET utilization and that activities on capacity development need to be designed to suit the specific national circumstances and to involve all relevant stakeholders. A programme with general application in both developing and industrialized countries should cover areas such as policies, financing, R&D, awareness-raising, information and professional formation.

Policies

Planning and legislation

A well-designed and sustained policy framework is key to market development and scale up for RETs, and should include awareness-raising with policy-makers to better understand the current energy market 'distortions', their consequences and what realistic contributions RETs can make in the short and long-term. This will need to be combined with enhancing the capacity of energy planners and analysts – for example, to assess the full costs of different supply options, to design policy intervention matching the reformed energy sectors and to ensure that RETs are considered in a realistic manner in the range of supply options. The legislative framework for the energy sector needs to be part of the policy development focus, as it is generally recognized as one of specific barriers to RET promotion in many countries. Policy-makers also need to pay attention to the opportunities for creating new jobs and SMEs by introducing RETs.

Entrepreneurs and business development

The increased market orientation of the energy sector has been extensively discussed, along with the potential for larger-scale RET contributions. A crucial component in such a development is strong private-sector participation, and RETs offer good opportunities especially for SMEs both in developing and in industrialized countries. But capabilities on both the technological and the business aspects of RETs need to be strengthened.

Finance

With the increased commercialization of energy markets and RET implementation, a key area of capacity development and increased awareness is with the international and national financial institutions. Chapter 8 of this book deals extensively with this issue, emphasizing the need to increase awareness and skills, for example in areas such as independent evaluation of RET projects, emphasizing long-term opportunities that may offset the up-front investment costs, and making financing of RETs include risk assessment and hedging strategies. The new opportunities offered through carbon funds related to the flexible mechanisms (emissions trading, clean development mechanism and joint implementation) under the Kyoto Protocol have the potential to become a new important source of finance for RETs, but both national entities responsible for carbon credits and trade and financial institutions need to build the capacity to make full use of these new opportunities.

R&D, awareness-raising, information and professional formation

Resource and technology data

Particularly in developing countries, but also in many industrialized countries, the lack of appropriate data on resources and technology performance is an important barrier to increased implementation of RETs. A recent major United Nations Environment Programme/Global Environment Facility (UNEP/GEF) project entitled Solar and Wind Resource Assessment (SWERA) is an important contribution to addressing the resource data issue, but evidently only covers a limited number of the countries needing better information.

Testing and licensing

An important contribution to the successful development of the Danish wind industry was the enforcement of strict testing and licensing procedures – still applicable – which helped ensure that quality of the developed turbines was high and in this way increased the credibility of a new technology. This experience has recently been transferred to India, where testing facilities have been established for wind turbines. Another example of recent approaches is the World Bank, which, in the area of PV for example, has issued a series of training programmes on quality for PV in partnership with the PV Global Approval programme.

Research and development

In a global world, the technically, socially and environmentally sound progress of RETs should be managed in global cooperation. To adopt know-how from advanced research institutes, to exchange information on progress, to adapt technologies to the national (natural) potentials, opportunities, needs and economic possibilities is still a lot of work. As R&D is crucial for the development of country-tailored businesses, an extended transfer of know-how is needed.

The listing of action areas is in no way comprehensive and will need to be combined with dedicated efforts on education and training, to ensure that the necessary expertise is available on R&D, as presented in the chapter dealing with this topic.

NOTE

Contributions to this chapter were made by Ram Shrestra, Youba Sokona, Ogunlade Davidson, Gisela Prasad, Daniel Bouille, Nicolas Di Sbroiavacca, Eric Usher, Sami Kamel, Joyti Painuly, Romeo Pacudan and Gordon Mackenzie.

REFERENCES AND FURTHER READING

Bouille, D. and McDade, S. (2002) 'Capacity Development', in *Energy for Sustainable Development: A Policy Agenda*. Johansson, T. B. and Goldemberg, J., New York, United Nations Development Programme

Department for International Development (DFID) (2002) *Energy for the Poor*, London, DFID

CIDA (1996) *Technical Cooperation to Capacity Development: Changing Perspectives in CIDA (Canadian International Development Agency)*, February

Energy Sector Management Assistance Programme (ESMAP) (1999) *A Review of the Renewable Energy Activities of the UNDP/World Bank Energy Sector Management Assistance Programme 1993–1998*, World Bank, Washington, DC

G8 Renewable Energy Task Force (2002) *Final Report*

German Advisory Council on Global Change (2003) *World in Transition – Towards Sustainable Energy Systems*

Hartley, N. (2003) 'Does the UK Have the Right Institutions for Energy Policy Making?', paper given at British Institute of Energy Economics, Energy Economics Conference, 25–26 September 2003, Oxford

International Energy Agency (IEA), Photovoltaic Power Systems Programme (2003a) *PV for Rural Electrification in Developing Countries – A Guide to Capacity Building Requirements*, Paris, IEA

IEA, Photovoltaic Power Systems Programme (2003b) *Institutional Framework and Financial Instruments for PV Deployments in Developing Countries*, Paris, IEA

Martinot, E., Chaurey, A., Lew, D., Moreira, J. B. and Wamukonya, N. (2002) 'Renewable Energy Markets in Developing Countries', *Annual Review of Energy and the Environment*, vol 27, pp309–348

Martinot, E., Ramankutty, R. and Rittner, F. (2000) *The GEF Solar PV Portfolio: Emerging Experience and Lessons*, Washington, DC, Global Environment Facility

Performance and Innovation Unit (2002) *The Energy Review*, London, Performance and Innovation Unit

Sawin, Janet (2004) *National Policy Instruments: Learning from Variety – Time to Concentrate on Success*, Thematic Background Paper Renewables 2004, (www.renewables2004.de), Bonn

Sonntag-O'Brien, V. and Usher, E. (2004) *Mobilising Finance for Renewable Energies*, Thematic Background Paper Renewables 2004, (www.renewables2004.de), Bonn

United Nations Development Programme (UNDP) (2000) *World Energy Assessment*, New York, UNDP

UNDP (2002) Special Issue: Technical Cooperation, *Development Policy Journal*, New York, UNDP

Water, Energy, Health, Agriculture and Biodiversity (WEHAB) Working Group (2002) *A Framework for Action on Energy*, New York, WEHAB

Index